H. G. Natke · C. Cempel

Model-Aided Diagnosis of Mechanical Systems

Springer
Berlin
Heidelberg
New York
Barcelona
Budapest
Hong Kong
London
Milano
Paris
Santa Clara
Singapore
Tokyo

H. G. Natke · C. Cempel

Model-Aided Diagnosis of Mechanical Systems

Fundamentals, Detection, Localization, Assessment

With 83 Figures

Springer

Professor Dr. rer. nat. Dr. hc. Hans Günther Natke
Universität Hannover
Curt-Risch-Institut für Dynamik, Schall- und Meßtechnik
Appelstr. 9A
D - 30167 Hannover

Professor Dr. habil. Czeslav Cempel
Technical University Poznan
Applied Mechanics Institute
3 Piotrowo Street
PL - 60-965 Poznan

ISBN 3-540-61065-0 Springer-Verlag Berlin Heidelberg New York

Library of Congress Cataloging-in-Publication Data
Natke, Hans G.:
Model-aided diagnosis of mechanical systems : fundamentals, detection, localization, assessment / H. G. Natke ; C. Cempel. - Berlin ; Heidelberg ; New York ; Barcelona ; Budapest ; Hong Kong London ; Milano ; Paris ; Santa Clara ; Singapore ; Tokyo : Springer 1997

ISBN 3-540-61065-0
ISBN 0-387-61065-0

NE: Cempel, Czeslaw:

Production: PRODUserv Springer Produktions-Gesellschaft, Berlin
Typesetting: Lewis & Leins, Berlin
SPIN: 10529674 68/3020 - 5 4 3 2 1 0 - Printed on acid-free paper

Preface

The Task of the Book. This book is concerned with the diagnosis of the damage and faults of mechanical systems in the diverse fields of engineering. It takes into consideration the various stages of the life cycle of a system from a diagnosis-orientated point of view and from a methodological one. In Fig. 1 it is indicated that analysis, testing and diagnosis is linked with all the stages of a system's life, with an enormous effect on costs, reliability and safety. What the title of the book promises is seen in Fig. 2. The contents are focussed on those system functions in which the system exists.

Holistic modelling (see Fig. 2) is the key to the methodology discussed in this book. The holistic model includes the life behaviour (slow-time coordinate) as well as the classical dynamics (expressed with the fast-time coordinate). Therefore the holistic model encompasses the system dynamics and its evolution. The possible interactions of a holistic model with the various stages of system design and development are shown in Fig. 2. The impact is shown in the lower part of Fig. 2 by the reduction of system life cycle costs, and the extension of the system life span. This is the main benefit of holistic modelling and diagnosis.

Fig. 3 gives an overview, indicating the importance of diagnostics and of model-based diagnostics in particular. The various models are shown which lead to the holistic model for diagnosis. The goal is, finally, to perform a system condition assessment in the past and in the present through an adjusted mathematical model (adaptive model), and for the future by the use of prediction with the help of adaptive models.

Model-supported diagnosis guarantees

- safety
- serviceability
- the reduction of costs, and
- it increases the lifetime.

Additionally, it can be stated that

- structural safety will be provided through adaptive (with respect to loading and state condition) modelling using measured data and the resulting prediction
- the holistic viewpoint permits easy design including new materials
- the uncertainties in general are minimized
- the assessment is made objectively.

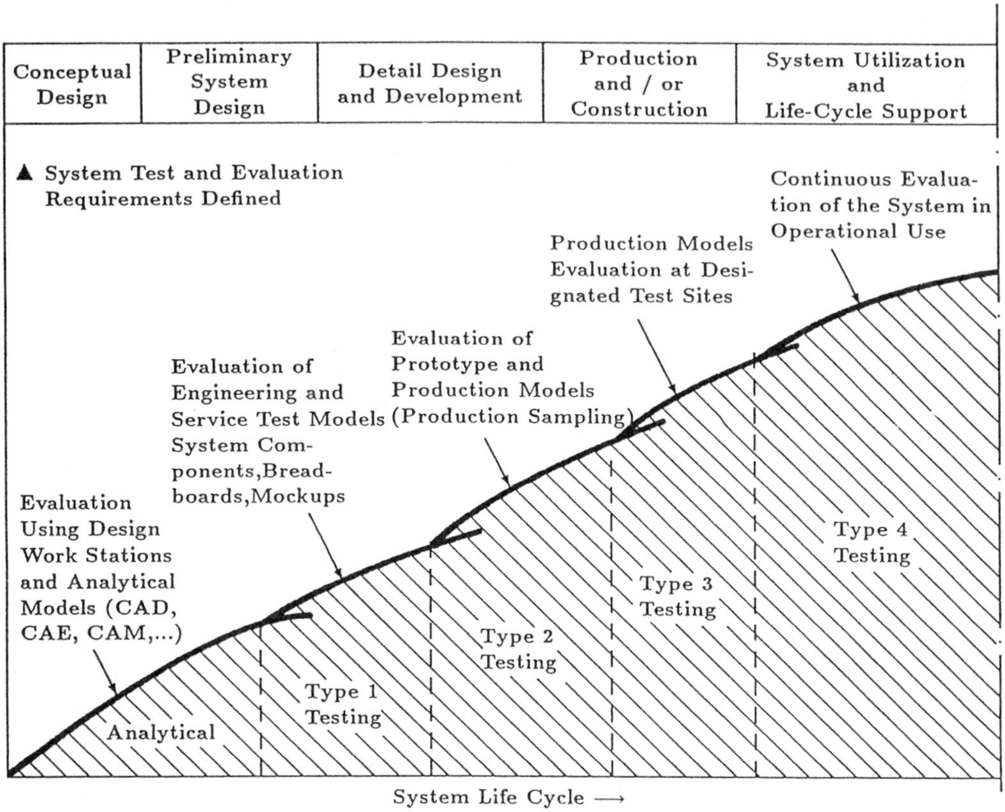

Fig. 0.1. Stages of system evaluation and testing during the life cycle (modified from Blanchard 1989)

Contents. The first chapter introduces the subject, and describes its objectives and scope.

In Chapt. 2 the foundation of system properties, modelling, monitoring, damage, symptoms, models and methods of damage detection, localization etc., that are common in classical diagnostics using signature analysis are reviewed.

Chapter 3 provides the fundamental guidelines for measurements and signal processing as a fundamental aspect when applying the contents of Chapt. 4.

Chapter 4 contains the principal part of the book: the knowledge base in the form of verified and validated mathematical models which are adjusted to the states of the system at current life times: using adaptive models. These models describe the current state of the system, permit a comparison with previous states, and therefore serve the purposes of fault detection, localization, and the cause-finding of faults and their assessment, predictions due to future forcings and trend predictions.

The diagnostic decision-making based on the validated models is discussed in Chapt. 5. It is based on thresholds, and on standard deviations

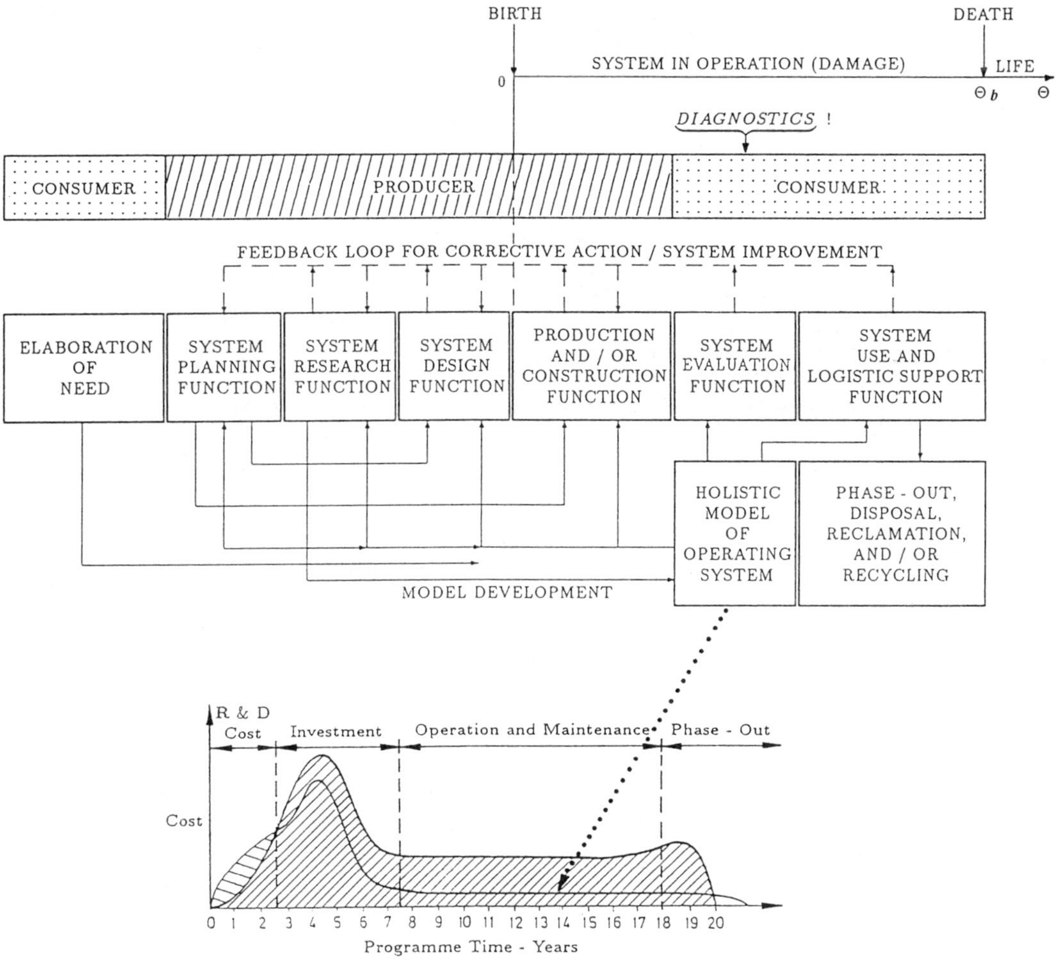

Fig. 0.2. The various system stages and the effect of the holistic model

of global and local quantities coming from the mathematical models, both in deterministic and probabilistic decision-making. Chapter 5 also contains a brief discussion of the assessment to be made with the resulting consequences. This chapter is tailored especially for the purpose of the book.

Finally, some methods are illustrated by examples in Chapter 6.

The book should provide the foundation for the diagnosis, so that the reader will be able

1. to develop a formulation for his particular application, and
2. to find stimulation for further applications and investigations.

It will not give recipes for model aided diagnosis for special cases, but possible solutions are discussed.

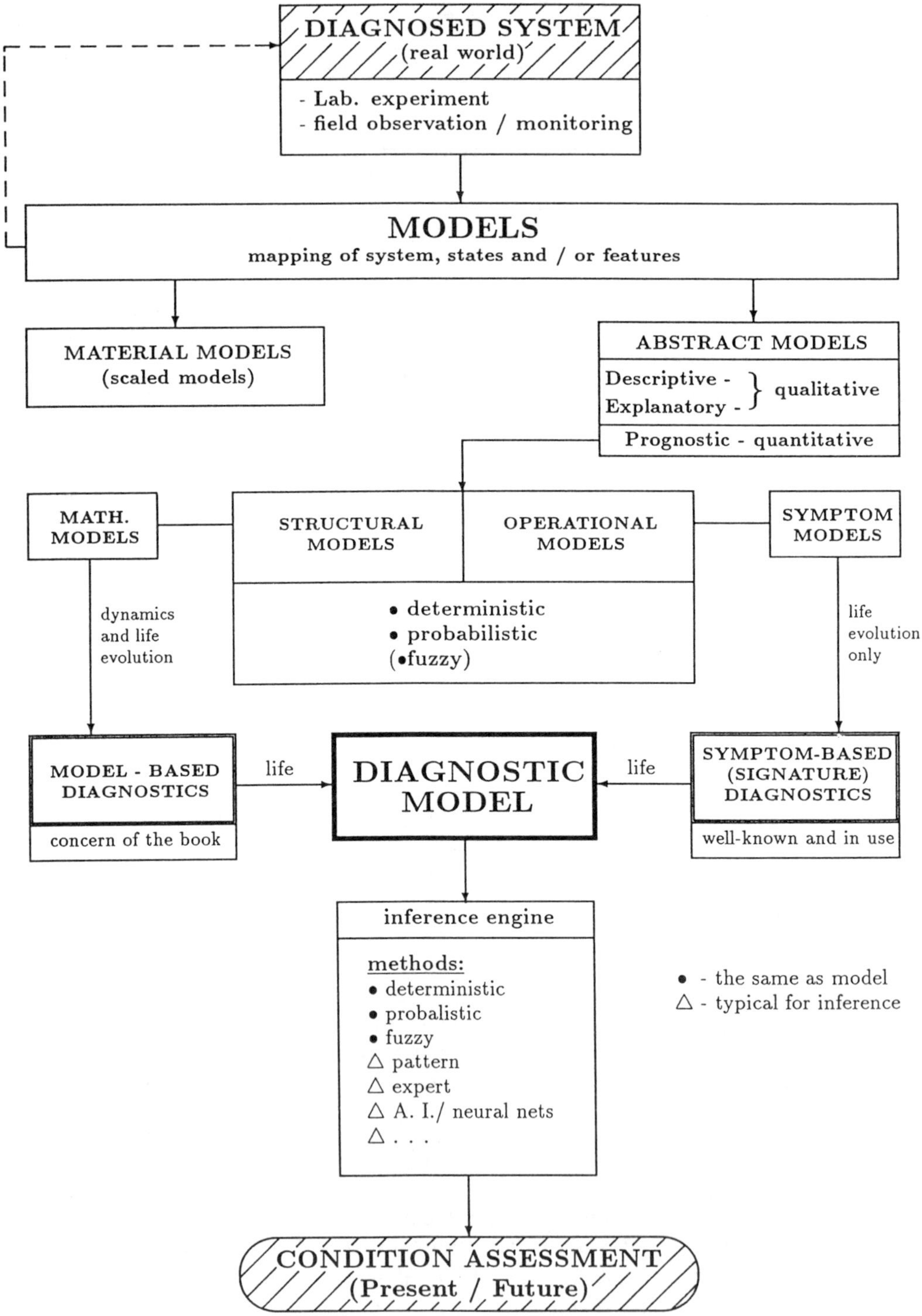

Fig. 0.3. The various models for a holistic diagnosis

Requirements on the Reader. Although Chapt. 2 repeats some known facts of modelling and classical diagnostics, the reader should have some expertise in at least one of the following topics and some basic knowledge in all the others:

- design, properties and operation of the system under diagnostic considerations
- statistics, stochastics, estimation theory, and reliability
- signal analysis and signature analysis
- classical diagnostics (using signature analysis)
- mathematics including matrix theory.

Acknowledgement. The authors would like to express their gratitude to their colleagues and friends Yakov Ben-Haim (Haifa, Israel) and James T. P. Yao (College Station, TX, USA) for their critical and helpful discussions on various aspects of the subject.

Contents

CHAPTER 1

Introduction

1.1 The Intention

The word "system" is used here for a technical construction (object) in its operational environment from, for instance, civil, mechanical, automotive, coastal, ship, or aerospace engineering. The forces acting on the system during its utilization affect the system's properties. An example is the stiffness degradation through a crack due to cyclic forcing, and the hardening effect of high static forces at the beginning of a fatigue test. If the system property changes reduce safety, lifetime or other aspects of performance, and comfort, then failures (defects, faults, damage) can occur. Because the safety effects of failures can be catastrophic, early failure detection, localization and assessment are necessary during the system's life.

From the diagnostic point of view, the utilisation stage within the system life cycle is characterized by monitoring, usually starting with a weak point observation, continuing by considering some alert thresholds and by exceeding them, beginning with measurements. This is a condition-dependent routine measurement in order to detect system changes, whatever they are. This will be explained and introduced later. If system changes are detected, and if they are significant, then these changes have to be localized. The causes have to be found out, their effects have to be assessed, and diagnosis of failures has to be performed as the basis of maintenance and actions.

The book deals with the diagnosis of failures in systems, based on *reliable* mathematical models using observations of the system. Here, diagnosis means failure detection, localization, and the assessment of its severity as well as the forecasting of its evolution. The prediction of the system condition is included, too.

1.2 The Concept

Figure 1.1 shows the concept of model-aided diagnosis in principle[1]. System analysis yields the prior mathematical model, which generally is uncertain due to the approximations included and due to unawareness.

[1] Maintenance [3] is outside the scope of this book.

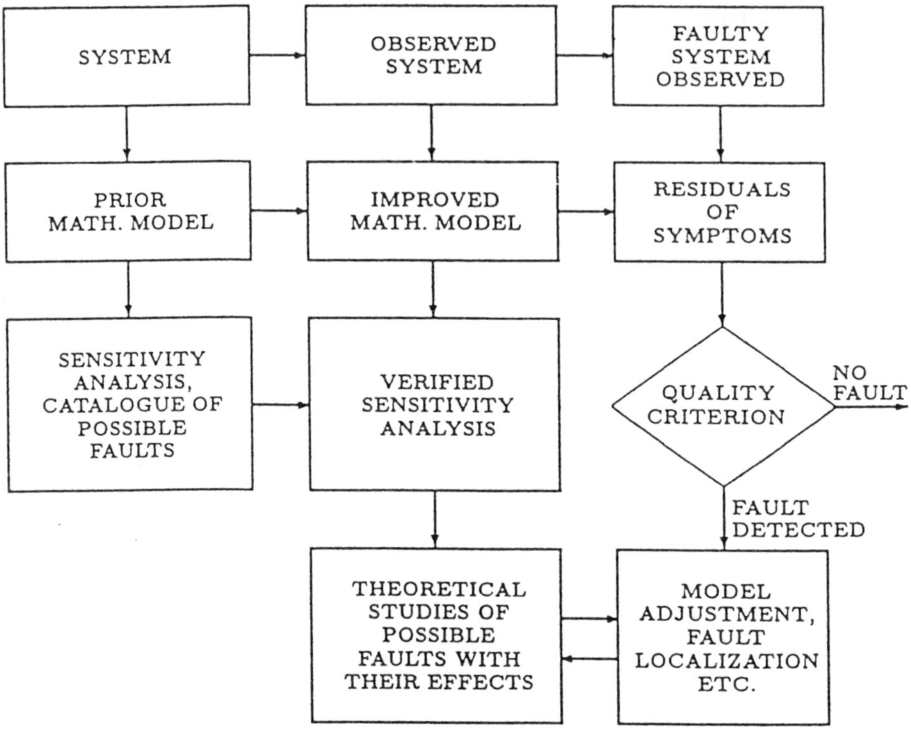

Fig. 1.1. The concept in principle

This model can be used to study possible physically based, assumed faults and their effects on the system. The result is then a catalogue of possible faults with their causes. This prior knowledge is essential in diagnosis. In addition, related symptoms (which are sensitive quantities for failure detection), features, and patterns, can be investigated and taken as a knowledge base for later diagnostic purposes. After this first step, system identification (ID) is applied in order to verify (in the sense of confirmation, checking and not finding the truth) and validate the prior mathematical model so that it is usable. Measurements of the unfaulty system performed at a predetermined life time θ_0 are used to correct this prior mathematical model, which will be used for verification and validation, or, if necessary, for improving the above-mentioned catalogue of possible faults. This verified and validated mathematical model is now the reference knowledgebase for comparison with mathematical models describing future modifications of the system observed due to faults. The third step includes the measurement of the state-modified system due to faults (as an indirect measurement of the system changes). If these measurements show significant changes in the symptoms chosen, a state-dependent mathematical model can be obtained by applying system identification, which is adjusted to the current system state, and which also results in a verified and validated mathematical model. The interesting quantities of this state-dependent model, when compared with the corresponding quantities of the reference model, will give feature residuals for the decision concerning the existence of significant modifications, which is equivalent to failure/damage detection. The next step to be done is diagnosis, which means fault localization, assessment and finding out possible (distinguishable) causes, before deciding on actions involving further system operations.

At every stage of the system's life θ an adjusted mathematical model has to be performed which represents the faulty system in evolution and serves for simulations and predictions as a basis for decisions. If the various life times are denoted by θ_i, $i = 0, 1, \ldots$, and the corresponding models with $\mathcal{M}(\theta_i) =: \mathcal{M}_i$, then it can easily be seen that the models $\mathcal{M}(\theta_i)$ represent time-invariant approximations at times θ_i of a time-variant model $\mathcal{M}$ of the system under consideration. In addition, the known models $\mathcal{M}(\theta_i)$, $i = 1(1)N_1$ at times till θ_{N_1}, enable one to set up an extrapolated model $\mathcal{M}_{N_1+1}$ for approximations of fault evolution and for trend predictions.

As can be concluded from the paragraph above, system identification [4–7] is the key to model-based fault detection, diagnosis and further decisions. Figure 1.2 [8] indicates an extended (due to model updating) system identification methodology for damage detection, localization, assessment and decision. The upper part in Fig. 1.2 represents the input/output relationship, including a possible modification of the system due to failures. Measurements of the system outputs, sometimes also of the system inputs, and perhaps combined with additional laboratory test results, give the most informative data set available for the (existing) system, provided that the measurements are performed professionally. Measurements are generally incomplete for various reasons (e.g., a sensor cannot be fastened at a necessary point or it fails, or a natural mode is missed in the dynamic response due to the existing excitation), and the measured values are erroneous. Therefore excitation and measurement problems, data acquisition, processing and reduction including error minimization play an important role. Incompleteness of the measured data base can be reduced, for example, by prior knowledge from system analysis. Combining theoretical with experimental analysis (tests, measurements) by the application of system identification will result in a verified and validated mathematical model. This means that a model with known confidence and sufficiently small errors is available. This updated mathematical model describing the system dynamics at the desired stage of the life time is the best available knowledge-base for failure localization and diagnosis, and consequently for decision-making purposes. In special cases, predictions of system conditions can include limit states.

Here system identification is a part of Test And Computer Aided Modelling (TACAM) using estimation methods and, of course, prior knowledge from system analysis. The mathematical models are uncertain. The uncertainties exist in the inputs, in the model structure as well as in its parameters if the model is structured[2]. Any distortion and environmental conditions not modelled will contribute to these uncertainties. In addi-

[2] The structure of a mathematical model refers to the type of equation written.

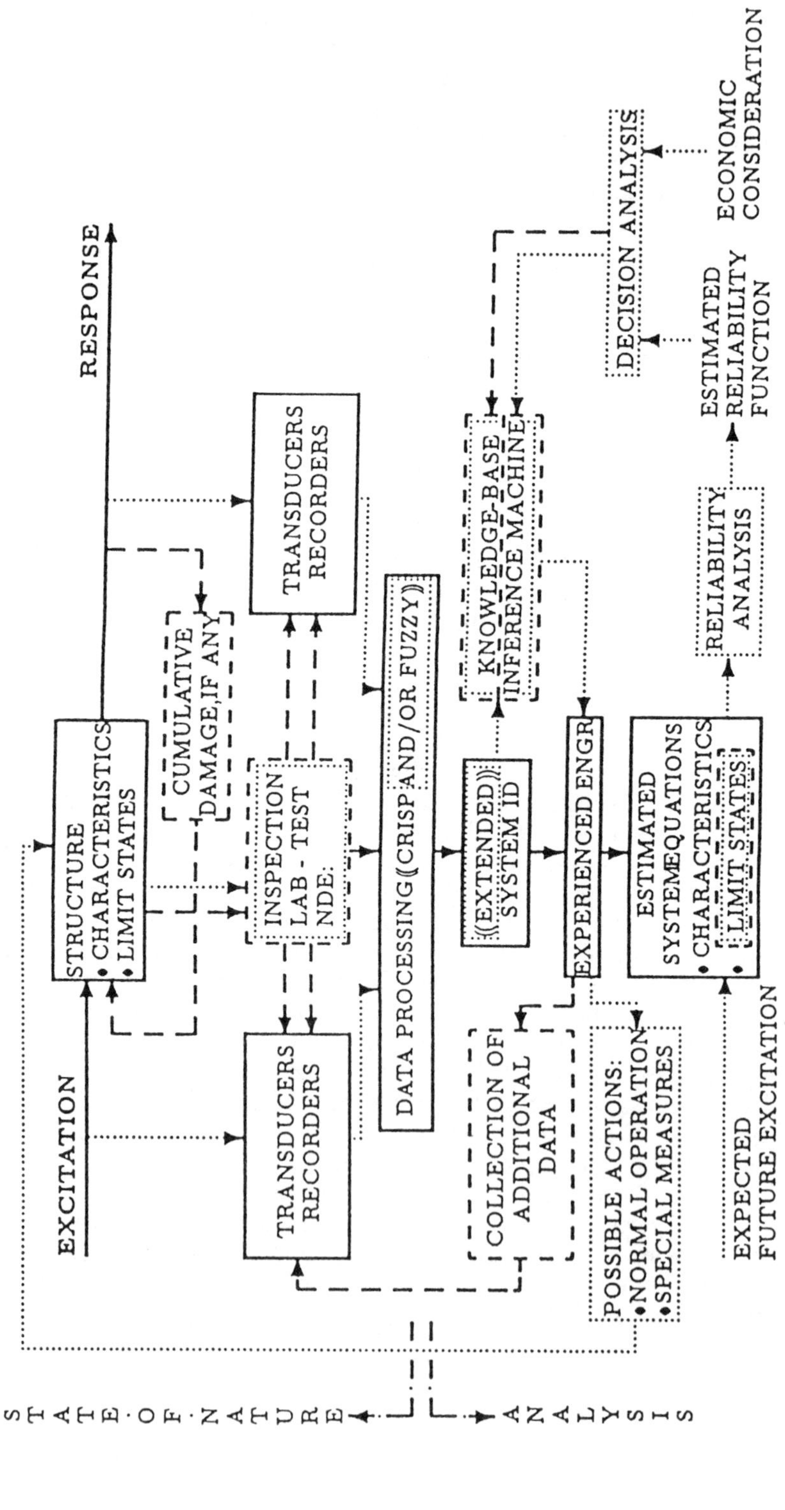

Fig. 1.2. Extended system identification methodology

tion, the test and measurement equipment can change the system under observation. Therefore this modified system has to be modelled in order to avoid systematic errors (bias). Uncertainties also have to be modelled either deterministically and/or stochastically in order to complete the available information. This also completes the knowledge base.

TACAM can be summarized as follows: it

- is based on direct and/or indirect measurements, where the latter means an enlargement of the observation space,
- can be handled by a minimum set of measurements (and no more pick-ups than in classical monitoring)
- provides an insight into the dynamic process and into the system properties,
- already includes the detection and localization of faults, and
- has the capability of performing cause-finding and assessment.

Decision analysis is another important field in this context. Decisions can be formulated deterministically or probabilistically. Multi-hypothesis decisions as well as adaptation of the testing procedure play an important role.

The major advantages of model-based diagnosis are the following:

- The utility of limited measurements can often be greatly extended when supplemented by prior knowledge from theoretical analysis of the system.
- It allows one to probe the dynamic process and the physical properties of the system (e.g. simulations in order to find countermeasures), unlike classical signal evaluation of unstructured (nonparametric) models, such as spectral densities.
- For complicated systems, usable models provide the best knowledge base for the localization and assessment of anomalous behaviour.

1.3 Modelling

Modelling is an addition to the tests/measurements fundamental to model-supported diagnosis. Therefore a brief summary should remind us of some important facts to be considered.

1.3.1 Modelling of Systems

Mathematical modelling of the dynamic system behaviour can start with a momentary (time $t = const.$) model, i.e. a static model. Time dependency will be introduced by mechanical principles. Fig. 1.3 shows the principal steps in modelling. The physical model contains the physically relevant characteristics resulting from the given forces (type, direction etc.) and the environmental effects with respect to the goal of the analysis. The mathematical formulation of the physical model is the next step, which often only can be made approximately. An example is the complex behaviour

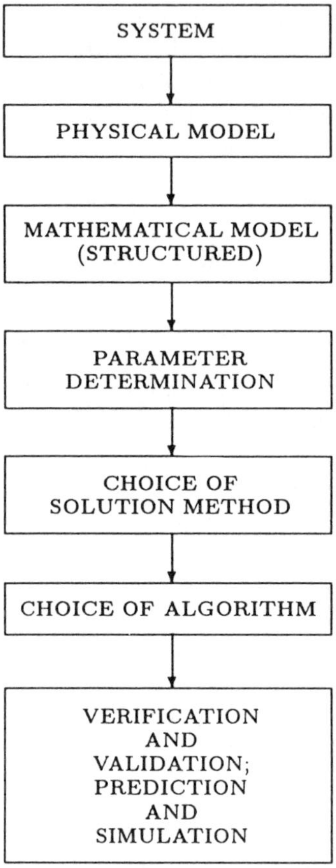

Fig. 1.3. System analysis, steps of modelling

of dissipative processes, which are often modelled as viscous damping due to the simplicity and convenience of this representation. In system analysis the models are generally structured; that means they are parametrical. Parameter determination uses drawings and material lists and takes into account the type of connections. Then the solution method and the algorithms applied are chosen, and with predictions and simulations the model has to be verified and validated.

Modelling itself is a hierarchical process, which is briefly described in Fig. 1.4 and concerns the third block of Fig. 1.3. After the physical model of the system has been established, the construction of the mathematical model starts with the choice of the general class of model to be developed. This requires consideration of boundary conditions, the level of detail required for the intended application, and the number of input and output variables to be used. Then the model structure is defined, which means specifying the type of functional relationship among the variables, whether integral or differential equations, spatially discrete or continuous, etc. These relationships concern the equations of motion, the measurement

MODELLING PROCESS:

1. CHOICE OF MODEL CLASS (MODEL FRAME):
   FRAME DEFINITION: - BOUNDARIES OF SYSTEM
   - LEVEL OF DETAILS OF DESCRIPTION (GOAL, UNCERTAINTY)
   - NO. OF INPUT, OUTPUT VARIABLES (CONTINUOUS, DISCRETE)

2. DETERMINATION OF MODEL STRUCTURE:
   FUNCTIONAL RELATIONSHIP BETWEEN VARIABLES CONCERNING:
   - EQUATIONS OF MOTION
   - MEASURING (OBSERVATION)
   - RESTRICTIONS

3. DEFINING MODEL PARAMETERS

4. PARAMETER DETERMINATION

Fig. 1.4. Modelling as a hierarchical process

equations, and possibly some constraints. The next level in the modelling process is the definition and determination of the parameters required in the model structure which has been chosen. The result, the mathematical model of the system, should be physically interpretable.

The types of mathematical models are manifold. The mathematical model as a partial differential equation (PDE) for field descriptions with an infinite number of degrees-of-freedom (DOF) has the advantage that it can describe not-closed systems, such as the half space. Spatially discretized models, which lead to ordinary systems of differential equations (ODE) with a finite number of DOF, can only describe finite domains. Consequently, closed systems are not represented, except that special boundary conditions defining additional submodels are introduced. It is only rarely that analytical solutions exist. Consquently, numerical procedures have to be chosen. Additionally, not all systems can be modelled analytically. Therefore a priori discretization (spatial, temporal) will be performed. Spatially discretized models lead to ODE for single-DOF (SDOF) or multi-DOF (MDOF) models, which can be time-continuous or time-discrete. Each numerical handling implies a discretization. Therefore dealing with discretized models leads to no loss of generality in practice. With principal (generalized) coordinates in mind, an MDOF model can be transformed (modal transformation) under particular pre-assumptions into a set of SDOF models. In the following it is mainly spatially discretized models that will be considered.

The classical equations of motion (number n) are formulated as ODE of second order, while in the state space domain the equations are of first order (but with $2n$ equations). Differential equations of motion can be

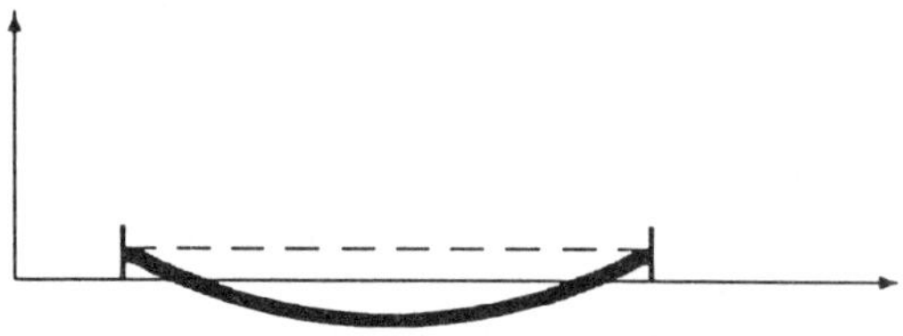

Fig. 1.5. Incomplete mode shape description using two nodes

expressed as integral-differential equations (IDE) or as integral equations (IE). Sometimes the latter formulations are advantageous for numerical reasons [9].

Models should be as simple as possible and as complex as necessary. Some practical suggestions on the procedure of modelling are discussed in the following.

- The order of a spatially discrete dynamic model must be chosen taking into account the types of external forces (e.g., duration, frequency content), and the dynamic responses to be represented. If the dynamic response is thought of as decomposed into eigenmodes, the model must reflect the necessary number of effective modes with their shapes, nodal points or lines, and curvatures, the latter, for example, for a beam in bending is related to the local bending moment, which is dominated by certain modes, while other modes have little effect on the stress-related failures of the beam.
- Within spatial discretization, the nodes and coordinates have to be chosen
  - from a static point of view (e.g. with respect to stress peaks)
  - from a dynamic point of view (band-limited model $\leadsto$ limited number of DOF)
  - taking into consideration the description of deflections and curvatures (see Fig. 1.5)
  - so that nodes of the theoretical analysis coincide with measuring points (in order to avoid later interpolation and to minimize expenditure)
  - so that numerical requirements are fulfilled (e.g. for integration with regard to discontinuities).
- Inertia forces and related distributions can generally be calculated without any difficulty by using the available routines. The spatially discretized results should be consistent with the restoring force model. For example, in finite element models the inertia and stiffness matrices should be consistent (based on the same displacement field). A common error is to use a lumped-mass matrix together with a finite element stiffness matrix.
- Modelling of dissipative forces is critical. Only tests can provide a realistic description. Fittings and joints are common sources of friction processes. For small displacements the connections often can be modelled as rigid from a dynamic point of view. For slightly larger displace-

ment amplitudes the friction can sometimes be linearized and modelled with viscous and other types of damping representations[3].

- Modelling of external forces has to take into account the force direction which influences the physical model (type of stressing of the system elements, e.g. as a plate or a disk).
- The choice of coordinates of in the equations of motion defines the coupling of the equations. Strong coupling increases the computational expenditure and can introduce numerical difficulties (counter-measure: introduction of principal coordinates for diagonalizing the matrices).
- Model inaccuracies (structure, parameters) restrict the accuracy of the prediction. Consequently, one should determine the number of significant digits which correspond to the given accuracy of the model.
- Submodelling[4] is useful in many cases. However, it is difficult to obtain the correct boundary conditions; often a coarse total model is needed in order to find out valid submodel boundary conditions.

Models of linear systems are distinguished by various frequency ranges. The low frequency range is characterized by a modal type response, i.e. the dynamic response is determined by the first eigenmodes of the system. In the intermediate frequency range the expansion of the dynamic response requires a very large number of modes. It is therefore charactereized by a high modal density. Asymptotic methods can be used in the high frequency range, and transfer functions are often used for the mathematical description. The external excitation determines the frequency range to be considered. When the excitation is limited to a specific frequency interval a band-limited mathematical model can be used. The actual frequency intervals corresponding to these ranges are system specific. For example, the high frequency range of eigenvibrations of a radar tower is generally below the low frequency range for the vibrations of a diesel engine.

In addition, propagating and standing waves have to be distinguished, dependent on whether a finite wave propagation velocity is important in the (damage) process under consideration. The type of model must be chosen to fit the purpose. For example, the d'Alembert solution represents propagating waves, while the Bernoulli solution can only describe standing waves (except in special cases).

Non-linear behaviour is much more difficult to model than linear behaviour, and it is sometimes important in damage processes. The model structure in general is unknown. We will restrict ourselves to linear and linearized models. However, piecewise linearization around the working point results in a model that cannot be used to describe any non-linear behaviour.

---

[3] Sometimes, friction is used for making rigid joints, i.e. high-strength bolts.

[4] Submodelling refers to the decomposition of the system into subsystems. The subsystems may be macro-elements, elements etc.

	SUBJECT AREA	EXCITATION INPUT	STATE OF ART: SYSTEM -CHARACTERISTICS -LIMIT STATES	RESPONSE OUTPUT
DIRECT PROBLEM	STRUCTURAL ANALYSIS	GIVEN (STATIC)	GIVEN OR ASSUMED	TO BE FOUND
	STRUCTURAL DYNAMICS	GIVEN (DYNAMIC)	GIVEN OR ASSUMED	TO BE FOUND
	RANDOM VIBRATION	STATISTICS AVAILABLE	GIVEN OR ASSUMED (DETERMINISTIC OR STOCHASTIC)	STATISTICS TO BE FOUND
INDIRECT PROBLEM	STRUCTURAL DESIGN	SPECIFIED	CONFIGURATION TO BE FOUND	LIMITS SPECIFIED
	RELIABILITY-BASED DESIGN	STATISTICS GIVEN	CONFIGURATION TO BE FOUND	LIMITS (PROBABILITY) SPECIFIED
	SYSTEM IDENTIFICATION	RECORD (DATA) AVAILABLE NOT AVAILABLE	CHARACTERISTICS TO BE ESTIMATED	RECORD (DATA) PARTIALLY AVAILABLE
AIM	DAMAGE ASSESSMENT SAFETY EVALUATION	RECORD (DATA) AVAILABLE	RESULTS OF INSPECTION AVAILABLE (DAMAGE OR LIMIT STATES TO BE ESTIMATED)	RECORD (DATA) PARTIALLY AVAILABLE

Fig. 1.6. Aim, overview and classification of modelling

The dynamics of linear time-invariant models ($\theta = \theta_i = const.$) can be represented by spatially discrete models in the standard form, suppressing the indices with respect to life time θ,

$$\mathbf{M}\ddot{\mathbf{u}}(t) + \mathbf{B}\dot{\mathbf{u}}(t) + \mathbf{K}\mathbf{u}(t) = \mathbf{N}\mathbf{f}(t), \tag{1.3.1}$$

where $\mathbf{u}(t)$ is the displacement vector and $\mathbf{f}(t)$ is the input vector. The quadratic inertia, damping and stiffness matrices $\mathbf{M}$, $\mathbf{B}$, $\mathbf{K}$ are constant on the time scale of the dynamics of the system, but may evolve during the lifetime of the system, representing the damage and ageing processes. $\mathbf{N}$ is a rectangular input matrix specifying the components (spatial coordinates) upon which non-zero external forces act.

The model quality depends on the knowledge and experience possessed by the system analyst, both of the system to be modelled and of the related process.

The aims of system analysis and the corresponding quantities are described in Fig. 1.6 based on [10]. Input/output relations are classified for various subject areas of application. Direct problems are distinguished from indirect or inverse problems. In direct problems the model and the input quantities are given, and the dynamic response is sought. In inverse problems the output and input quantities are given and the model is desired. The latter is called system identification or the design problem. The input problem, which is not indicated in this figure, is the inverse problem of finding the input which produces a specified output when parts of the model are known. Diagnosis is classified as an inverse problem with emphasis on the identification of system modifications.

An important difference between static and dynamic models should be pointed out, because it plays a very large role in diagnostics. Static models should, in general, predict stress distributions including the maxima. Therefore these models are highly detailed, and if a finite element (FE) model is used it consists of many static DOF (typically, several orders of magnitude higher than that of dynamic models). Dynamic models are usually based on energy considerations, which involves integral, scalar quantities. In consequence, such a dynamic model is generally unable to represent stress peaks, which can be decisive in diagnostics.

The resulting model is, as already mentioned, uncertain with regard to its structure and its parameters. This means that possible errors are unknown in location and quantization. Each model must therefore be verified and validated [11]. Verification means the reconstruction of the used data within the applied quality criterion (e.g within the least squares sense). Validation concerns the homomorphy between system and model with respect to the investigated behaviour. Verification, for example, can be based on the process data[5], while validation includes the model structure and the parameters. Verification and validation can be done using measurements, which reduces the uncertainties and results in a mathematical model with sufficiently small errors: TACAM.

1.3.2 Modelling of Uncertainties

The diagnosis of faults must be performed in the face of systematic errors and uncertainties. The latter may occur in the domains of

1. model structure
2. model parameters
3. the type of failures which can occur
4. fault evolution, and
5. measurements.

The analyst usually has prior information which characterizes these uncertainties. Model-based formulation and evaluation by a diagnostic algorithm requires explicit mathematical representation of these uncertainties. In other words, mathematical modelling of the uncertainties is as important as the modelling of the system dynamics.

Three fundamentally different approaches can be used in the modelling of uncertainty. The choice of an approach depends primarily on the nature and extent of the available information.

Probabilistic models. Point processes, stochastic processes, and time series fall into this category. Here many references are available, such as [5, 12, 13]. The realization of a probabilistic model typically requires a substantial

[5] The process is defined only by the input and output quantities.

amount of either theoretical or empirical information, i.e. knowledge of the stationarity or ergodicity of the process, knowledge of the probability density functions, etc. Careless use of such models can lead to erroneous results, especially in the diagnosis of rare events.

However, probabilistic methods, stochastical procedures and estimators are well established, well-known within the engineering community and accepted, although approximations are used and assumptions often have to be made which cannot be proven easily. Several probabilistic models for decision-making will be presented in Chapt. 5.

Fuzzy-set models. Since the theory of fuzzy sets was presented by Zadeh (1965) [14] it has been well developed and applied to solve many practical problems (e.g. [15–17]). In particular, the application of fuzzy sets is useful in cases where explicit computation cannot be easily made and statistical data are insufficient for the formulation of probabilistic models. Mathematical fuzziness means multivaluedness or multivalence. It is expressed by the membership functions $m_A : X \longrightarrow [0, 1]$. With $x \epsilon X$ the membership value measures the elementhood (or degree) to which element x belongs to set A. Although such methods have been investigated in diagnostics recently (e.g. see [18, 19]) more work needs to be done in this direction. Several promising approaches using fuzzy logic will be presented and discussed in Chap. 5.

Convex models. Another non-probabilistic methodology for representing uncertainty is based on convex models. A convex model is a set of functions with chosen properties, so that each function represents an allowed realization of an unknown or uncertain event such as a specific failure or an uncertain input. The analyst's prior information on the range of variation of the uncertain event is invested in specifying the structure of the set. In other words, the structure of a set specifies the properties shared by the functions belonging to the set.

Convex models are usually reserved for those situations where uncertainties are characterized by only fragmentary information. An extensive discussion of convex modelling is found in [20]. Three convex models are briefly described in order to introduce this method.

The simplest is the uniform-bound convex model: let $\eta(t)$ represent the element of the set to be considered (e.g. amplitude, geometrical imperfection), then it is defined as

$$\mathcal{F}_{\mathrm{UB}} = \{\eta(t) : \ |\eta(t)| \leq \widehat{\eta}\} \tag{1.3.2}$$

The quantity $\widehat{\eta}$ is the defined bound. The convex model $\mathcal{F}_{\mathrm{UB}}$ is the set of all $\eta(t)$ not exceeding $\widehat{\eta}$.

In a more detailed analysis the bound $\widehat{\eta}$ may vary in a range from φ_1 to φ_2. Then an envelope-function, $\widehat{\eta}(\varphi, z)$ could be defined:

$$\widehat{\eta}(t) = \begin{cases} 0 & \varphi \notin [\varphi_1, \varphi_2] \\ \overline{\eta} & \varphi \in [\varphi_1, \varphi_2] \end{cases} \tag{1.3.3}$$

where $\overline{\eta}$ is a constant. This is the envelope-bound convex model:

$$\mathcal{F}_{\text{EB}} = \{\eta(\varphi, z) : \; |\eta(\varphi, z)| \leq \widehat{\eta}(\varphi, z)\} . \tag{1.3.4}$$

A brief discussion of the convex-model representation of uncertainties in the input function, $f(t)$ in Eq.(1.3.1), will conclude this discussion. Consider, for example, uncertain ground motion during an earthquake. Many convex models are available for this purpose [21]. The integral-energy-bound model will now be introduced. The input function, $f(t)$, deviates from a nominal input vector, $f^0(t)$, where the time-integral of the energy of the deviation is bounded:

$$\mathcal{F}_{\text{IEB}} = \left\{ f(t) : \int_0^t \left[f(\tau) - f^0(\tau)\right]^T \left[f(\tau) - f^0(\tau)\right] \, d\tau \leq \rho^2(t) \right\} . \tag{1.3.5}$$

This set represents the uncertainty in the realization of the input. It includes strongly transient functions, since the integral is bounded, but not the instantaneous magnitude of the functions. Furthermore, one can constrain the uncertain inputs to a certain time interval by choosing the energy-bound $\rho^2(t)$, so that it vanishes outside this interval. In practical applications, the determination of $\bar{\eta}$ and $\rho^2(t)$ is an important consideration.

This brief discussion of convex models may be summarized by explaining how convexity comes into play.

One factor contributing to the tendency towards convexity of uncertainty-sets is the dearth of information about the uncertain phenomenon. When only very limited information about the phenomenon is available, such as envelope-bounds or sparse spectral information, and we ask for the set of *all* the functions consistent with this information, the result is that such sets are often convex. An additional cause of convex uncertainty-sets is discussed in Sect. 2.1 of [20].

Convex models are only mentioned as an interesting tool. However, they are not used in the following. They are intended to provide stimulation for further research and possible application.

1.3.3 Holistic Modelling

In addition to the general modelling principles, which take the existing uncertainties into account, model adjustment at life time θ_i, $\mathcal{M}(\theta_i)$, with time-invariant models (1.3.1) is fundamental for model-based diagnosis. However, the background is the holistic model, which will be summarized briefly (Fig. 1.7).

L may be the model operator defined in an appropriate domain. It is applied to the total dynamic response $x(r, \theta, t)$, defined in a corresponding domain, with r the spatial coordinate, θ the life time (slow time), and t the time of dynamics (fast time). The equation of motion then follows as

$$L[x](r, \theta, t) = f(r, \theta, t), \tag{1.3.6}$$

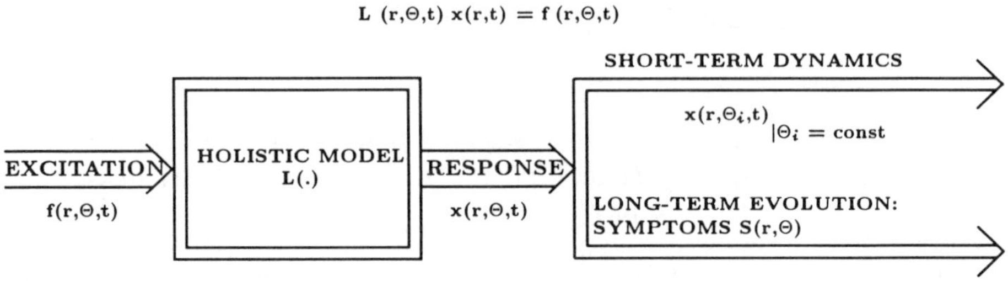

Fig. 1.7. Holistic model and its responses

where $f(.)$ designates the external force.

As expressed above, the short-term dynamic response can be considered for $\theta = \theta_i = const.$, resulting in $x(r, \theta_i, t) =: x_i(r, t)$ as the current (with respect to θ) dynamic response. Symptoms are sensitive quantities for the detection and observation of system modifications and their evolution. Symptoms can be performed in the deterministic case by applying the operator ϕ to the total dynamic response, and in the case of stochastic models with the expectation operator $E_t\{.\}$, applied to a sample dependent on t:

$$S(r, \theta) = \begin{cases} \phi[x(r, \theta, t)] \text{ in the deterministic case,} \\ E_t\{\phi[x(r, \theta, t)]\} \text{ in the case of statistical models.} \end{cases} \tag{1.3.7}$$

1.4 Summary

The concept of model-supported diagnosis is mainly based on adjusted mathematical models. The adjustment is performed to various state conditions given by measurements. Such a verified and validated mathematical model with sufficiently small errors is the best available knowledge base to the system. System identification methods thus have to be applied. The following are essential: prior knowledge of system analysis, and the introduction of the slow-time coordinate (life time) and the fast-time coordinate, the latter used for the description of the classical dynamics of the system. The distinction between the evolution and the dynamics of the system permits the modelling of the system at various discrete life times through time-invariant models, which means with constant coefficients of the ODEs within a limited interval of the related discrete life time.

CHAPTER 2

Review of Tools and Concepts of Diagnosis

Symptom-based diagnosis is described in [22–24]. Symptoms are sensitive quantities with respect to a fault, and they are used for fault detection and also for diagnostic purposes. Classical diagnostic methods are signal-based and signature-supported. Simple signature analysis employs data-reduction methods, for example counting the number of peaks of a spectral function. In general, signature analysis involves scalar information. Multiple features are applied independently. Extended signature analysis uses vectorial information and patterns. A set of features will be evaluated and used for decision and assessment.

In the following, damage and faults will be described with consideration of their effects on the models used (parameters, structures). The resulting symptoms will be derived and then discussed. Finally, damage initiation and evolution in operating systems are considered.

2.1 Damage, Faults and their Descriptions

The conceptions of damage and faults will be introduced and discussed. Evolving faults lead to system modifications, and consequently to model modifications. These model modifications and their effects on the dynamic variables are investigated in detail, which is fundamental in diagnosis.

2.1.1 General Discussion

Damage is the result of a defect, failure or fault. These words will be used more or less equivalently[1]. Damage is (life) time-dependent; it may concern the total system or parts of it. Additionally, one has to distinguish between repairable and non-repairable faults. What is damage? It is a reduction of the ability to fulfil a pre-specified operation/function. This definition includes safety requirements as well as comfort conditions.

The various types of physically elementary faults and defects are

[1] However, defined differences do exist:

- defect: any absence of a characteristic of an item from the requirements,
- failure: termination of the ability of an item to perform a required function. After failure the item has a fault.
- fault: the inability of an item to perform a required function,
- damage: injury or harm which reduces the value or serviceability.

- holes or voids,
- cracks,
- fractures,
- surface discontinuities,
- mass reduction by abrasion and (all types of) erosion,
- mass increase (e.g. by pollution),
- material ageing etc.

Various causes of faults are

- the lack and/or reduction of strength due to
  - external forces including unexpected forces, e.g. century earthquake
  - environmental forces including, for example, chemical effects, and/or
  - human errors with respect to design, detail construction and construction work (→ initial failures)
- defects due to operation (e.g. wear, abrasion)
- sedimentation, and
- vandalism.

Many causes not explicitly enumerated are contained in the above list, such as resonance, self-excitation (as flutter), improper handling due to partial ability (e.g. colour-blindness), accidents etc. It should be noted that a fault can cause a feedback effect on the input quantity (→ interaction problems). In addition, one has to take into account special effects, such as the time-dependent domino effect. For example, if one essential element in a truss fails, then this may lead to the failure of other members. Another example is an earthquake, which not only causes buildings to collapse due to its wave propagation from the epicentre, but which also may cause electrical installations and gas pipes to fail and result in fire and subsequent explosions.

The forcing of the system which induces stresses is essential here. Limiting stresses and stress concentrations can cause damage. In consequence, a detailed analysis (e.g. FE calculations, multi-body (MB) models, fracture analysis) leads to an assessment of the resulting stresses. Satisfying the design criteria concerning the strength and strain properties, structural elements, and the total system with respect to external forces and to force paths will protect the system to a great extent against failures from design loading.

The above-mentioned faults will modify the initial mathematical model describing the undamaged system. The following models of system faults will be considered:

- model parameter modifications including changes of boundary conditions[2];
- model structure modifications, such as the introduction of additional DOF, or of non-linearities.

---

[2] Continuous models include the boundary conditions in the Green's functions [9] and discrete models in the flexibility influence matrix.

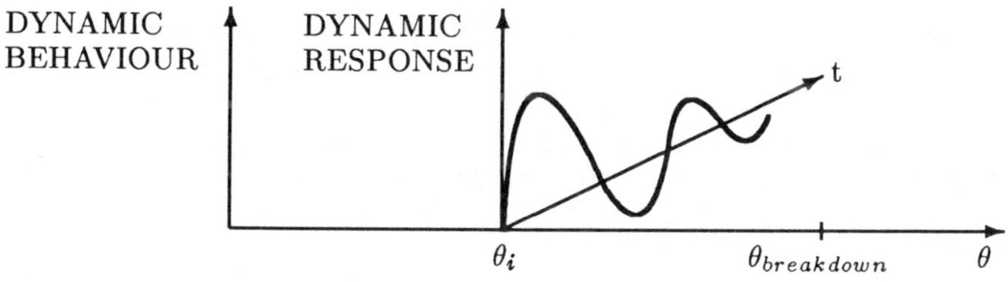

Fig. 2.1. Slow (θ) and fast (t) time coordinates

A crack, for example, which opens and closes during a cyclic motion leads to non-linear behaviour: both sides of the crack may come into contact during motion, inducing impacts and friction. The crack results in additional DOF in the related FE-model.

These system modifications will result in modified energies, eigenquantities, stress and acoustic levels etc. compared with, for instance, the non-modified system. This can be seen, for example, by looking at modified state conditions and by comparing the state vector at the life time θ_{i+1}, $\mathbf{x}^T(t, \theta_{i+1}) = \{\mathbf{u}^T(t, \theta_{i+1}), \dot{\mathbf{u}}^T(t, \theta_{i+1})\}^T$, with the state vector determined at time θ_i, $\mathbf{x}(t, \theta_i)^T = \{\mathbf{u}^T(t, \theta_i), \dot{\mathbf{u}}^T(t, \theta_i)\}^T$. A comparison of states or symptoms at times θ_{i+1} and θ_i is only rational if a significant (e.g. abrupt) change occurred during this time interval. Slowly varying changes must be detected by comparing the corresponding quantities with those of a chosen reference model describing a system condition much younger than $\theta_{i+1} - \theta_i$. The coordinate θ for the life time is often called the slow time coordinate, compared with the fast time coordinate t (see Figure 2.1).

Here the term 'comparison' is chosen instead of the term 'correlation', because the transformation into the parity domain[3] can be done deterministically as usual.

In the following only such faults will be considered which affect the mathematical model as follows:

1. Parameter modifications which also include modifications of boundary conditions by
   - stiffness reduction
   - inertia increase/decrease,
   - damping modification (often an increase; however, pre-stressed concrete behaves in a very complicated way [25]).

2. Model structure modifications by
   - an increase of the number of DOF within a finite frequency interval (see the example in [26]),
   - modification of the type of damping,
   - the change from linear to non-linear behaviour, generally connected with additional forcings, e.g. impacts occuring through the closing of cracks.

[3] Generation of residuals.

In this way, damage and faults are defined in the parameter domain and in the model structure domain. However, detected parameter changes must be (re-)transformed into the fault domain in order to determine the type of damage, and to distinguish between different faults, which means knowing their relation to the real construction. This subject is discussed in Chap. 5. Figure 2.2 illustrates the three domains of faults.

In the (real and existing) system, faults may appear as mentioned in the Figure. These faults are mapped in the related model by parameter and/or model structure changes. Finally, the introduced diagnostic model is based on symptom measurements (monitoring), and it results in the condition assessment of the system.

2.1.2 Parameter Modifications: Preliminary Examples

The effect of failure-related parameter modifications on the model will be studied by sensitivity investigations using linearized differential error analysis, and non-linearized error analysis by including higher order terms in the corresponding Taylor series, or by recalculations. The results of the sensitivity analysis are amplification factors and functions which indicate the extent to which modified physical parameters, for example stiffness coefficients, influence dynamic variables, for example the dynamic response of the system. Therefore sensitivity analysis can also be used to reduce the number of parameters to be considered in modification investigations by neglecting those of less importance. Additionally, the quantities of sensitivity analysis are used in the formulation of the system identification procedures (model adjustment to different states) discussed in Chap. 4. They are also included in the learning phase of diagnosis.

The following examples illustrate the sensitivity of model quantities due to parameter modifications. The next subsection contains a more general perspective.

Eigenmode sensitivity due to stiffness modifications along a beam. It is easy to derive linearized sensitivity quantities for a uniform cantilever, especially for the eigenvalues with respect to stiffness modifications ΔB along the entire length l_0. The normalized span coordinate is designated by $\xi = x/l_0$. We are now interested in normal mode modifications with respect to a bending stiffness modification. Asymptotically, the normal mode of order k can be expressed analytically (time independent) by [9]

$$\hat{w}(\xi) = A_k \sin \lambda_k \xi, \tag{2.1.1}$$

where A_k is an integration constant, $\lambda_k^4 = \omega_k^2 \mu_0 l_0^4 / B_0$ the eigenvalue with the eigenfrequency ω_k, μ_0 the mass per unit length and B_0 the bending stiffness of the nominal beam. The variation of the modal displacement due to a bending stiffness modification ΔB thus follows:

$$\Delta\hat{w}_k(\xi) \doteq \frac{dw_k}{dB_0}\Delta B \doteq -\,[A_k \lambda_k \xi \Delta B \cos(\lambda_k \xi)]/(4B_0). \tag{2.1.2}$$

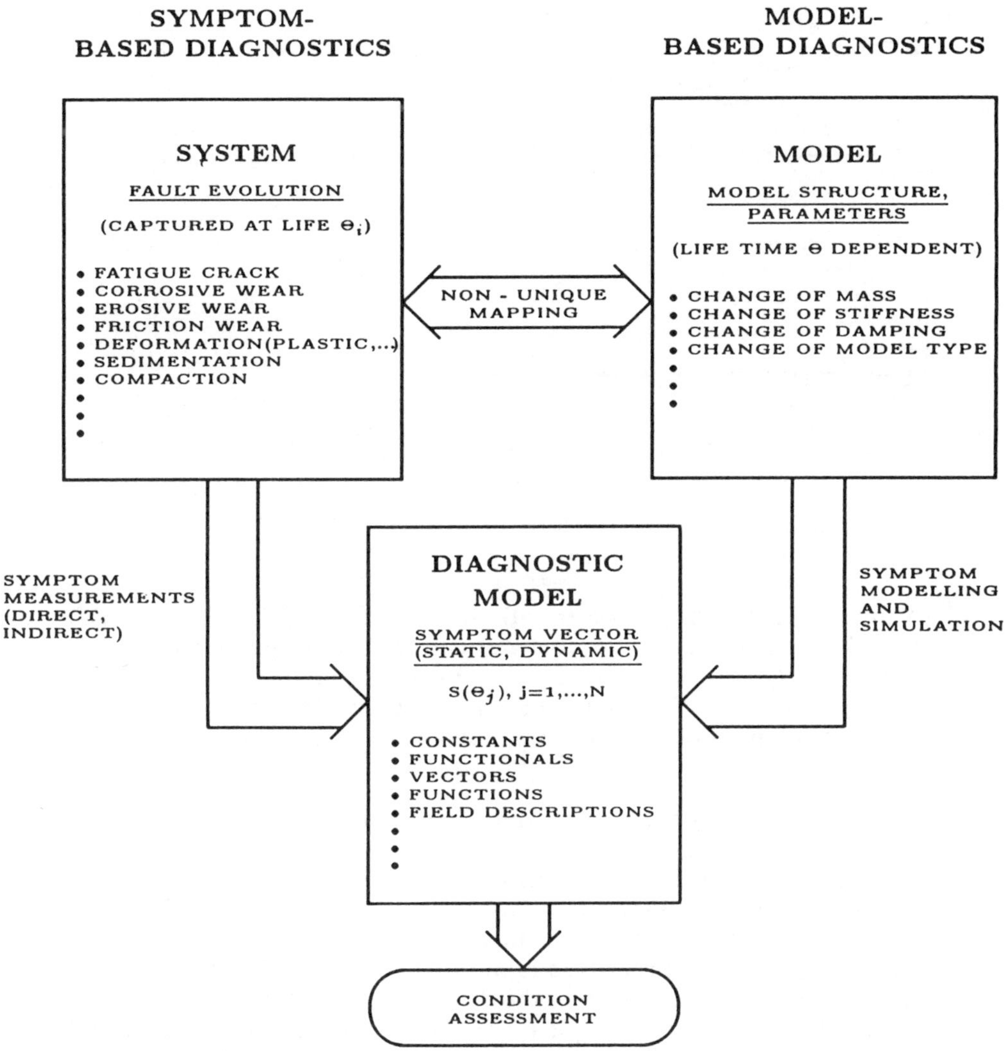

Fig. 2.2. The domains of faults: the faults are mapped differently in diagnostics

The modification of the normal mode is asymptotically proportional to
- the eigenvalue λ_k (increasing sensitivity with increasing k)
- the related coordinate ξ
- the ratio $\Delta B/B_0$.

The maximum modification of $\hat{w}_k(\xi)$ along ξ is given by $\Delta\hat{w}'_k(\xi) = 0$, where the prime denotes differentiation with respect to ξ. Comparing Eqs. (2.1.1) and (2.1.2) reveals that $\Delta\hat{w}_k(\xi)$ is proportional to $\Delta\hat{w}'_k(\xi)$, $\Delta\hat{w}_k(\xi) \sim \hat{w}'_k(\xi)$. Consequently, $\Delta\hat{w}'_k(\xi) \sim \Delta\hat{w}_k''(\xi) = 0$: the maximum modification of $\hat{w}_k(\xi)$ is asymptotically connected with the bending moment (the second

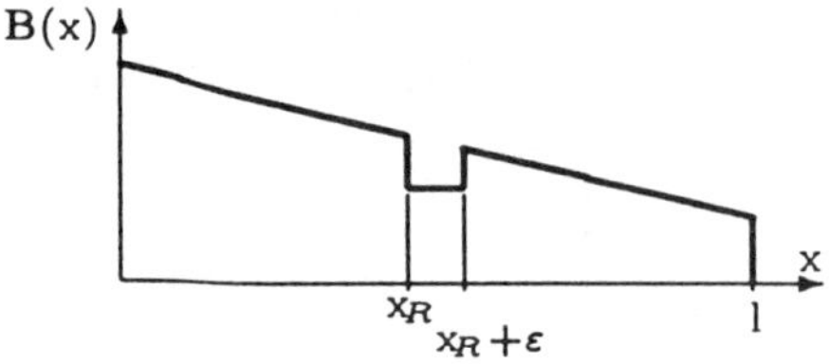

Fig. 2.3. Local stiffness reduction of a beam

Fig. 2.4. One-sided elastically supported beam

derivative) equal to zero, however with the shear force $\hat{Q}_k(\xi) \sim \hat{w}_k'''(\xi) \neq 0$. It also follows that $\max_\xi \hat{w}_k'(\xi)$ does **not** correspond to the maximum amplitude $\hat{w}_k(\xi)$ $(\hat{w}_k'(\xi) = 0)$. We obtain not only quantitative information about the sensitivity of potential quantities (to be measured), but also an insight into the physical relationships involved.

Local stiffness modifications of a beam. The local stiffness modification of a beam is illustrated in Fig. 2.3.

The effect of this stiffness reduction can be modelled in different ways. The beam can be partitioned into two parts at the point of stiffness reduction and connected with a torsional spring with an equivalent spring constant [27]. Another procedure is the integration or summation of the flexibilities under consideration of the discontinuities at the corresponding reduced stiffness interval. The use of FE models (MDOF) is also practical.

If static unit point loads are taken, the optimum location for maximum deflection is the free end of the beam. Loading the free end is the best choice for detecting the presence of a local stiffness reduction. However, a single load is insufficient for localizing this reduction and assessing its severity. One may employ hierarchical adaptive procedures for seeking the location and degree of stiffness reduction, for example as described in [28].

Boundary condition modification of a one-sided elastically supported beam. The Green's function (flexibility influence function) of the Bernoulli beam shown in Fig. 2.4 is

$$G(x, \xi) = G_{BC}(x, \xi) + \int_0^{min(x,\xi)} \frac{(x-\eta)(\xi-\eta)}{B(\eta)} d\eta. \tag{2.1.3}$$

$B(x)$ is the stiffness distribution. It is the deflection along the elastic axis[4] at point x due to a vertical force at point ξ. G_{BC} describes the influence function with respect to the boundary conditions [9]:

$$G_{BC} = \frac{x\xi}{k_2} + \frac{1}{k_1}. \tag{2.1.4}$$

The differential modification of G_{BC} for small changes in the spring coefficients is

$$\Delta G_{BC}(x, \xi) = -\frac{x\xi}{k_2}\frac{\Delta k_2}{k_2} - \frac{1}{k_1}\frac{\Delta k_1}{k_1}. \tag{2.1.5}$$

For a rotational stiffness modification, Δk_2, a unit force at ξ produces a deflection at x which is proportional to Δk_2 and to the product $x\xi$. On the other hand, a translatory stiffness modification, Δk_1, produces a deflection which is independent of its position.

Reaction forces of a beam supported at both ends. Shear forces of a Bernoulli beam are proportional to the third spatial derivative of the displacement[5]. Consequently, if the stiffness reduction is modelled by an equivalent spring, then this spring will produce a discontinuity of the shear force distribution. Additionally, it will show the location of discontinuity very clearly. So the shear force distribution can serve as a symptom for the localization of such a defect. The shear forces will directly affect the reaction forces of a beam supported at its ends. Following the procedure of looking for the shear forces of a bending beam as a sensitive feature (symptom), we can then consider the reaction forces at the boundaries [29] as sensitive quantities for boundary condition modifications.

It can be concluded from the relationship between natural modes and boundary conditions, as described in [30], that only the eigenquantities of low-order modes are sensitive to boundary influences. This statement is used in the first example of this subsection, in [26] and in [31].

Eigenfrequency shifts for various structural modifications. Figures 2.5 and 2.6 show examples of eigenfrequency shifts.

Figure 2.5 shows the percentage eigenfrequency changes of the first five modes of a simply supported beam with the support at two different locations and with location distances a and a'. It can be noted that some shifts are positive and some are negative, and that the shifts are rather large.

Figure 2.6 shows a cantilevered beam which is supported at one end and additionally attached by a rotational spring, k_2. The related spring constant k_2 is taken as being parametrical. Fig. 2.6 gives the eigenfrequency shifts

[4] It is defined as the axis where vertical forces induce only bending moments.

[5] It is known that the differentiation with respect to the spatial coordinate can be substituted by integration, taking the equation of motion into consideration [9].

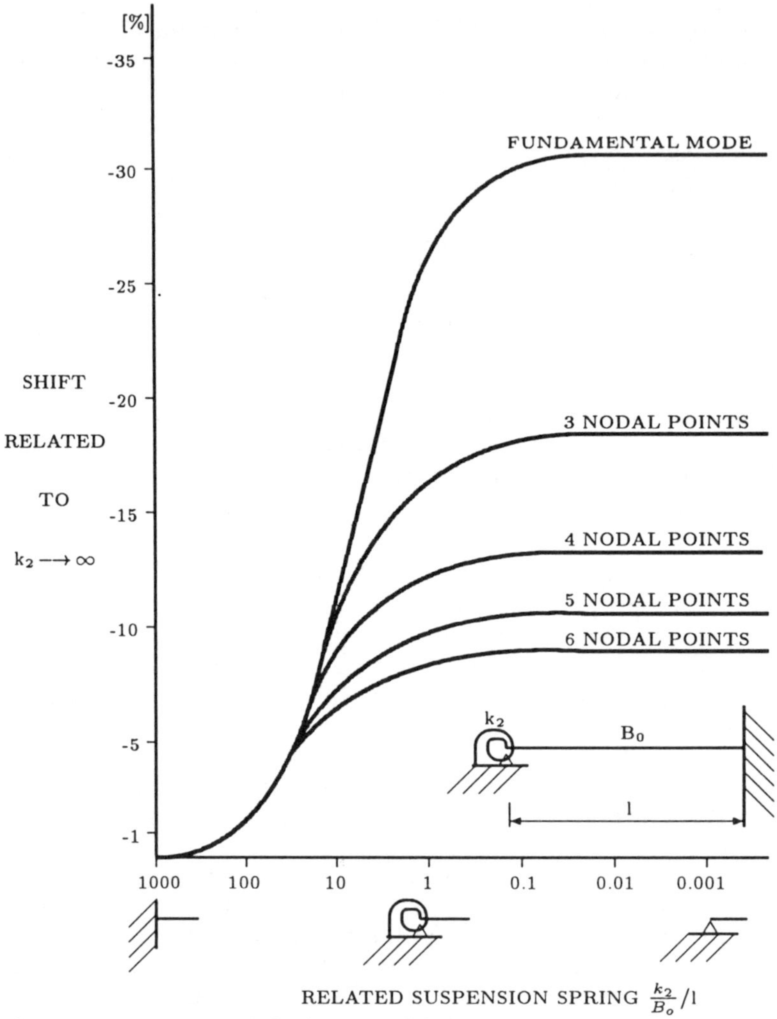

Fig. 2.5. Eigenfrequency shifts due to modified support distances

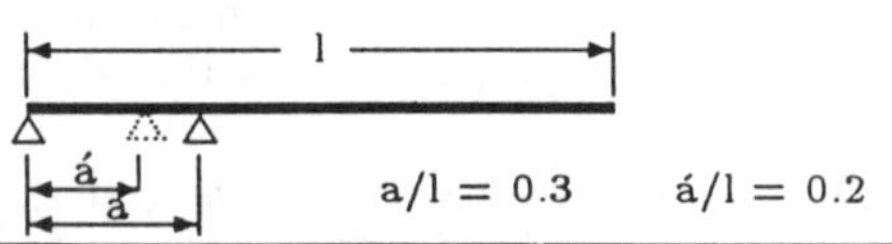

Δf_1	Δf_2	Δf_3	Δf_4	Δf_5
-23%	-21%	-15%	+15%	+16%

Fig. 2.6. Eigenfrequency shifts in percentages of a beam with bending conditions as indicated

relative to the eigenfrequency related to $k_2 \to \infty$. The shifts for the five modes shown are relatively large. The curves show that the percentage changes of the eigenfrequencies of the lower modes are greater.

Eigenfrequencies are not always as sensitive as indicated in these examples. This can be seen by taking the Rayleigh quotient and by modifying the stiffness and/or the mass distribution locally. The displacements expressed by the corresponding eigenfunctions/vectors, which will not change greatly, or expressed by admissible functions/vectors (taken from the unmodified model), yield slightly modified generalized (integral) stiffnesses and masses.

2.1.3 Parameter Modifications: General Analysis

Several general problems are now discussed that arise from the sensitivity of modal quantities and dynamic responses due to model parameter modifications. Finally, subsystem modelling will be applied.

Modal sensitivity. Here the eigenvalue problem of the associated undamped system will be considered. The inertia changes are denoted by $\Delta\mathbf{M}$ and the stiffness changes by $\Delta\mathbf{K}$. $\mathbf{M}$ and $\mathbf{K}$ are the unmodified inertia and stiffness matrices. $\hat{\mathbf{U}}_0$ is the corresponding modal matrix with column vectors equal to the eigenvectors $\hat{\mathbf{u}}_{0i}$, and $\mathbf{\Lambda}_0$ the diagonal matrix of eigenvalues λ_{0i}. The corresponding changes of the modal quantities are denoted by $\Delta\hat{\mathbf{U}}$ and $\Delta\mathbf{\Lambda}$. Then the eigenvalue problem of the modified system reads

$$-(\mathbf{M} + \Delta\mathbf{M})(\hat{\mathbf{U}}_0 + \Delta\hat{\mathbf{U}})(\mathbf{\Lambda}_0 + \Delta\mathbf{\Lambda}) + (\mathbf{K} + \Delta\mathbf{K})(\hat{\mathbf{U}}_0 + \Delta\hat{\mathbf{U}}) = \mathbf{0}. \quad (2.1.6)$$

Now the generalized mass matrix normalized as

$$\mathbf{M}_g := \hat{\mathbf{U}}_0^T\mathbf{M}\hat{\mathbf{U}}_0 = \mathrm{diag}(m_{gi}) = \mathbf{I} \quad (2.1.7)$$

is introduced, where $\mathbf{I}$ is the unity matrix, and the matrix of generalized stiffnesses which follows with (2.1.7) from the matrix eigenvalue problem of the unchanged undamped model as

$$\mathbf{K}_g := \hat{\mathbf{U}}_0^T\mathbf{K}\hat{\mathbf{U}}_0 = \mathrm{diag}(k_{gi}) = \mathbf{\Lambda}_0. \quad (2.1.8)$$

The eigenvectors are linearly independent and form a basis set, so that the eigenvector changes $\Delta\hat{\mathbf{U}}$ can be expressed as a linear combination of the nominal eigenvectors with a matrix $\mathbf{A}$:

$$\Delta\hat{\mathbf{U}} := \hat{\mathbf{U}}_0\mathbf{A}. \quad (2.1.9)$$

With the decomposed eigenvector modifications with respect to the unmodified normal modes (2.1.9) the linearized Eq. (2.1.6) yields

$$\Delta\mathbf{\Lambda} \doteq \hat{\mathbf{U}}_0^T\Delta\mathbf{K}\hat{\mathbf{U}}_0 + \mathbf{\Lambda}_0\mathbf{A} - (\mathbf{A} + \hat{\mathbf{U}}_0^T\Delta\mathbf{M}\hat{\mathbf{U}}_0)\mathbf{\Lambda}_0. \quad (2.1.10)$$

The introduction of the generalized matrices of the parameter modifications,

$$\Delta \mathbf{K}_g := \hat{\mathbf{U}}_0^T \Delta \mathbf{K} \hat{\mathbf{U}}_0 =: [\Delta k_{gik}], \tag{2.1.11}$$

$$\Delta \mathbf{M}_g := \hat{\mathbf{U}}_0^T \Delta \mathbf{M} \hat{\mathbf{U}}_0 =: [\Delta m_{gik}], \tag{2.1.12}$$

Equation (2.1.10) leads (the main diagonal elements of $\mathbf{\Lambda}_0 \mathbf{A}$ and $\mathbf{A}\mathbf{\Lambda}_0$ are equal) to:

$$\Delta\lambda_{0i} \doteq \Delta k_{gii} - \lambda_{0i} \Delta m_{gii}. \tag{2.1.13}$$

Equation (2.1.13) indicates that, to first order parameter modifications, the eigenvalue changes are independent of the eigenvector changes, and depend only on the nominal modal quantities and on the parameter modifications.

Taking into account that $\mathbf{\Lambda}$ is a diagonal matrix, the non-diagonal elements of Eq. (2.1.10) result for distinct eigenvalues in

$$a_{ik} \doteq \frac{\Delta k_{gik} - \lambda_{0k} \Delta m_{gik}}{\lambda_{ok} - \lambda_{0i}}, \quad i \neq k, \tag{2.1.14}$$

and

$$a_{ii} \doteq \Delta m_{gii}/2, \tag{2.1.15}$$

while employing the generalized orthogonality properties (2.1.7) and (2.1.8). Therefore the modifications $\Delta\mathbf{\Lambda}$ and $\Delta\hat{\mathbf{U}}$ are expressed in a linearized way by the generalized stiffness and inertia modifications.

The effects of modified boundary conditions are included in the above formulations. However, an additive formulation with flexibility coefficients $\mathbf{G} = \mathbf{K}^{-1}$, if existent, is preferable, as can be seen from what has already been stated in the examples.

Sensitivity investigations with respect to initial conditions are discussed in [5]. Sensitivity analysis for the eigensolutions of the model of the damped system can be dealt with easily in the state space [5].

Sensitivity of the stiffness and the flexibility matrices with respect to modal quantity changes. The inverse problem of how, for example, the prior stiffness or the flexibility matrix are affected by modal quantity modifications, is considered as follows: the spectral decompositions of the flexibility and stiffness matrices as given in [5] read

$$\mathbf{G} = \sum_{i=1}^{n} \frac{1}{\lambda_{0i} m_{gi}} \hat{\mathbf{u}}_{0i} \hat{\mathbf{u}}_{0i}^T, \tag{2.1.16}$$

$$\mathbf{K} = \sum_{i=1}^{n} \lambda_{0i} m_{gi} \mathbf{M} \hat{\mathbf{u}}_{0i} \hat{\mathbf{u}}_{0i}^T \mathbf{M}, \quad \mathbf{M} = \mathbf{M}^T. \tag{2.1.17}$$

In the flexibility formulation the dyadic products are weighted with $1/(\lambda_{0i}m_{gi}) = 1/(\omega_{0i}^2 m_{gi}) = 1/k_{gi}$, which are decreasing factors with increasing index i and correspondingly numbered eigenvalues. The reciprocal statement holds true for the stiffness formulation. However, one has to pay attention to the normalization, and therefore to the variation, of m_{gi}. If the normalization, for instance, is changed compared with (2.1.7) to

$$\hat{\mathbf{u}}_{0i}^N := \frac{1}{|\sqrt{\lambda_{0i}m_{gi}}|}\hat{\mathbf{u}}_{0i} \tag{2.1.18}$$

it follows

$$\mathbf{G} = \sum_{i=1}^{n} \hat{\mathbf{u}}_{0i}^N \hat{\mathbf{u}}_{0i}^{NT} : \tag{2.1.19}$$

this is a sum of dyadic products explicitly without any weighting. Therefore a check must be made to determine whether the lower modes essentially determine **G**, while the higher modes determine **K**. However, if one looks at the decomposition of the kinetic energy of a free vibrating system approximated by a sufficient number of DOF, it is known that the higher terms are negligible. Therefore it seems to be preferable for local stiffness modifications to look for higher modes. And inversely, if higher modes are important one should choose the stiffness formulation. The corresponding statements concerning the lower modes hold true for the flexibility formulation.

The decomposition of the flexibility matrix of the modified system results from the generalized orthogonalization and leads to the reciprocal sensitivity formulation, as the parameter modifications depend on the eigenquantity changes:

$$(\mathbf{K} + \Delta\mathbf{K})^{-1} =: \mathbf{G} + \Delta\mathbf{G} = (\hat{\mathbf{U}}_0 + \Delta\hat{\mathbf{U}})(\mathbf{K}_g + \Delta\mathbf{K}_g)^{-1}(\hat{\mathbf{U}}_0 + \Delta\hat{\mathbf{U}})^T. \tag{2.1.20}$$

In Eq. (2.1.20) the modified generalized matrix of stiffnesses is no longer diagonal; in consequence, no explicit sum can be given. When the generalized stiffness matrix $\mathbf{K}_g = \mathbf{\Lambda}_0$ (see Eq. (2.1.8)) is substituted, it follows in a first order approximation that

$$(\mathbf{\Lambda}_0 + \Delta\mathbf{K}_g)^{-1} = [\mathbf{\Lambda}_0(\mathbf{I} + \mathbf{\Lambda}_0^{-1}\Delta\mathbf{K}_g)]^{-1} \doteq \mathbf{\Lambda}_0^{-1} - \mathbf{\Lambda}_0^{-1}\Delta\mathbf{K}_g\mathbf{\Lambda}_0^{-1}, \tag{2.1.21}$$

and

$$\begin{aligned} \mathbf{G} + \Delta\mathbf{G} &= (\hat{\mathbf{U}}_0 + \Delta\hat{\mathbf{U}})(\mathbf{\Lambda}_0^{-1} - \mathbf{\Lambda}_0^{-1}\Delta\mathbf{K}_g\mathbf{\Lambda}_0^{-1})(\hat{\mathbf{U}}_0 + \Delta\hat{\mathbf{U}})^T, \\ \Delta\mathbf{G} &= \Delta\mathbf{G}^T \doteq \Delta\hat{\mathbf{U}}\mathbf{\Lambda}_0^{-1}\hat{\mathbf{U}}_0^T + \hat{\mathbf{U}}_0\mathbf{\Lambda}_0^{-1}\Delta\hat{\mathbf{U}}^T. \end{aligned} \tag{2.1.22}$$

In experimental analysis Eq. (2.1.22) serves for the modification (design) problem.

The equation can be applied directly for the eigenquantities of hysteretically damped systems using a complex stiffness matrix $\mathbf{K} + j\mathbf{D}$.

Another treatment of the sensitivity of parameter matrices with respect to modal quantity changes uses the modified modal quantities

$$\left.\begin{array}{lcl}\hat{\mathbf{U}}_m & := & \hat{\mathbf{U}}_0 + \Delta\hat{\mathbf{U}}, \\ \mathbf{\Lambda}_m & := & \mathbf{\Lambda}_0 + \Delta\mathbf{\Lambda}\end{array}\right\} \tag{2.1.23}$$

with the modified stiffness matrix

$$\mathbf{K}_m := \mathbf{K} + \Delta\mathbf{K} \tag{2.1.24}$$

assuming $\Delta\mathbf{M} \equiv \mathbf{0}$. The matrix eigenvalue problem (2.1.6) now reads

$$-\mathbf{M}\hat{\mathbf{U}}_m\mathbf{\Lambda}_m + \mathbf{K}_m\hat{\mathbf{U}}_m = \mathbf{0}, \tag{2.1.25}$$

and with (2.1.24)

$$\Delta\mathbf{K}\hat{\mathbf{U}}_m = -\mathbf{K}\hat{\mathbf{U}}_m + \mathbf{M}\hat{\mathbf{U}}_m\Delta\mathbf{\Lambda}_m. \text{ [6]} \tag{2.1.26}$$

This is called [32] the modal residual matrix **R**, indicating modifications in the stiffness matrix (2.1.24) due to modified (measured and therefore known) modal quantities. Right-hand multiplication of (2.1.26) with $\hat{\mathbf{U}}_m^{-1} = \hat{\mathbf{U}}_m^T$ for symmetric parameter matrices directly yields

$$\Delta\mathbf{K} = -\mathbf{K} + \mathbf{M}\hat{\mathbf{U}}_m\mathbf{\Lambda}_m\hat{\mathbf{U}}_m^T. \tag{2.1.27}$$

Dynamic response sensitivity. Sensitivity investigations of the forced responses of the model of the damped system can be handled more easily in the frequency domain than in the time domain. The Laplace transform is chosen with the Laplacian variable s. With zero initial conditions and the Laplace transforms

$$\mathbf{U}(s) := \mathscr{L}\{\mathbf{u}(t)\}, \ \ \mathbf{P}(s) := \mathscr{L}\{\mathbf{f}(t)\}, \tag{2.1.28}$$

the equation of motion (1.3.1) reads

$$(s^2\mathbf{M} + s\mathbf{B} + \mathbf{K})\mathbf{U}(s) = \mathbf{N}\mathbf{P}(s). \tag{2.1.29}$$

With the dynamic stiffness matrix

$$\mathbf{S}(s) := s^2\mathbf{M} + s\mathbf{B} + \mathbf{K} \tag{2.1.30}$$

Eq. (2.1.29) converts to

$$\mathbf{S}(s)\mathbf{U}(s) = \mathbf{N}\mathbf{P}(s). \tag{2.1.31}$$

[6] See Sect. 4.2.3; the modal residual vector is identical to the equation error of the matrix eigenvalue problem written in the modal vectors.

A modification in the dynamic stiffness matrix $\Delta\mathbf{S}(s)$ and the nominal, unmodified external excitation $\mathbf{P}(s)$ gives the equation of motion

$$[\mathbf{S}(s) + \Delta\mathbf{S}(s)][\mathbf{U}(s) + \Delta\mathbf{U}(s)] = \mathbf{NP}(s) \qquad (2.1.32)$$

or in linearized form

$$\Delta\mathbf{S}(s)\mathbf{U}(s) \doteq -\mathbf{S}(s)\Delta\mathbf{U}(s) =: \Delta\mathbf{P}(s), \qquad (2.1.33)$$

which can be interpreted as an additional force. This relationship can be used for determining the modified dynamic response,

$$\Delta\mathbf{U}(s) \doteq -\mathbf{S}(s)^{-1}\Delta\mathbf{S}(s)\mathbf{U}(s), \qquad (2.1.34)$$

which requires only the knowledge of the unmodified model and the dynamic stiffness modification. Alternatively, $\Delta\mathbf{S}(s)$ can be determined if linearly independent dynamic responses due to corresponding force vectors [33] (see localization of modifications, Sects. 2.2.6 and 2.1.4) are known. Equation (2.1.34) states that the additional force (2.1.33) is weighted by the transfer matrix

$$\mathbf{H}(s) := \mathbf{S}(s)^{-1} \qquad (2.1.35)$$

and that Equation (2.1.34) can be used to estimate the change (e.g. by taking the norm) $\Delta\mathbf{U}(s)$ for an assumed $\Delta\mathbf{S}(s)$ only with knowledge of the unmodified model. If the changes due to $\Delta\mathbf{S}(s)$ are significantly observable, this quantity can serve as a symptom.

The above equations can also be written as

$$[\mathbf{S}(s) + \Delta\mathbf{S}(s)]\mathbf{U}(s) \doteq \mathbf{NP}(s) - \mathbf{S}(s)\Delta\mathbf{U}(s) =: \mathbf{NP}(s) + \Delta\mathbf{P}(s) \qquad (2.1.36)$$

which verifies the statement that the modified system produces an additional force. We shall return to these equations for fault detection and localization in Sect. 2.2.6.

Subsystem modelling. Subsystem modelling is useful for various purposes: to reduce the orders of the models to be analyzed, to control the degree of coarseness of the model parametrization, to investigate submodel (component) sensitivity to parameter modifications, and to localize faults in the corresponding submodels. Let us continue with the associated undamped model and assume a stiffness matrix decomposition in the form

$$\mathbf{K} = \mathbf{K}(\mathbf{a}_K) = \sum_{i=1}^{I} a_{Ki}\mathbf{K}_i, \qquad (2.1.37)$$

where the vector $\mathbf{a}_K$ contains the real dimensionless design parameters a_{Ki}. We require the sensitivity of the eigenvalues λ_{0r} and eigenvectors $\hat{\mathbf{u}}_{0r}$ of the eigenvalue problem

$$(-\lambda_{0r}\mathbf{M} + \mathbf{K})\hat{\mathbf{u}}_{0r} = \mathbf{0}, \quad r = 1(1)n, \qquad (2.1.38)$$

with respect to variations of the real factors a_{Ki}. By inserting (2.1.37) into the matrix eigenvalue problem (2.1.38), it follows by differentiation that

$$(-\tfrac{\partial \lambda_{0r}}{\partial a_{Ki}}\mathbf{M} + \mathbf{K}_i)\hat{\mathbf{u}}_{0r} + (-\lambda_{0r}\mathbf{M} + \mathbf{K})\tfrac{\partial \hat{\mathbf{u}}_{0r}}{\partial a_{Ki}} = \mathbf{0}. \tag{2.1.39}$$

Assuming symmetric matrices, normalizing the eigenvectors as $m_{gr} = 1$, and taking into account Eq. (2.1.38), left-hand multiplication of the above equation with the transposed eigenvector of $\hat{\mathbf{u}}_{0r}$ produces

$$\tfrac{\partial \lambda_{0r}}{\partial a_{Ki}} = \hat{\mathbf{u}}_{0r}{}^T\mathbf{K}_i\hat{\mathbf{u}}_{0r}. \tag{2.1.40}$$

The differential change of the rth eigenvalue with respect to the ith submodel of the stiffnesses depends only on the generalized submodel stiffness related to the unmodified eigenvector $\hat{\mathbf{u}}_{0r}$. Here the first order modification of the eigenvalue is independent of the eigenvector modifications (in agreement with Eq. (2.1.13)).

Sensitivity of modal quantities to finite parameter changes. Finite modifications take into account higher order terms in the corresponding expansions; see, for instance, [34]. In this context it should be mentioned that recalculations with finite modifications can, of course, use the results of the eigenvibrations of the unmodified model in order to reduce the computational expenditure. The matrix eigenvalue problem (2.1.38) for the unmodified system, and that of the modified system,

$$[-\omega_{Mi}^2(\mathbf{M} + \Delta\mathbf{M}) + \mathbf{K} + \Delta\mathbf{K}]\hat{\mathbf{u}}_{Mi} = \mathbf{0}, \tag{2.1.41}$$

are taken as being solved with the transformation

$$\hat{\mathbf{u}}_{Mi} = \hat{\mathbf{U}}_0\hat{\mathbf{q}}_{Mi}. \tag{2.1.42}$$

By employing the congruence transformation with $\hat{\mathbf{U}}_0$ and taking into account the generalized orthogonal properties of the eigenvectors, and using the results of the unmodified model one obtains

$$[-\omega_{Mi}^2(\mathbf{I} + \hat{\mathbf{U}}_0^T\Delta\mathbf{M}\hat{\mathbf{U}}_0) + \mathbf{\Lambda}_0 + \hat{\mathbf{U}}_0^T\Delta\mathbf{K}\hat{\mathbf{U}}_0]\hat{\mathbf{q}}_{Mi} = \mathbf{0}. \tag{2.1.43}$$

If the modifications appear only at a few points, then $\mathbf{I}$ and $\mathbf{\Lambda}_0$ will dominate in the above equation, so that an iteration starting with $\omega_{Mi} \doteq \omega_{0i}$, $\hat{\mathbf{q}}_{Mi} \doteq \mathbf{e}_i$ will often converge in a few steps.

Selective sensitivity of forced response. First, the sensitivity of the forced response to parameter modifications will be discussed. Now consideration is given to the very closely related problem of choosing the force so that the response is sensitive to a very small set of parameters, and at the same time insensitive to the remaining parameters [35].

The dynamic response in the image domain follows from Eq. (2.1.31) to

$$\mathbf{U}(s) = \mathbf{H}(s)\mathbf{N}\mathbf{P}(s), \tag{2.1.44}$$

with the use of the transfer matrix (2.1.35). For the sake of simplicity it is assumed that all the components of the dynamic response are available (observable). The dynamic stiffness matrix (2.1.30), decomposed corresponding to pre-given subsystems, is represented as in Eq. (2.1.37):

$$\mathbf{S}(s) = \sum_{i=1}^{S} a_{Si}\mathbf{S}_i. \tag{2.1.45}$$

Equation (2.1.45) expresses the assumption that all three parameter matrices are linear in the design parameters a_{Si}.

The output sensitivity to the i-th model parameter, a_{Si}, is defined as:

$$\zeta(a_{Si}) = \left\| \frac{\partial \mathbf{U}(s)}{\partial a_{Si}} \right\|^2 \tag{2.1.46}$$

where $\|.\|$ is the Euclidian norm for vectors. When $\zeta(a_{Si})$ is large, the Laplace-transformed output is sensitive to variations of the model parameter a_{Si}.

Let $\mathcal{T}$ be a set of parameter indices. The condition for selective sensitivity to the parameter indexed in $\mathcal{T}$ is:

$$\zeta(a_{Si}) = \begin{cases} 0 & if \quad i \notin \mathcal{T}, \\ \text{not zero} & if \quad i \in \mathcal{T}. \end{cases} \tag{2.1.47}$$

The construction of a selectively sensitive input vector is now based on the following equations. The partial derivative in Eq. (2.1.46) follows from Eq. (2.1.44) to

$$\frac{\partial \mathbf{U}(s)}{\partial a_{Si}} = -\mathbf{H}\mathbf{S}^{-1}\mathbf{H}\mathbf{N}\mathbf{P}. \tag{2.1.48}$$

So let us find a vector ϕ satisfying

$$\mathbf{S}\phi = \begin{cases} 0 & if \quad i \notin \mathcal{T}, \\ \text{not zero} & if \quad i \in \mathcal{T}. \end{cases} \tag{2.1.49}$$

This is the first selectivity equation. Solutions of the above equation are independent of the design parameters. Solutions of (2.1.49) will often exist, even for small index sets $\mathcal{T}$, because the submodel matrices $\mathbf{S}_i$ are very sparse. Having found a ϕ, we choose the input $\mathbf{P}_s$, as a solution of

$$\left.\begin{aligned} \mathbf{N}\mathbf{P}_s &= \mathbf{S}\phi \\ &= \textstyle\sum_{i\in\mathcal{T}} a_{Si}\mathbf{S}_i\phi. \end{aligned}\right\} \tag{2.1.50}$$

The input vector resulting from the latter two equations will satisfy the sensitivity requirement of Eq. (2.1.47). These relations are necessary and sufficient. The strong requirements (2.1.47) and (2.1.49) can be weakened by substituting the zero by a small positive number.

Additional remarks. In addition to the modifications of the eigenquantities and dynamic variables, all the model properties should be considered, such as reciprocity (which can be lost during life time), linearity etc. in order to find out the effects and the extent of the parameter modifications. It must be mentioned here that the choice of a suitable quantity for the intended purpose is also important; for example velocity (impedance, mobility) or acceleration (apparent mass, inertance) instead of displacement will be appropriate if the higher frequency range is considered.

Further knowledge can be obtained by going deeper into the properties of, for example, the sensitivity of eigenquantities with respect to wave length and mode order [26]. In [26] it is shown that the mode shape sensitivity for a beam approaches its maximum when the dimensionless fault size is greater than 1/2. Knowledge of excitation characteristics (shape, time-dependency, frequency content, modal decomposition, etc.) also leads to extended statements, as is also used in adaptive testing [28].

2.1.4 Model Structure Modifications

Evolving model structure modifications, for example due to system degradation are decisive changes, because they can concern the choice of model classes. Model structure modifications are alterations of the dynamic stiffness matrix in general, of the type of damping, of the model order, and of the properties of the parameter matrices, and can lead to changes from a linear to a non-linear structure. An example of the latter is the viscously damped behaviour for relatively small displacement amplitudes to larger ones due to stiffness reduction, which implies frictional behaviour. Nonlinearities will be mentioned briefly in the following.

Dynamic stiffness matrix. Modifications $\Delta\mathbf{S}(s)$ of the dynamic stiffness matrix $\mathbf{S}(s)$, Eq. (2.1.30), have already been discussed briefly in Sect. 2.1.3, in the paragraph on dynamic response sensitivity. If linearly independent dynamic responses $[U_1(s), \ldots, U_n(s)] =: \mathbf{\Xi}(s)$ are measured due to suitable forces $[P_1(s), \ldots, P_n(s)] =: \mathbf{\Pi}(s)$ then, suppressing the argument s in Eq. (2.1.33), it results in

$$\mathbf{S}\,\Delta\mathbf{\Xi} + \Delta\mathbf{S}\,\mathbf{\Xi} \doteq \mathbf{0},$$

and it follows that

$$\Delta\mathbf{S} \doteq -\mathbf{S}\,\Delta\mathbf{\Xi}\,\mathbf{\Xi}^{-1}.$$

Non-linearized, it holds true that

$$\Delta \mathbf{S} = -\mathbf{S}\, \Delta\mathbf{\Xi}\, \mathbf{\Xi}^{-1}(\mathbf{I} + \Delta\mathbf{\Xi}\, \mathbf{\Xi}^{-1})^{-1}.$$

Therefore modifications of the dynamic stiffness matrix can be detected. As already mentioned, significant modifications in $\mathbf{S}(s)$ can serve as a symptom.

Type of damping. Our knowledge of damping forces is smaller than that of the other forces in mechanics. It is difficult to model them theoretically. In general, the damping must be determined by tests, [5]. In consequence, the damping forces are often modelled by equivalent viscous or hysteretic damping forces.

Damping ratio estimates are relatively erroneous, [5], compared with, for example, eigenfrequency estimates. When one looks at the damping estimation from the free response data of an SDOF model, $e^{-\delta t}(A \cos \omega_D t + B \sin \omega_D t)$, it can be seen that through the *ln*-operation and for models with small damping ratios, small (multiplicative) errors can substantially distort the damping coefficient. When dynamic response measurements are taken, a long measuring time is needed in order to achieve small standard deviations for damping ratio estimates. However, for various reasons the data are often not available for the time required. The inner damping of a continuum that is proportional to the strain velocity $\dot{\epsilon}$ compared with the external damping, e.g. the common viscous damping, is described for a bar, for instance in [9]. The differences in the modelled damping effects can be substantial.

Consideration is given to the viscously damped system (1.3.1) in the time domain and the corresponding hysteretically damped system with the complex stiffness matrix $(\mathbf{K} + j\mathbf{D})$, the latter being restricted to harmonic excitation. With the use of the ease-hypothesis that the generalized damping matrices (with respect to the eigenvectors of the associated undamped model) are diagonal matrices, the following abbreviations will be introduced:

$$b_{gr} := \hat{\mathbf{u}}_{0r}{}^T \mathbf{B} \hat{\mathbf{u}}_{0r}, \;\; \mathbf{B}_E = diag(b_{gr}), \tag{2.1.51}$$

$$d_{gr} := \hat{\mathbf{u}}_{0r}{}^T \mathbf{D} \hat{\mathbf{u}}_{0r}, \;\; \mathbf{D}_E = diag(d_{gr}). \tag{2.1.52}$$

Each damping ratio is related to one DOF, r, (proportional damping)

$$D_r := \frac{b_{gr}}{2\sqrt{k_{gr} m_{gr}}} =: \frac{\delta_r}{\omega_{0r}}, \tag{2.1.53}$$

$$g_r := \frac{d_{gr}}{\omega_{0r}^2} \doteq 2D_r, \;\; D_r^2 \ll 1, \;\; r = 1(1)n. \tag{2.1.54}$$

In a first approximation it is $2D_r \doteq g_r$.

Let us now consider the particular proportional damping

$$\mathbf{B} = \alpha\mathbf{M} + \beta\mathbf{K}, \tag{2.1.55}$$

with real constants α and β. In generalized coordinates, suppressing the index r, it follows that

$$b_g = \alpha m_g + \beta k_g = 2D\omega_0 m_g. \tag{2.1.56}$$

Damping only proportional to $\mathbf{K}$ ($\alpha = 0,\ \beta \neq 0$) will result in a damping ratio proportional to ω_0,

$$D_K = \frac{1}{2}\beta\frac{k_g}{\omega_0 m_g} = \frac{1}{2}\omega_0\beta. \tag{2.1.57}$$

This type of damping increases with increasing eigenfrequency. b_g only proportional to the inertia matrix $\mathbf{M}$ ($\alpha \neq 0,\ \beta = 0$) results in

$$D_M = \frac{1}{2}\frac{\alpha}{\omega_0} \tag{2.1.58}$$

which means a damping reduction with increasing ω_0. For structural damping we obtain:

$$D_{SK} = \frac{1}{2}\beta,\quad \alpha = 0, \tag{2.1.59}$$

independent of ω_0, and

$$D_{SM} = \frac{1}{2}\frac{\alpha}{\omega_0^2},\quad \beta = 0, \tag{2.1.60}$$

dependent on ω_0^{-2}.

As can be seen, the type of damping differs substantially.

Number of DOF. Another model structure modification is the change of the number in DOF within a (frequency-) bandlimited model. An example is the modelling of a crack in a plate. The linear model for a plate with a crack leads to modified boundary conditions for the open crack, and these are handled by introducing additional nodes in the FE model, which means enlarging the orders of the related matrices, and therefore introducing additional DOF (modes). The result is demonstrated in Table 2.1 and Fig. 2.7 [26].

The general formulation is as follows: the dynamic response of the non-modified system with n DOF is assumed to be modelled by

$$\mathbf{S}_n(s)\mathbf{U}_n(s) = \mathbf{P}_n(s). \tag{2.1.61}$$

Table 2.1. Natural frequencies [Hz] of the plate shown in Fig. 2.7

PLATE CONDITION/DOF	I	II	IIIa	IIIb	IV
UNIFORM (NO FAULT)	49.8	95.8	-	115.7	163.2
WITH CRACK	49.7	87.8	112.1	115.9	163.3
CRACK + HOLE	49.7	87.8	112.1	115.9	163.4

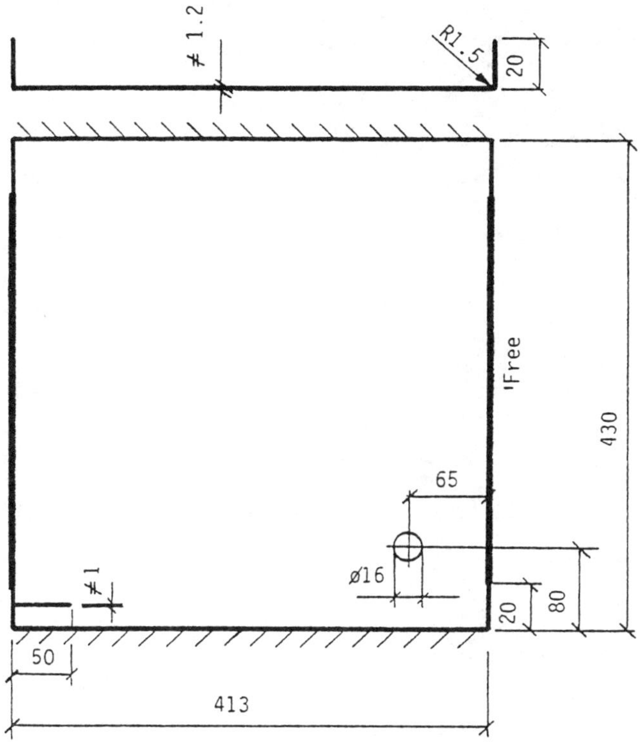

Fig. 2.7. Eigenmodes of a plate with a hole and a crack. **a)** plate dimensions, **b)** additional mode IIIa (see Table 2.1) due to the crack, **c)** nodal line of mode IIIb (see Table 2.1) of the uncracked plate, **d)** nodal line of mode IIIb of the cracked plate

With an additional DOF the equation of motion can be written in re-ordered form in the following way, while suppressing the argument s,

$$\begin{bmatrix} & & & S_{1,n+1} \\ & & & S_{2,n+1} \\ & \mathbf{S}_n & & \cdot \\ & & & \cdot \\ & & & \cdot \\ & & & S_{n,n+1} \\ S_{n+1,1} & S_{n+1,2} \ldots & S_{n+1,n} & S_{n+1,n+1} \end{bmatrix} \begin{Bmatrix} U_{(n+1)1} \\ U_{(n+1)2} \\ \cdot \\ \cdot \\ \cdot \\ U_{(n+1)n} \\ U_{(n+1)n+1} \end{Bmatrix} = \begin{Bmatrix} \\ \mathbf{P}_n \\ \\ \\ \\ 0 \end{Bmatrix} \qquad (2.1.62)$$

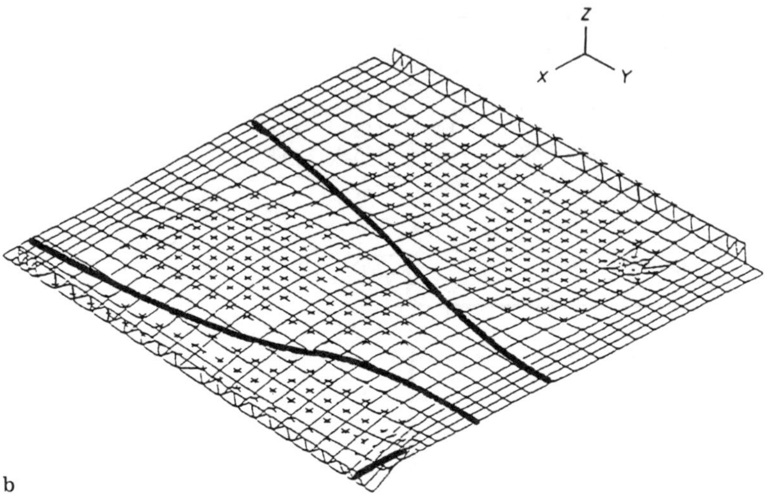

b

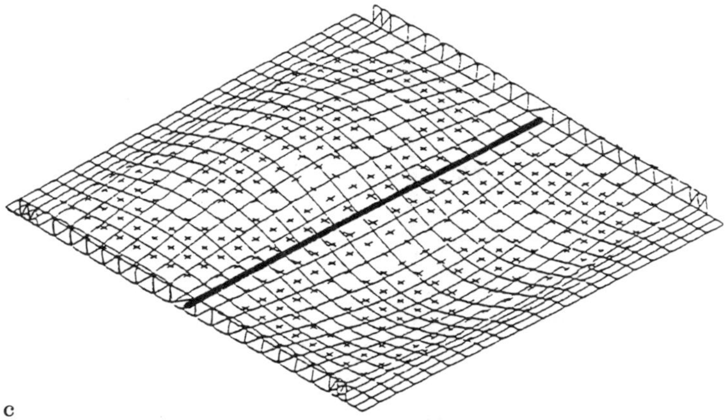

c

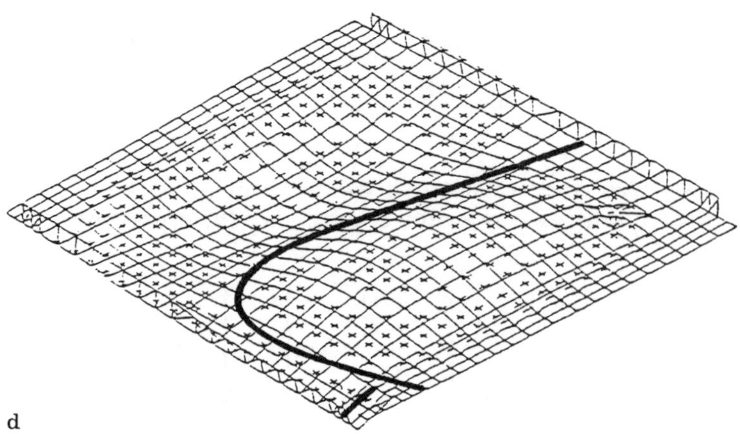

d

Fig. 2.7. (continued)

with the same excitation as before. Briefly, (2.1.62) should be written with $\mathbf{U}_{(n+1)} =: (\mathbf{U}'^{T}_{n}, U_{n+1})^T$, where the prime denotes the membership to the model with $n+1$ DOF, as

$$\begin{bmatrix} \mathbf{S}_n(s) & \mathbf{s}_{n,n+1}(s) \\ {}^{'}\mathbf{s}_{n+1,n}(s) & S_{n+1,n+1}(s) \end{bmatrix} \begin{Bmatrix} \mathbf{U}'_n(s) \\ U_{n+1}(s) \end{Bmatrix} = \begin{Bmatrix} \mathbf{P}_n(s) \\ 0 \end{Bmatrix}. \tag{2.1.63}$$

The second row of (2.1.63) leads to the response

$$U_{n+1}(s) = -\frac{1}{S_{n+1,n+1}(s)}\mathbf{s}_{n+1,n}(s)\mathbf{U}'_n(s) \tag{2.1.64}$$

which can be substituted into the first row of Eq. (2.1.63):

$$\mathbf{S}_n(s)\mathbf{U}'_n(s) - \frac{\mathbf{s}_{n,n+1}(s)\mathbf{s}_{n+1,n}(s)}{S_{n+1,n+1}(s)}\mathbf{U}'_n(s) = \mathbf{P}_n(s). \tag{2.1.65}$$

The result is an equation for $\mathbf{U}'_n(s)$ which can be expressed by the non-modified result and a deviation,

$$\mathbf{U}'_n(s) = \mathbf{U}_n(s) + \Delta\mathbf{U}_n(s): \tag{2.1.66}$$

$$[\mathbf{S}_n(s) - \frac{\mathbf{s}_{n,n+1}(s)\mathbf{s}_{n+1,n}(s)}{S_{n+1,n+1}(s)}]\Delta\mathbf{U}_n(s) = \frac{\mathbf{s}_{n,n+1}(s)\mathbf{s}_{n+1,n}(s)}{S_{n+1,n+1}(s)}\mathbf{U}_n(s). \tag{2.1.67}$$

Dependent on rank considerations Eq. (2.1.67) can be solved.

As can be seen, the coupling terms in Eq. (2.1.67) are very important, so that in the case of modal vibrations simple frequency criteria are generally insufficient for determining modification effects [36].

If $n-1$ DOF now have to be considered instead of n DOF, corresponding partitioning leads to the desired result:

$$\begin{Bmatrix} \mathbf{U}'_{n-1} \\ U_n \end{Bmatrix}, \begin{Bmatrix} \mathbf{P}_{n-1} \\ P_n \end{Bmatrix}, \tag{2.1.68}$$

$$\mathbf{U}_{n-1} = \mathbf{U}'_{n-1} + \Delta\mathbf{U}_{n-1}, \tag{2.1.69}$$

it follows

$$\Delta\mathbf{U}_{n-1}(s) = \mathbf{S}^{-1}_{n-1}(s)\mathbf{P}_{n-1}(s) \tag{2.1.70}$$
$$-[\mathbf{S}_{n-1}(s) - \frac{\mathbf{s}_{n-1,n}(s)\mathbf{s}_{n,n-1}(s)}{S_{n,n}}]^{-1}[\mathbf{P}_{n-1}(s) - \frac{\mathbf{s}_{n-1,n}(s)}{S_{n,n}(s)}P_n(s)],$$

if the the matrix $\mathbf{S}_{n-1}(s)$ is regular.

If more than ± 1 DOF has to be changed, a corresponding submodel formulation has to be performed, [36]. This formulation also clearly shows the influence of the coupling terms. The transformation into generalized coordinates means that the investigations are clearly arranged.

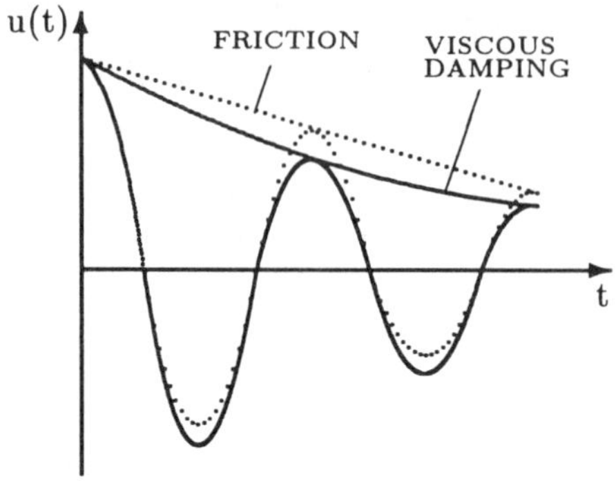

Fig. 2.8. Comparison of a viscous damping force with dry friction

Nonlinearities. Model structure modification can occur for the finite deflections by a change from linear external damping to dry friction (non-linear behaviour, e.g. fittings), as already mentioned. This can be detected by the envelope observation of the related free vibrations which change from an exponential function to a straight line, Fig. 2.8. If it is assumed that the unmodified system under consideration behaves linearly, then this assumption requires, for example, that the excitation level is small enough so that the resulting deflections are also sufficiently small. System modifications, for example stiffness reduction, may now introduce non-linear behaviour by enlarged deflections. This non-linear behaviour must first be detected. This detection can be done directly by looking for non-linear characteristics, or indirectly, by looking for violated assumptions of the linear system [37] (for details see Sect. 2.2.3). The question arises whether the observed non-linearity also has to be modelled for the purpose under consideration.

If one does not know the nonlinear behaviour of the system, and if it is contained non-negligibly in the measured response, and if one models and estimates the parameters of a structured *linear* model, then sometimes the estimates are not physically interpretable (e.g. negative mass). These unrealistic parameter estimates can serve for detection of nonlinearities [38].

2.2 Symptoms

In the previous section, modifications of the modal parameters and dynamic responses were investigated that are due to parameter and model structure modifications, the latter stemming from system modifications caused by failures. However, state modifications are not only the result of parameter and structure modifications, of course, and initial conditions for the free vibrations and the forcing are also essential.

2.2.1
Introductory Remarks: Symptoms, Discriminants, Features

Detection and diagnosis of failures are based on measurements, through the intermediary of models. Measurements should be reduced to a minimum number of measurants which are most sensitive and informative with respect to the expected system modification ($\rightsquigarrow$ monitoring). Directly or indirectly measurable quantities which are sensitive to failures are called symptoms. Additionally, symptoms should be sensitive to the damage evolution (related to θ), but should be insensitive to distortions. A discriminant is a symptom which is sensitive to a particular fault, and therefore discriminants allow us to distinguish between various faults. Features are special arrangements of symptoms which enable us to distinguish between several faults (with respect to a class). Patterns are established by features in order to characterize different system conditions. If a symptom is related to a subsystem it will serve for the localization of a modification to the system.

The sensitivity results of the previous section are used to find out symptoms, discriminants and features. The basic and simple quantity is the symptom, which should have the following properties:

- directly (e.g. strain) or indirectly (e.g. stress as a model-based reconstruction) measurable,
- functional relationship to a damage measure,
- high sensitivity to a fault/damage as a local property, but robust towards unknown disturbances as a global property (contradictory requirements which need optimization),
- distinguishability of various terms in the model, which also includes fault separation[7],
- the absolute value is a non-decreasing function of time, unless the system is repaired etc.,
- permit trend estimation.

2.2.2
Symptoms of Linear Systems

The following system-related characteristics can serve as symptoms:

- constants (e.g. cross-section measures independent of the fast time coordinate t, but they can be dependent on the life time, e.g. due to corrosion),
- functionals (scalars like an eigenfrequency which, for example, can be expressed as a Rayleigh quotient),
- vectors (discretized function or an assembly of scalar symptoms),
- functions (like eigenfunctions),
- field descriptions (multi-dimensional, for example the velocity field of a continuum) by direct or indirect (model-supported) measured quantities.

[7] Sometimes called isolation of faults.

Examples of scalars are

- maximum response $\max_t x(t)$,
- input energy,
- rms-value $x_{rms} := \mid \sqrt{\frac{1}{T}\int_0^T x^2(t)dt} \mid$ for periodic signals,
- average of the absolute signal $x(t)$: $x_{av} := \frac{1}{T}\int_0^T \mid x(t) \mid dt$,
- form (shape) factor x_{rms}/x_{av},
- crest factor x_{peak}/x_{rms},
- impulse factor x_{peak}/x_{av},
- variance σ_x^2 (total power, see [5]),
- 4th root of Kurtosis β, with $\beta := \frac{1}{T}\int_{-\infty}^{\infty} x^4(t)dt / \left(\frac{1}{T}\int_{-\infty}^{\infty} x^2(t)dt\right)^2$,
- Rice frequency $f_r = \dot{x}_{rms}/(2\pi x_{rms})$, where $\dot{x}_{rms}$ stands for the rms-value of the velocity signal (it is connected with the power spectral density and not with the probability density function (pdf) as variance etc.).

A hinge mechanism will be considered as an example of a function leading to a symptom. The signal considered with the i-th rotation (kinematics) may be written as [24]

$$\Phi_i(t, \theta) = c_0(t) + v_i(t, \theta) + n_i(t, \theta), \quad (i-1)T < t \leq iT, \quad i = 1, 2, \ldots, \tag{2.2.1}$$

with

T	the rotation period
θ	life time (slow time coordinate),
t	time of dynamics (fast coordinate),
$c_0(t)$	nominal signal of the mechanism,
$v_i(t, \theta)$	fault signal of the mechanism,
$n_i(t, \theta)$	noise signal independent of the nominal and fault signals.

Taking into account the periodicity of the motion one obtains the total symptom

$$\Phi(t, \theta) = \sum_{i=1}^{\infty} \Phi_i(t, \theta) * \delta(t - iT), \tag{2.2.2}$$

where the symbol $*$ denotes the convolution and $\delta(t)$ is the Dirac function. The total corresponding quantities of (2.2.1) are designated by the same symbols, but without indices. The fault intensity of the mechanism may then be measured by the expectation of the fault signal, which can be chosen as a symptom

$$F(\theta) = E_t\{v(t, \theta)\}.$$

The ratio (see Eq. (2.2.1) combined with (2.2.2))

$$SNR_0 = \frac{F(\theta)}{E_t\{\Phi(t, \theta)\} - F(\theta)} = \frac{E_t\{v(t, \theta)\}}{E_t\{c(t) + n(t, \theta)\}}$$

then describes the fault signal dependent on θ related to the remaining signals dependent on θ. The information content here is the fault signal

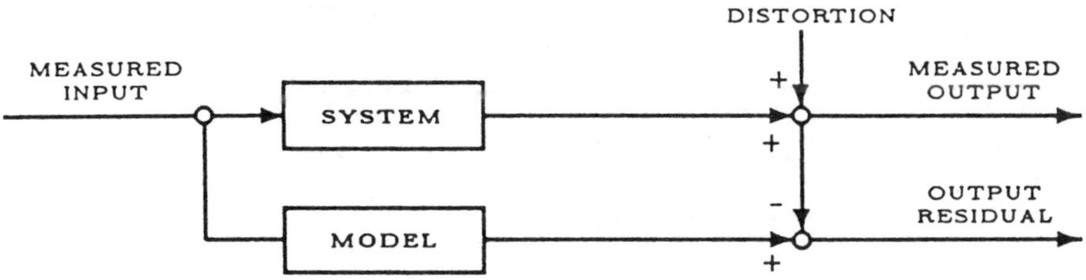

Fig. 2.9. Output residual

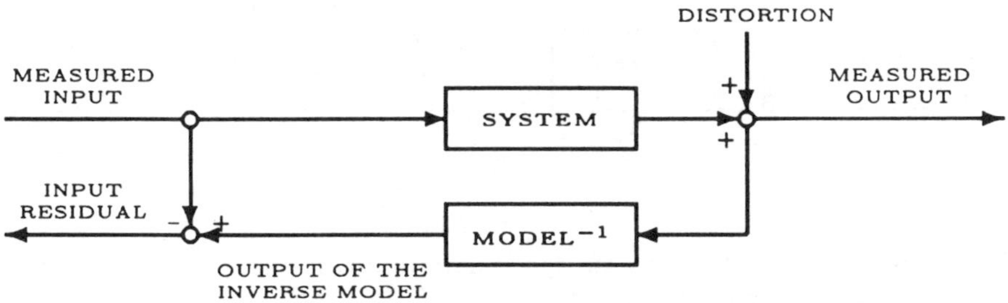

Fig. 2.10. Input residual

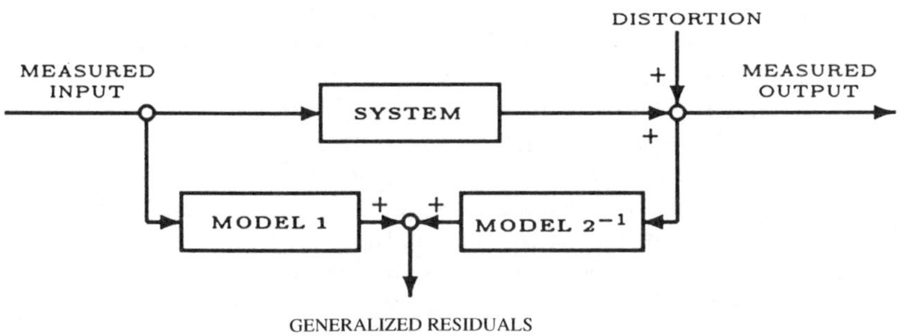

Fig. 2.11. Generalized residual

at the point of generation, and therefore it may be called a signal to noise ratio, because the remaining signals in the denominator play the role of distortions. Generally, the symptoms are measured at locations not identical to the locations of the fault generation. Then the spatial coordinate and the dynamic behaviour of the system must be included, the latter described by impulse response functions in the time domain or by frequency response functions in the frequency domain.

It should be mentioned that we distinguish between global and local symptoms. Global symptoms are the norms of residuals to be generated (see Sect. 4.2.3), i.e. the output, input and generalized residuals (see Figs. 2.9 to 2.11), with respect to the dynamic responses. The components

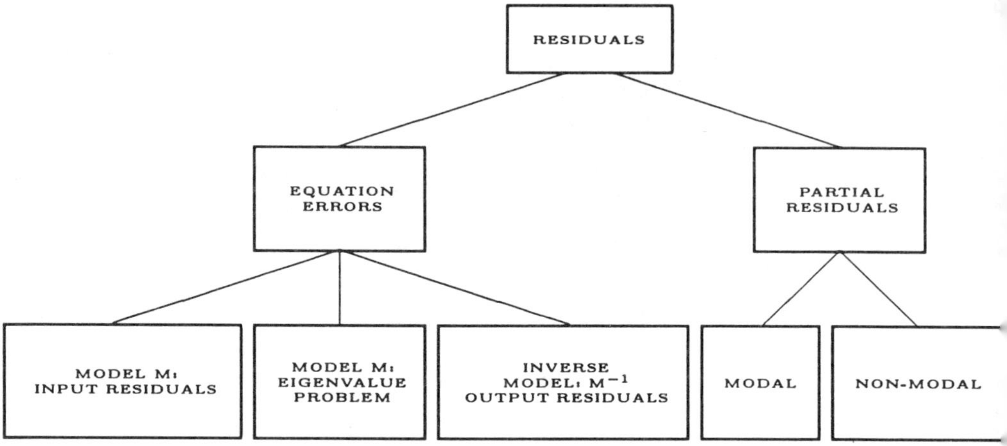

Fig. 2.12. Classification of residuals

of the residual vectors generated from dynamic responses are also global residuals, because they only implicitly contain the information of local parameter modifications. Additionally, we can introduce the equation error, which in the case of dynamic response problems is equal to the input (force) residual. In the case of the eigenvalue problem we can distinguish between the equation error (zero force residual) and partial residuals, such as the differences between eigenfrequencies of the damaged and undamaged system, or such as the eigenfrequency residuals **and** the corresponding differences between the eigenvectors, and in addition between the generalized masses. The fulfilment of the generalized orthogonality properties etc. can also be taken into account. Fig. 2.12 gives an overview of possible residuals. The non-modal residuals are important in engineering, because the engineer is mostly interested in, for example, stress distributions and acoustic levels. One obtains local residuals, for example, by taking the elementwise differences of the stiffness matrices at life times θ_i and θ_{i+1}. Flexibility and inertia matrixes can be taken instead of stiffness matrices. Here the reader is referred to Sects. 2.1.3 and 2.2.6, and Chap. 4.

Of course, one can combine global with local residuals (see Chap. 4). One can also choose special indicators, like the MAC[8] [39] etc. Local symptoms with their local information contents will serve simultaneously for fault detection and localization. The product of the residuals of the modal vector components are also a suitable symptom for fault detection [26].

Finally, in very general terms we can state the following: the measured signal is designated by $x(r, t, \theta)$. The operator to transform $x(r, t, \theta)$ into a symptom, briefly called the symptom operator, may be designated by $\Psi\{x(r, t, \theta)\}$ (e.g. Kurtosis). If the symptom is a statistical variable, we

[8] Modal Assurance Criterion: it is the cosine of the angle between, for example, a measured and calculated eigenvector.

have to apply the expectation operator E. If the process is ergodic, we can substitute the expectation by time averaging with the probability one:

$$E\{\Psi\{s(r,t,\theta)\}\} = \lim_{T\to\infty} \frac{1}{T} \int_0^T \Psi\{s(r,t,\theta)\}dt. \tag{2.2.3}$$

The discriminant of the k-th fault can then be described by the operator Ξ_k:

$$d_k(r,\theta) := E\{\Xi_k\{\Psi\{x(r,t,\theta)\}\}\}, \tag{2.2.4}$$

or directly from the signal:

$$d_k(r,\theta) := E\{\Xi_k\{x(r,t,\theta)\}\}.$$

As an example of a discriminant let us consider a truss. The bars serve for the decomposition (2.1.37) of the stiffness matrix, the matrices $\mathbf{K}_i$ in (2.1.37) representing the undamaged bars. After some time stiffness reduction in a particular bar can be detected and located in the factors a_{Ki} if the stiffness matrix (2.1.37) is adjusted to the damaged state. The adjustment factors a_{Ki} can serve as discriminants in this case, where localization is identical with designating a fault in a particular bar.

2.2.3 Symptoms of Non-linear Systems

Various types of non-linearities are shown in Fig. 2.13. Nonlinear behaviour can result from modifications of a linearly behaving system. The symptoms discussed in the previous section can also be applied for non-linear systems. However, non-linear effects differ fundamentally from linear behaviour. Linearized models cannot describe the non-linear behaviour. Consequently, symptoms including nonlinear effects compared with the corresponding quantities of a linear model lead to deviations from the linear behaviour, so they can serve for detection: violated assumptions of linearity. Therefore, as already stated, the detection of non-linear behaviour can be performed indirectly by tests, if one assumes a linear model and looks for violated assumptions. Here artificial harmonic excitation with different levels of response amplitudes is most effective. Characteristics which can be chosen are

- superposition,
- reciprocity,
- parameter independency of sample (initial conditions, damping),
- Nyquist plots (their geometry),
- ...

The distortions are sometimes informative (pattern recognition) when one looks, for example, at distorted Nyquist plots [40] (Fig. 2.14); here the isochrones ($\omega =$ const.) for simple cases indicate the type of non-linearity. Non-linearities affect the dynamic response most in the resonance neighbourhood [41, 42].

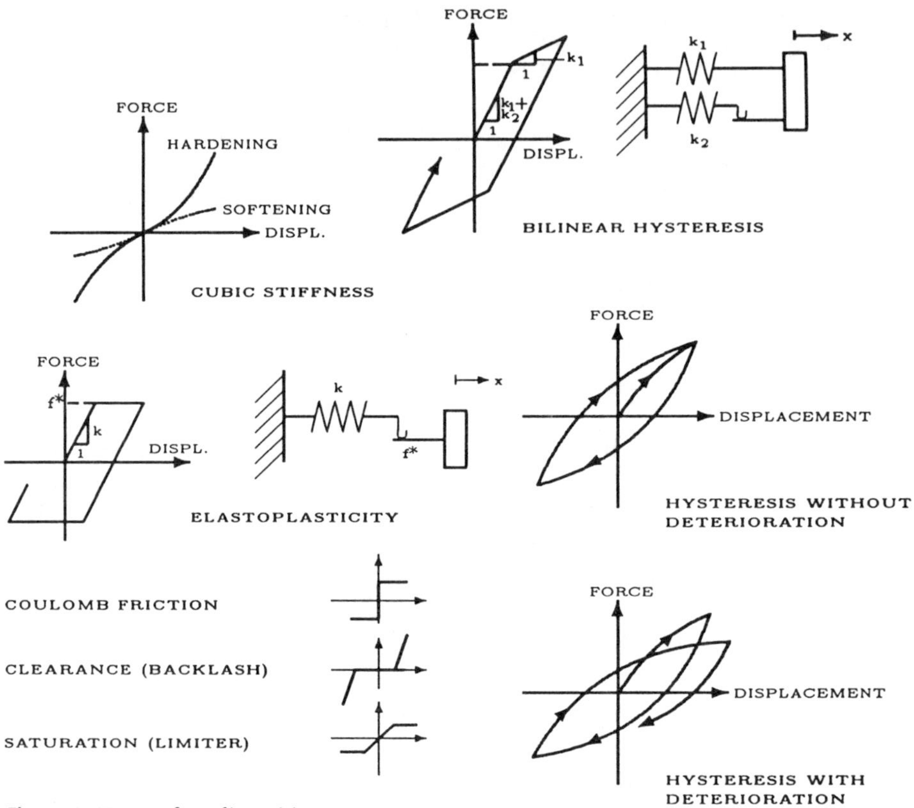

Fig. 2.13. Types of nonlinearities

Direct methods for detection are based on such symptoms as

- signals due to special excitation and filtering [5],
- indicator functions (e.g. SIG-function [43]),
- Hilbert transforms [44]
- high order correlation functions [45],
- multispectral density functions [46],
- dispersion functions [47],
- histogram measures [48] (pattern classification),
- NARMA, NARMAX models [45], [49] to [53],
- polynomial fits [54],
- ...

Before going into details of the itemized methods, the following problems should be mentioned in this context:

- It is hard to distinguish between the bias in the process model and in the noise model.
- Does the detected non-linear behaviour require and also permit a non-linear model?

STIFFNESS	DAMPING	POLAR PLOT OF DISPLACEMENT RESPONSE	PATTERN OF ISOCHRONES (ω = const)
LINEAR	HYSTERETIC	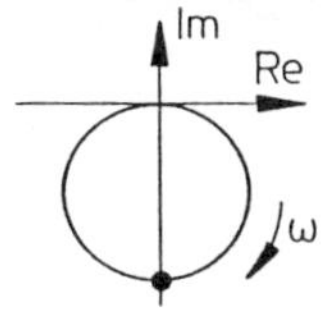	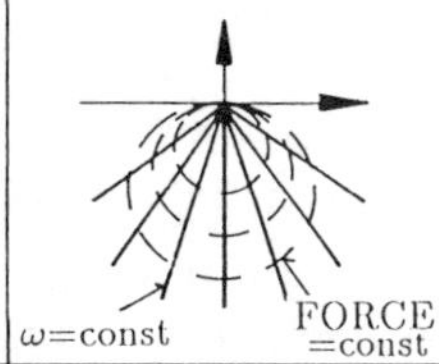
LINEAR	HYSTERETIC+ COULOMB TYPE	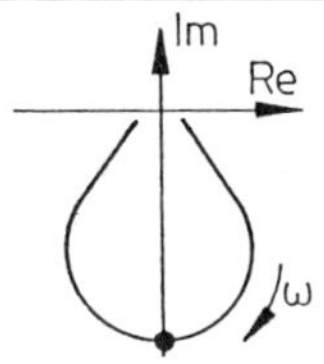	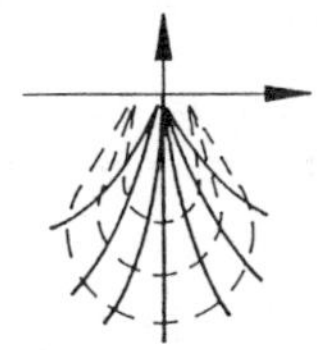
LINEAR	VISCOUS QUADRATIC	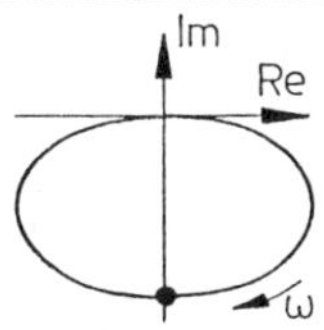	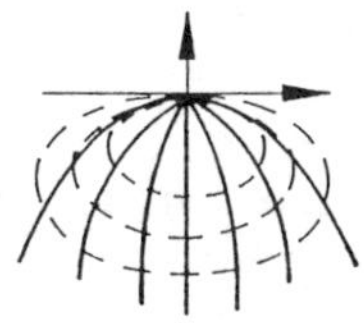
HARDENING	HYSTERETIC	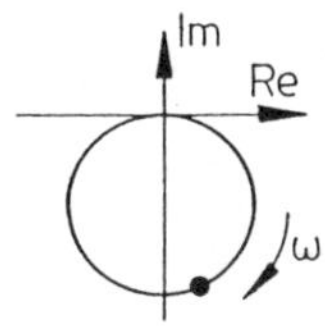	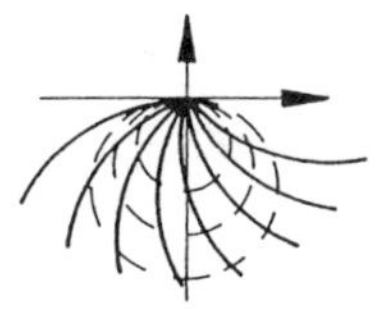
SOFTENING	HYSTERETIC	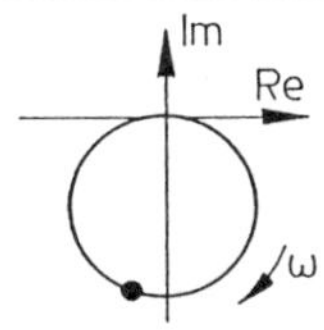	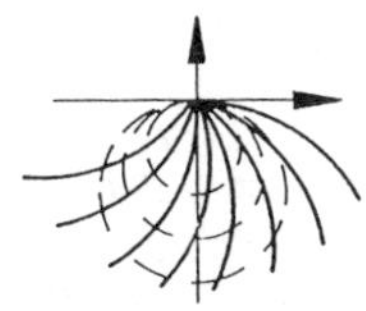

Fig. 2.14. Recognizing nonlinearities from the patterns of isochrones

- Detection in general does not include structure (of the model) identification.
- The choice of inputs and outputs must reveal the structure patterns (complete set of information with respect to the initial conditions: sufficiently large deflection amplitudes without system destruction).

Note: Although, for example, an SDOF model with cubic stiffness and initial harmonic excitation will respond with sub-harmonics for free vibrations with particular initial conditions [55]. In self-excited vibrations of systems with dry friction a stable or unstable vibration can occur.

We now come to the details of the direct methods itemized above.

The *SIG-function* [43] as a direct measure is defined by

$$SIG := \frac{\int_0^T [x(t) - x_f(t)]^2 dt}{\int_0^T x^2(t) dt} \tag{2.2.5}$$

where $x(t)$ is the measured and $x_f(t)$ the filtered, linearized dynamic response within a time interval $[0, T]$.

The *Hilbert transform* is defined as follows. The one-sided Fourier transform is defined by:

$$\mathcal{F}\{f(x)\} = \int_0^\infty e^{-jyx} f(x) dx =: F(y) = F_1(y) + jF_2(y) \tag{2.2.6}$$

with

$$f(x) = \begin{cases} 0 & \text{for} \quad x < 0 \\ \neq 0 & \text{for} \quad x > 0. \end{cases} \tag{2.2.7}$$

With the unit step function $1(x)$ Eq. (2.2.7) can be rewritten in the form

$$f(x) = 1(x) f(x).$$

Assuming $f(x) \epsilon L^2(0, \infty)$, the Fourier transform leads for $f(x)1(x)$ to the Hilbert transform

$$F_2(y) = -\frac{1}{\pi} VP \int_{-\infty}^{\infty} \frac{F_1(\eta)}{y - \eta} d\eta =: \mathcal{H}\{F_1(y)\}, \tag{2.2.8}$$

and to the inverse Hilbert transform

$$F_1(y) = \frac{1}{\pi} VP \int_{-\infty}^{\infty} \frac{F_2(\eta)}{y - \eta} d\eta =: \mathcal{H}^{-1}\{F_2(y)\}; \tag{2.2.9}$$

in addition, the Parceval theorem holds true,

$$\int_{-\infty}^{\infty} F_1^2(y) dy = \int_{-\infty}^{\infty} F_2^2(y) dy. \tag{2.2.10}$$

The abbreviation VP stands for

$$VP \int_{-\infty}^{\infty} \ldots = \lim_{\epsilon \to \infty} \left(\int_{-\infty}^{y-\epsilon} \ldots + \int_{y+\epsilon}^{\infty} \ldots \right). \tag{2.2.11}$$

The real functions $F_1(y), F_2(y)$ are named conjugate functions. It is necessary and sufficient for the existence of the Hilbert transform that the components of $F(y)$ are conjugate functions. The Hilbert transform, for example, exists for causal linear functions. In consequence, for the unit impulse response function for a causal SDOF model $g(t)$ it holds true that

$$\mathcal{F}\{g(t)\} = F(j\omega) = F_1(j\omega) + F_2(j\omega), \tag{2.2.12}$$

and it follows

$$\left.\begin{array}{lllll} \Re[\{F(j\omega)\}] & = & \Re F(j\omega) & = & F_1(j\omega), \\ \Im[\mathcal{H}\{F(j\omega)\}] & = & \Im F(j\omega) & = & F_2(j\omega). \end{array}\right\} \tag{2.2.13}$$

If these equations do not hold true it is concluded that the system behaviour is non-linear [56]. This statement is valid if unit response functions for causal non-linear systems do not exist as proven in [44].

Equation (2.2.13) can be written as

$$H(y) = \mathcal{H}\{F(j\omega)\} = -\frac{1}{\pi} VP \int_{-\infty}^{\infty} \frac{F(j\omega)}{\omega - y} d\omega \tag{2.2.14}$$

with

$$H(y) = F(j\omega) \mid_{j\omega = y}, \tag{2.2.15}$$

if the system behaves linearly, otherwise the inequality holds. Fig. 2.15 from [57] shows the patterns (frequency response functions) of three types of non-linearities for two excitation levels. The differences between low and high levels are small. Figure 2.16 from [57] contains the Fourier and Hilbert transforms of various non-linear SDOF models.

It is noted that new developments in the description (detection) of linear and non-linear behaviour apply wavelets[9] and the Wigner distribution [46], [58], [59].

For a stationary signal with zero mean a *high order correlation function* is, for example

$$\Phi_{uu^2}(\tau) = E\{u(t+\tau)u^2(t)\}. \tag{2.2.16}$$

For an ergodic process the correlation function of order n is defined by

$$R_{x_1 \ldots x_n}(\tau_1, \ldots, \tau_n) = \lim_{T \to \infty} \frac{1}{T} \int_{-\infty}^{\infty} x_1(t-\tau_1) \ldots x_n(t-\tau_n) dt. \tag{2.2.17}$$

The high order correlation function gives a characterization of a nonlinear process which serves for model structure identification [4].

Bispectral power density functions $S_\omega(\omega_1, \omega_2)$ are bi-dimensional Fourier transforms of the corresponding higher order correlation functions [45].

Example: A non-linear process is given by $y(t) = u^2(t)$, $u(t)$ stationary Gaussian and mean-free, $E\{u(t)\} = 0$. The cross-correlation function is

$$\Phi_{yu}(\tau) = \int_{-\infty}^{\infty} \int_{-\infty}^{\infty} yu \; p(y, u; \tau) dy du = \int_{-\infty}^{\infty} u^3 p(u, \tau) du = 0.$$

[9] The disadvantage of the Fourier transform is the lack of a localization property: local modification of a signal changes the transformed signal everywhere. The application of wavelets does not have this disadvantage.

This cross-correlation function gives no indication of the quadratic relationship.

Therefore, more information is needed than is contained in the joint probability density function (pdf), and this is the conditional pdf. In order to overcome the lack of information on the usual correlation function (see the last example) the following *dispersion function* [47] is introduced:

$$\Theta_{yu}(t_1, t_2) = E_{u(t_2)}[\{E_{y(t_1)}[y(t_1) \mid u(t_2)] - E_{y(t_1)}[y(t_1)]\}]^2. \quad (2.2.18)$$

t_1, t_2 are independent variables. The dispersion function is the conditional mean of the random signal $y(t_1)$ given $u(t_2)$. The auto-dispersion function of the signal $u(t_1), u(t_2)$ is:

$$\Theta(t_1, t_2) = E_{u(t_2)}\{E_{u(t_1)}[u(t_1 \mid u(t_2] - E_{u(t_1)}[u(t_1)]^2\}. \quad (2.2.19)$$

Figure 2.17 shows some examples with the conditional pdfs and dispersion functions.

Histograms are well-known as approximations of pdfs and require no further discussion.

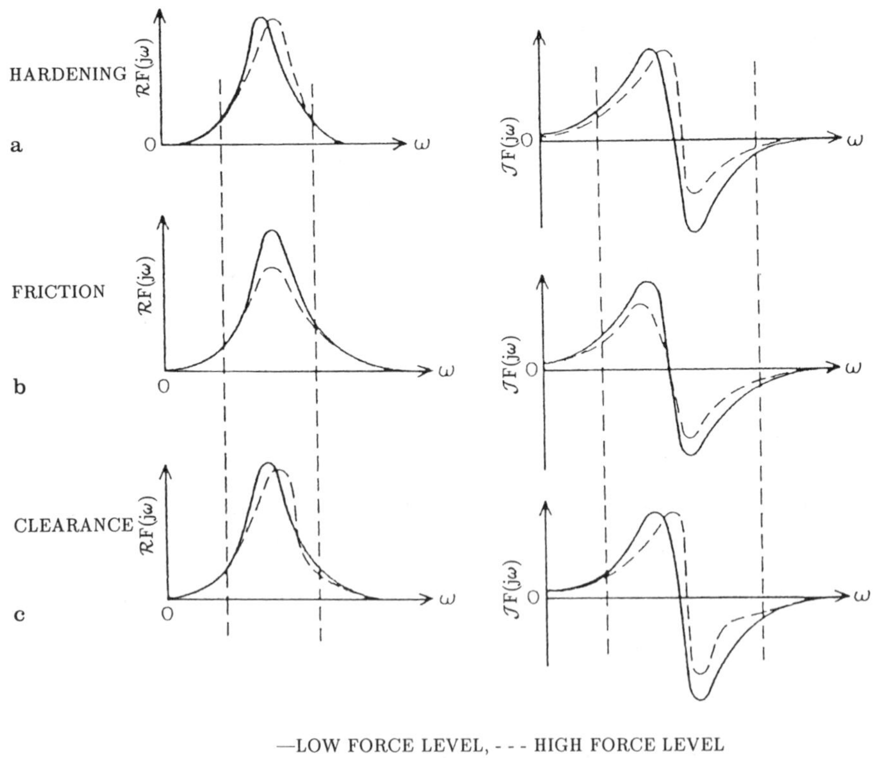

Fig. 2.15. Real and imaginary parts of nonlinear functions at two excitation levels for an SDOF system

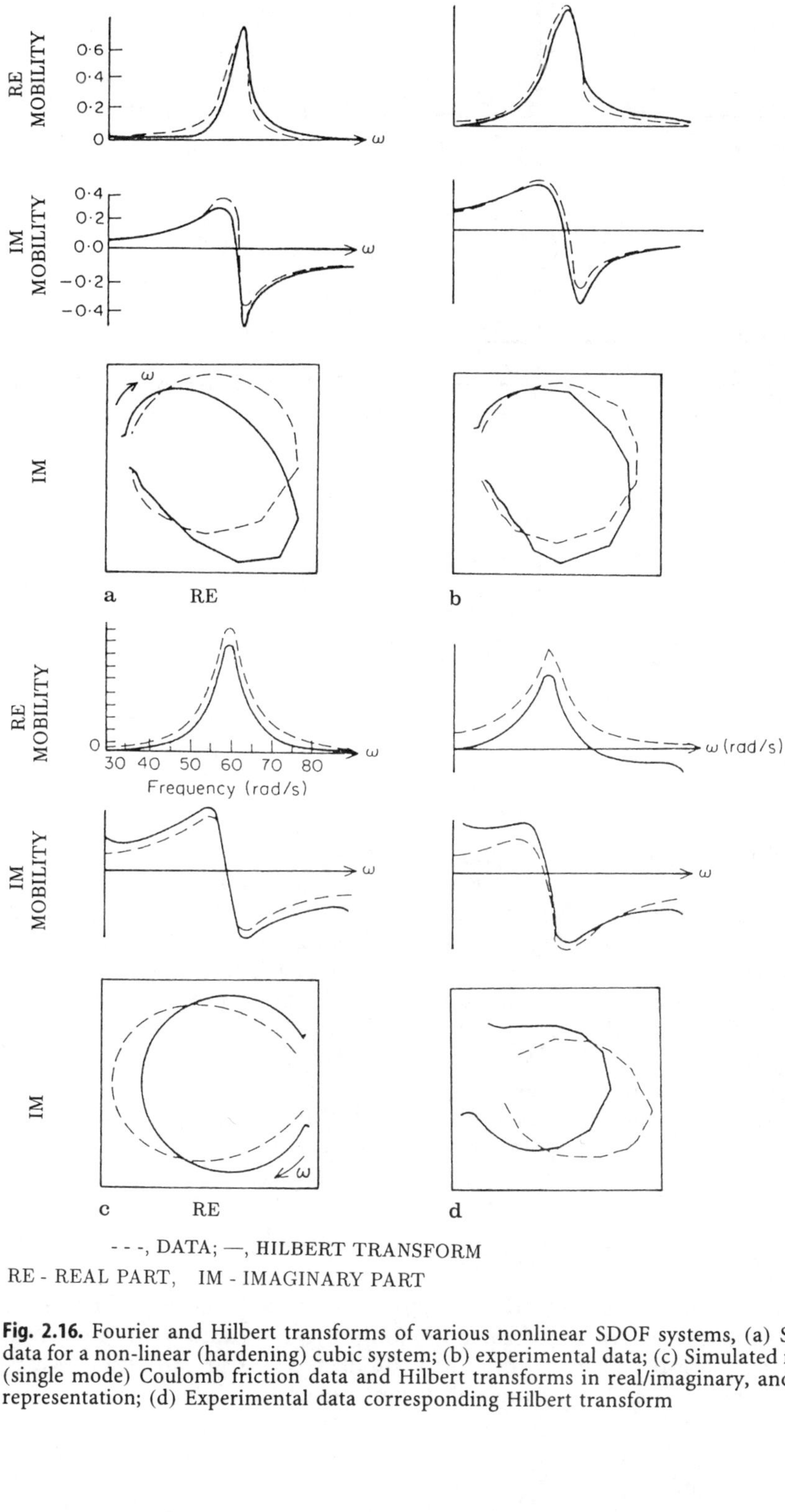

Fig. 2.16. Fourier and Hilbert transforms of various nonlinear SDOF systems, (a) Simulated data for a non-linear (hardening) cubic system; (b) experimental data; (c) Simulated nonlinear (single mode) Coulomb friction data and Hilbert transforms in real/imaginary, and Nyquist representation; (d) Experimental data corresponding Hilbert transform

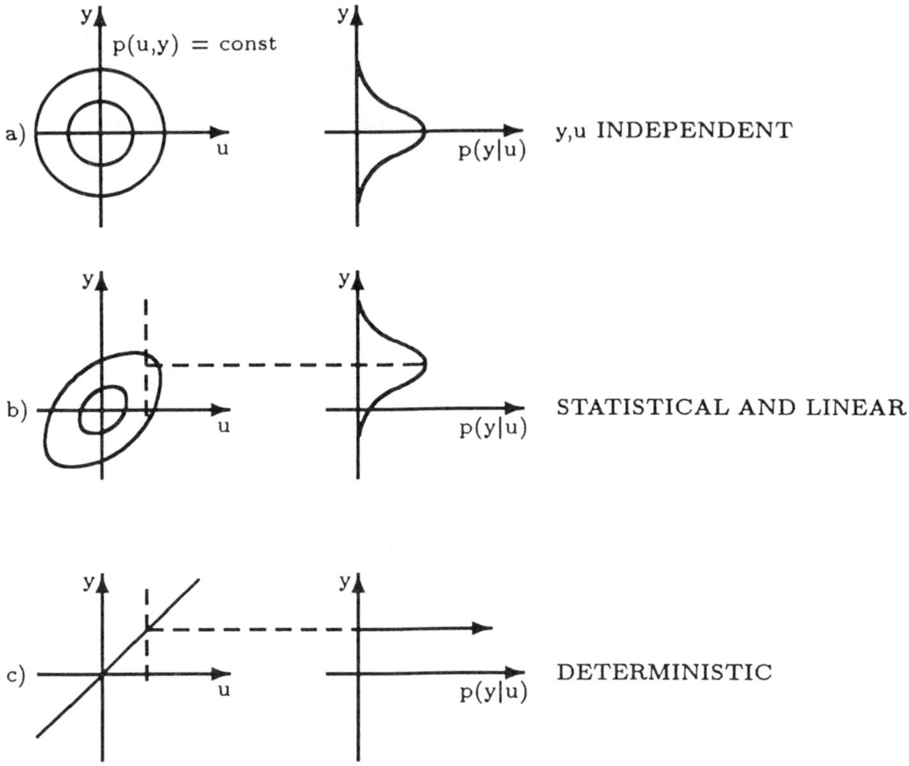

Fig. 2.17. Some conditional pdfs

Modelling of processes applying estimators is another, but expensive, method of detection. It is an advantage that the resulting mathematical model can be used for other purposes as well. The parametric model can be a differential equation or, when discretized, a difference equation: for non-linear models the latter is called a *non-linear ARMAX* model or briefly a NARMAX model. X stands for exogenous excitation. Further models are available, of course. If one restricts the class of models to *polynomials*, knowing that according to the Stone-Weierstrass theorem, polynomials approximate continuous single-valued functions in a finite interval uniformly, then their maximum power can be estimated simultaneously with the polynomial coefficients: structure identification [54], [60], [61].

2.2.4 Features and Patterns, Classification

Features. Symptoms assembled in a feature vector will serve as a basis for decision (Fig. 2.18). The features can belong to classes and form a pattern. A quality criterion (decision function) for decision is given, and this can be knowledge-based. The result will initiate several actions.

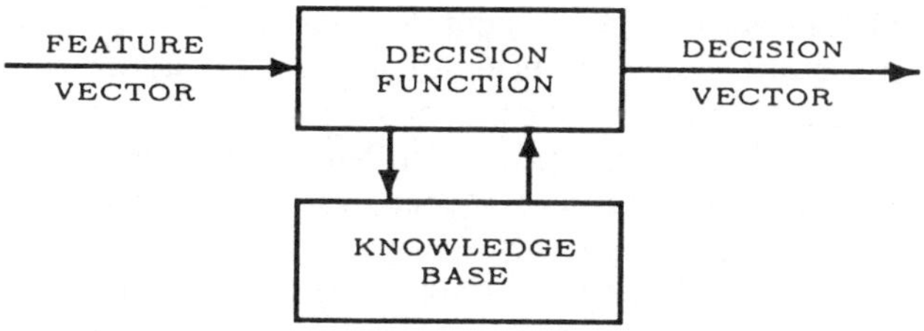

Fig. 2.18. Decision due to features

The starting point is a scalar or a signal (time function). The first step to be done is

- pre-processing: this can be done by amplification, filtering, linearization, or rectification and envelope calculation etc. Another type is digitizing, taking into account the sampling theorem (Shannon, [5]). It is possible that segmentation and digital filtering will be necessary (see Chap. 3).
- The second step is feature abstraction: the aim is to condense the information contained in a signal to a few significant characteristic quantities: the features (this is more than data reduction). A set of features will be assembled in a feature vector.

The following are examples of features :

- Eigenfrequency shifts $\Delta f_i / f_i$, f_i, the eigenfrequency of the test specimen, $i = 1, \ldots$, of a cantilever due to mass and stiffness modifications. The eigenfrequencies of the unmodified system are denoted as $f_{i\ unm}$.The residuals are $\Delta f_i := f_i - f_{i\ unm}$. It is assumed that local defects exist at the points x_k, $k = 1(1)K$. The characteristics d_{Bk} and $d_{\mu k}$ are introduced for the defects of stiffnesses and masses, respectively, at point k:

$$d_k = \begin{cases} 0 & \text{no defect exists} \\ 1 & \text{defect exists.} \end{cases}$$

  Each defect characterized by $d_{Bk}, d_{\mu k}$ is related to a frequency shift pattern $\Delta f_{Bik}, \Delta f_{\mu ik}$ for $k = 1(1)K$. With these patterns a coefficient matrix can be established which can be used for computing $d_{Bk}, d_{\mu k}$ as values between 0 and 1 [62] (see also [63]).
- Additional eigenfrequencies, as shown in Table 2.1 from the plate with a crack. The nodal lines also form very sensitive patterns (see Fig. 2.7).
- The participation coefficients of an expansion of a pattern $\{f\} = \mathbf{f}$, a $(n, 1)$–vector, with respect to an orthogonal basis $\{\Phi_\nu\}$, $\nu = 1(1)m$, can be chosen as features :

$$c_\nu = \{\Phi\}_\nu^T \{f\}, \ \{c_\nu\} = [\{\Phi_1\}, \ldots, \{\Phi_m\}]^T \{f\}.$$

  The inverse problem will approximate $\{f\}$:

$$\{f\}_{approx} = [\Phi_1, \ldots, \Phi_m]\{c_\nu\} = \sum_{i=1}^{m} c_\nu \Phi_\nu .$$

With $m = n$ the completeness of the vector basis is given.

- The Fourier coefficients of the Fourier series of a periodic signal can be chosen as features (interesting for rotating machines).
- Fourier and Laplace transforms, their discretized forms as well as the cepstrum [5] of a signal can be taken as features. Spectral amplitudes for constant bandwith (e.g. octave) may also serve as features.
- The construction of special quantities of a mathematical model for the extraction of features (e.g. model parameters) is most sophisticated.
- The stress distribution which varies dependent on crack sizes can serve as a feature.

Features must describe the modifications expected in a sufficiently significant way, and they must be statistically independent.

Patterns. What are patterns? The waveforms of time signals, as used in acoustics and seismic investigations are well-known patterns. In general, patterns consist of features which fulfil the following postulates:

P1: The information is contained in a representative sample.

P2: A (simple) pattern possesses features which are characteristic of their membership of a class.

P3: The features of patterns of one class compose a nearly compact set (subdomain) within the feature domain. These subdomains of features of different classes are separate.

The following are examples of patterns: a typical view of an areal of a city, a circuit diagram and, as already mentioned, the waveform of a signal, and the figure of a spectral function.

Features and patterns serve for the distinguishability (isolation) of faults and are based on the model-supported investigations (see the catalogue of possible faults in Fig. 3.1). The selective sensitivity already mentioned is another tool for fault isolation.

Classification. In consequence of P3 a classsification is now possible. Various sets of features with respect to different classes, each fulfilling P3, permit classification with a pre-chosen decision function (e.g. the Euclidian distance). Fig. 2.19 shows this classification for the feature domain of two dimensions (m_1, m_2). Classification is done here using a straight line, and for more than two dimensions by a hyper-plane.

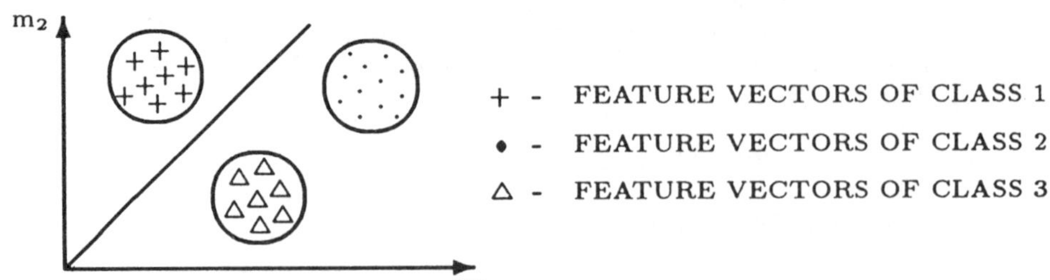

Fig. 2.19. Classification procedure for non-overlapping sets

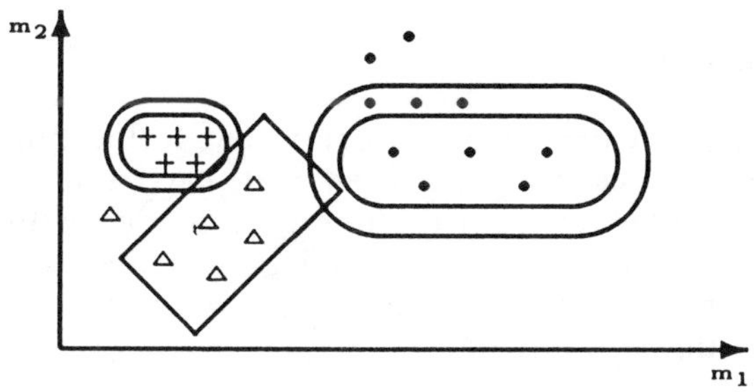

Fig. 2.20. Use of classes with tolerance domains

In addition to this deterministic procedure of classification, statistical procedures can, of course, be applied if, for instance, variances are included as tolerances (Fig. 2.20). Complicated model-supported classification can also be taken: a class of models and a subset of parameters are chosen as features. In addition, a learning classification can be introduced, where the classification is not fixed and the decision function will be adjusted optimally during application by, for example, external (e.g. time-dependent) information.

2.2.5 Remarks on the Decision Problem

Decision theory plays an important role, and it is possible to apply deterministically and statistically based procedures. If sufficient statistical data are not available, fuzzy logic can serve for decision-making. A real application in detail is, of course, problem-orientated and object-orientated and based on symptoms, features and patterns; functionality (operational-ability, serviceability) criteria, and safety assessment are mostly the basis for decisions (see Chap. 5). Holistic modelling combined with a method of making condition assessment by applying fuzzy sets is proposed in [64].

The damage detection of systems can use residuals. The first decision to be made is whether the detected modification is significant. The quality criterion applied can be deterministic (threshold, limit values) or statistical (by samples: the test sample in comparison with the learning, initial [design] sample; by patterns with tolerance values). If the standard deviation σ_x is used as a tolerance band within the statistical decision, then it is energy-based, because σ_x^2 is equal to the total integral over the power spectral density for mean-free samples [5] (p.105). The second decision will be model-based; it is a decision dependent on the modification assessment concerning further actions (see Fig. 1.2). In Chap. 5 the reader can find a more detailed illustration and discussion referring to the model-based diagnosis procedure.

2.2.6 Remarks on the Localization Problem

The localization of faults/damage in the system is equivalent to locating model modifications. The model representing the recent state can be modified due to faults/damage in comparison with the reference model (with respect to a previous state or to the unmodified model): this means a more or less qualitative determination, whereas the **estimation** of parameter value deviations is a quantitative approach.

Detection of faults. Global detection can be performed (see Sects. 2.1.3 and 2.2.2) using the Laplace transformed equation of motion (2.1.31) with respect to the life time $\theta = const.$ with zero initial conditions. The model describing the system at time θ_i will be denoted by the index i, and with $i-1$ the model describing the system at time $\theta_{i-1}, i = 1, \ldots$; the index 0 designates the initial state:

$$\mathbf{S}_i(s)\mathbf{U}_i(s) = \mathbf{NP}(s), \quad \mathbf{S}_{i-1}(s)\mathbf{U}_{i-1}(s) = \mathbf{NP}(s), \tag{2.2.20}$$

assuming that the external forces are unchanged during the system's lifetime. The modification of the dynamic stiffness matrix will then be expressed by

$$\mathbf{S}_i(s) - \mathbf{S}_{i-1}(s) =: \Delta\mathbf{S}_i(s). \tag{2.2.21}$$

It follows

$$\mathbf{S}_{i-1}(s)\mathbf{U}_i(s) = \mathbf{NP}(s) - \Delta\mathbf{S}_i(s)\mathbf{U}_i(s) : \tag{2.2.22}$$

the known (by measurement) and Laplace-transformed dynamic response $\mathbf{U}_i(s)$ multiplied by the known previous dynamic stiffness matrix will result in an additional (residual) force

$$\Delta\mathbf{P}_i(s) := -\Delta\mathbf{S}_i(s)\mathbf{U}_i(s). \tag{2.2.23}$$

The external force $\mathbf{NP}(s)$ is assumed to be known. If $\|\Delta\mathbf{P}_i(s)\|$ is greater than a pre-chosen threshold $P_s > 0$, then a significant modification is detected and further investigations are needed.This detection procedure is illustrated in Fig. 2.21; it is an input error method. If instead of $\Delta\mathbf{P}_i(s)$ one is looking for $\Delta\mathbf{U}_i(s) := \mathbf{U}_i(s) - \mathbf{U}_{i-1}(s)$, it is the output error method.

These procedures for detection use the dynamic stiffness matrix $\mathrm{S}(s)$ or the transfer matrix $\mathbf{H}(s)$, and input and output quantities: these are non-parametric models. In addition, adaptive excitation is suitable instead of that coming from the sytem's operation in order to enlarge a possible modification effect in $\Delta\mathbf{P}_i(s)$ or in $\Delta\mathbf{U}_i(s)$.

Sensitivity analysis has to be performed in order to determine symptoms of condition for the system under consideration. As shown in Sect. 2.1.3, modal quantities can be taken, and further symptoms are itemized in Sects. 2.2.2 and 2.2.3.

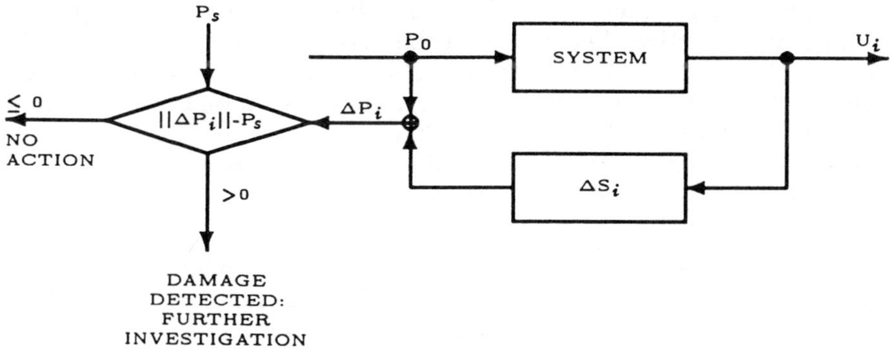

Fig. 2.21. Detection procedure based on a threshold P_s

Localization. Local residuals serve for detection as well as for localization, as has already been mentioned in Sect. 2.2.2[10]. In order to reduce the computational expenditure and the related numerical problems, one can introduce global damage discriminants [26], [33] as factors $a_{M\sigma}^{(i)}, a_{K\iota}^{(i)}, a_{G\iota}^{(i)}, a_{B\rho}^{(i)}$ of the physical parameter matrices associated with subsystems of the total system considered:

$$\mathbf{M}^{(i)} = \sum_{\sigma=1}^{S} a_{M\sigma}^{(i)} \mathbf{M}_{\sigma}^{(i-1)} \tag{2.2.24}$$

$$\mathbf{K}^{(i)} = \sum_{\iota=1}^{I} a_{K\iota}^{(i)} \mathbf{K}_{\iota}^{(i-1)}, \quad \mathbf{G}^{(i)} = \sum_{\iota=1}^{I} a_{G\iota}^{(i)} \mathbf{G}_{\iota}^{(i-1)}, \tag{2.2.25}$$

$$\mathbf{B}^{(i)} = \sum_{\rho=1}^{R} a_{B\rho}^{(i)} \mathbf{B}_{\rho}^{(i-1)}, \tag{2.2.26}$$

where the matrices $\mathbf{M}_{\sigma}^{(i-1)}, \mathbf{K}_{\iota}^{(i-1)}, \mathbf{G}_{\iota}^{(i-1)}, \mathbf{B}_{\rho}^{(i-1)}$ designate the submodels for the previous (reference) state,

$$\left.\begin{array}{c} \mathbf{M}^{(i-1)} := \sum_{\sigma=1}^{S} \mathbf{M}_{\sigma}^{(i-1)}, \\ \mathbf{K}^{(i-1)} := \sum_{\iota=1}^{I} \mathbf{K}_{\iota}^{(i-1)}, \ \mathbf{G}^{(i-1)} := \sum_{\iota=1}^{I} \mathbf{G}_{\iota}^{(i-1)}, \\ \mathbf{B}^{(i-1)} := \sum_{\rho=1}^{R} \mathbf{B}_{\rho}^{(i-1)}, \end{array}\right\} \tag{2.2.27}$$

and again the dependency on i is related to the life time θ_i. Deviations to be estimated of the a's from 1 (Eqs.(2.2.24) to (2.2.25) turn into Eq. (2.2.27) for the unmodified model) indicate modifications within the concerning

[10] Eigenfrequencies also have some local properties through their relationship to the modes which can be used (rank ordering [62], [63])

subsystems. If the submodelling for modification location is insufficient or insufficiently sensitive, then a hierarchical halving of the submodels can be done [26] together with an appropriate stopping rule, depending on the accuracy required for localization. The problem of how to choose these submodels (see Sect. 4.3.1) is assisted by

- prior knowledge of fault occurrence (experience), and
- theoretical fault investigations (see Fig. 1.1)
- sensitivity analysis.

Many other residuals can also be used. However, no attempt has been made to achieve completeness, and especially the choice of the residuals is dependent on the system, the fault and the purpose.

Problems. Consider the fact that the detection and localization are based on measurements which may be erroneous and incomplete, as already stated. The detection and localization results are therefore uncertain. If, for example, eigenquantities are chosen as symptoms corresponding to an N DOF model, with $N < n$, n the order of the prior model, the question arises of how to overcome this incompleteness for local detection by a comparison of the models, and if partial residuals are not taken. W.P. Rodden [65] proposes, as is also recommended in [66], a comparison of the spectral decomposition of the inverse inertia matrix and of the flexibility matrix of the measured eigenquantities with those of the prior mathematical model, taking only the corresponding N modal quantities. If one likes to work with regular matrices, then the missing eigenvectors can be substituted by arbitrarily linear independent vectors for comparison purposes only [67], or by order reduction [68].

The estimation of the a's by, for instance, the application of least squares, means handling inverse problems which, in general, are ill-conditioned for spatially discrete models. In particular, when computing with measured data, the measurement irregularities will generally be amplified. With the application of suitable algorithms [69], for example, the QR-method can avoid amplifying the inaccuracies. However, often no unique solution exists, so that regularization methods must be applied [70–72]. A very efficient procedure is the Bayesian approach, or if the confidence of the prior values cannot be estimated, the corresponding deterministic Tikhonov-Phillips method. The application of the pseudo-inverse combined with penalty terms regularizes the problem, which means that it gives a unique solution with satisfactory accuracy. These problems are discussed in Chap. 4.

Intensities. Stationary dynamic responses and eigenquantities contain some difficulties for the localization of modifications. Therefore, finally, structural intensities [73] should be mentioned which can lead to some new procedures in the future, although the methods have not been developed sufficiently yet. Only a simple example is therefore explained as a demonstration. Local energy transport by travelling waves can be described by the energy flux, as is well-known in acoustics, and they can

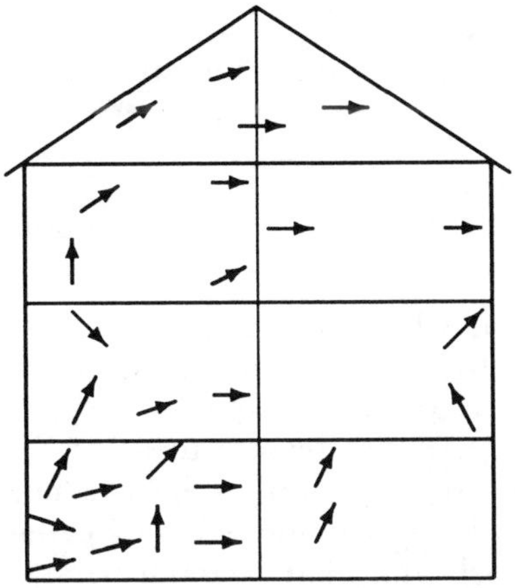

Fig. 2.22. Example of acoustic intensities

be measured by intensities which are vectorial quantities. They can also provide local information. Fig. 2.22 indicates an intensity distribution in a house due to a structure-borne vibration source. In consequence, intensity measurements are able to detect and localize sources (excitation) and sinks (damping) dependent on structural properties. FE modelling is used for determining energy fluxes [73] and finite differences [74].

The one-dimensional wave equation [9] will be considered:

$$m(x)\ddot{y}(x,t) - [S(x)y'(x,t)]' = 0, \tag{2.2.28}$$

with the stiffness distribution $S(x)$. The energy flux from x_1 to x_2 is obtained from the differentiated (with respect to time t) total energy contained in the interval $[x_1, x_2]$:

$$\frac{d}{dt}[E_{kin}(t) + E_{pot}(t)] = P(x,t)\ |_{x_1}^{x_2}, \tag{2.2.29}$$

where $P(x, t)$ is the instantaneous energy flux,

$$P(x,t) = -S(x)y'(x,t)\dot{y}(x,t). \tag{2.2.30}$$

As is done in acoustics, $P(x, t)\ |_{x_1}^{x_2}$ can be determined approximately by approximation of the derivatives:

$$x_0 = \frac{x_1 + x_2}{2}, \tag{2.2.31}$$

$$\dot{y}(x_0, t) \doteq \frac{\dot{y}(x_2, t) + \dot{y}(x_1, t)}{x_2 - x_1}, \tag{2.2.32}$$

$$y'(x_0, t) \doteq \frac{y(x_2, t) - y(x_1, t)}{x_2 - x_1}, \tag{2.2.33}$$

and this produces $P(x_0, t)$, as given in (2.2.30).

The average of the net energy flux in the positive x-direction (ascending wave) follows for the constant cross-section with $S(x) = S_0$ to [75]:

$$\overline{P(x_0 - ct)} = \frac{S_0}{2\pi c} \int_{-\infty}^{\infty} \frac{1}{sin2\omega\tau} Im[S_{12}(\omega)]d\omega \tag{2.2.34}$$

where $c^2 = S_0/(\rho A) = E/\rho$ for longitudinal vibrations (A is the cross-section area, ρ the material density, E is the Young's modulus), $\bar{\tau} := (x_2 - x_1)/c$, ω the variable frequency, and $S_{12}(\omega)$ the cross-power spectral density of the measured signals $\dot{y}(x_1, t)$, $\dot{y}(x_2, t)$ with respect to $\tau = \bar{\tau}/2$.

The advantage of one-dimensional wave propagation is the absence of dispersion. Dispersion makes the problems much more complicated for the bending of beams (Timoshenko model) and plates. For dispersive waves the superposition of the energy flux with respect to ascending and descending waves does not hold true, and therefore only approximations are available.

2.3 Damage Initiation and Evolution in Operating Systems

2.3.1 The Physics of Structural Wear and Damage

The wear of mechanical systems is directly linked to their operation and occurs in many ways, but it is always closely related to the dynamic phenomena, such as vibrations, acoustics including noise and ultrasound, and to the energy flow through the system, and dissipation inside the system.

Taking into account the main type of wear, one should first consider the fatigue phenomena which, if they occur throughout the machinery or structural part, cause a loss of integrity. If fatigue occurs on the surface, it can cause the pitting and/or spalling of moving parts (i.e. bearings, gears etc.) or fretting (fatigue corrosion) in inmovable structural joints, especially in a corrosive atmosphere. The energy dissipated inside the system due to these types of wear can be calculated by the equations given below (some are of a qualitative nature due to a broad meaning of coefficients)[24, 76].

Volumetric fatigue, according to Morrow's hypothesis [77], provides the following way of calculating the dissipated energy:

$$E_{d_1} = 3\left(\frac{\sigma_a^{1+b}}{k}\right)^{\frac{1}{b}} n, \quad n = \left[\int f(\theta)d\theta\right] \simeq [f \cdot \theta], \tag{2.3.1}$$

where the brackets [.] designate the Gaussian brackets, k, b are material constants, and σ_a is the alternative stress amplitude, n is the number of load cycles, and $f(\theta)$ is the instantaneous excitation frequency dependent on the life time θ of the component under consideration.

Surface fatigue energy can be calculated according to the Palmgren relationship [78] (Chap. 2) as energy dissipated due to pitting and spalling:

$$E_{d_2} = c_1 p^3 n, \tag{2.3.2}$$

where c_1 is again a material coefficient, and p the unit pressure in the mechanisms.

Corrosive fatigue - fretting, here the wearing energy is proportional to the volume V_0 of the removed material [78, 79] (Chap. 12)

$$E_{d_3} = c_2 V_0 = c_z \left[\frac{k_0\sqrt{p} - k_1 p}{f} - k_2 \hat{A} \right] n, \tag{2.3.3}$$

where c_z is a proportionality constant, k_0, k_1, k_2 are material constants, $\hat{A}$ is the vibration amplitude, and f the frequency .

The second important type of wear is adhesive and/or **abrasive wear** which proceeds in every rotating or sliding structural joint. The intensity of this type of wear depends on the lubrication quality, the unit pressure, as well as on the relative vibration amplitude. The energy dissipated by frictional wear is proportional to the volume of the abrasively removed material, which is governed by Archard's law [79, 80]

$$E_{d_4} = c_3 k_3 p \hat{V} R_e^{-1} \theta, \tag{2.3.4}$$

where c_3 is a proportionality constant, k_3 a material constant, R_e yields the stress of the material, and $\hat{V}$ is the sliding velocity amplitude.

Another type of wear is **erosion**. It is caused by such phenomena as cavitation, corrosion, impact of fluid stream, ions, or other particles etc. With the exception of erosion by radiation, it is mechanical vibration that intensifies the erosion, and sometimes the vibration initiates erosion, as in the case of cavitation. The dissipated energy for cavitational wear may be assessed by [78]

$$E_{d_5} = c_5 B \hat{V}^{b(\theta)} \theta. \tag{2.3.5}$$

Here again c_5 is a proportionality constant, B a material constant, and $b(\theta)$ is the cavitational exponent. The same type of energy equation can be invented for other types of erosion, but with different interpretations of $\hat{V}$ as the velocity amplitude of the flow of a liquid, a jet, of particles, etc.

The last type of wear in mechanical systems to be mentioned is **creep**, which is particularly important at higher temperatures and for complex mechanical loads: it is dependent on mean working stress and vibration with high frequencies. It may be stated here that the dissipated energy is proportional to the creep strain ε_{c_γ}, being highly dependent on the stress σ, the temperature T and the frequency f of ultrasound [81, 82]

$$E_{d_6} = c_6 \varepsilon_{c_\gamma} = c_4 \left\{ \frac{\sigma}{E} + e(T, f) \sigma^d \theta \right\}. \tag{2.3.6}$$

Here c_6 is a proportionality constant, $e(T, f)$ the creep function, E the Young's modulus, and the exponent d dependent on the material and temperature.

Having specified the main types of wear and associated dissipated energy, we can conclude that these energies are event-dependent, (n), as in fatigue processes, and life time-dependent (θ) for the other forms of wear. Hence, the the following can be written generally

$$E_d = \begin{cases} An & \text{event type wear} \\ B\theta & \text{life time type wear.} \end{cases} \tag{2.3.7}$$

In addition, it is advantageous to split the time-varying quantities into their mean values superimposed by the time-varying parts designated with the subscript a:

$$\left.\begin{array}{ll} \sigma = \sigma_m + \sigma_a & \text{stress} \\ p = p_m + p_a & \text{unit pressure} \\ V_s = V_m + V_a & \text{stream velocity.} \end{array}\right\} \tag{2.3.8}$$

The dynamic components defined above are a part of the internally dissipated power $N'_d(\sigma_a, p_a, V_m, \ldots)$ associated with the system operation. Hence, each energy of wear E_{di} can also be expressed in a more general sense:

$$E_{di} = E_{di}[\theta, n, N'_d(\sigma_a, p_a, V_a, \ldots), k_i, R_e, \ldots, V_m, p_m, \sigma_m, T, \ldots], \quad i = 1, \ldots, 6. \tag{2.3.9}$$

The variables in the above equation (except θ, n, $N_d(.)$) can be understood as attributes of the system behaviour dependent on the design quality Q, the manufacturing quality M, the use or load intensity L, the location of installation P, and the renewal quality R. The last relation may thus be rewritten as

$$E_{d_i} = E_{d_i}(\theta, n, N'_d, Q, M, L, P, R). \tag{2.3.10}$$

2.3.2 The Energy Model of System Damage and its Measure

Operating mechanical systems and their parts are open systems with respect to energy flux and dissipation (see the previous section). We can therefore treat them as dissipative systems in terms of General System Theory [83, 84]. Moreover, after a sufficiently (finite) operating time every mechanical system ends in breakdown as the result of wear, i.e. energy dissipation. From this simple observation it can be concluded that the energy dissipation capacity of each mechanical system is finite. This postulate is typical of an existing system. Hence let us designate those influence parameters enumerated in the previous section as components of the logistic vector $\Delta^T := \{Q, M, P, L, R\}$. One can thus write the limits of energy dissipation qualitatively as:

$$0 \leq E_d(\theta, V, \Delta) \leq E_{db}(\Delta). \tag{2.3.11}$$

For the new system (or its parts) we therefore assume the dissipated energy equal to zero, and the limit or breakdown value of it as $E_{db}(\cdot)$. E_{db} depends at least on the quality or the level of its design, manufacturing, load sequence and intensity, and maintenance operations (if any). This energy concept of damage and its link to the dissipative system would appear to be innovative.

Bearing in mind the variety of mechanical systems, their parts and their parametric dependencies on dissipated energy from Δ, we need a special dimensionless measure of system damage evolution or wear development. Following the authors' last works [85] to [87], the dimensionless measure of the total damage development or evolution in a system can be defined as

$$D(\theta, \Delta) := \frac{E_d(\theta, \Delta)}{E_{db}(\Delta)}, \tag{2.3.12}$$

with

$$\boxed{\text{new system}} = 0 \leq D \leq 1 = \boxed{\text{at breakdown}}.$$

What the particular forms of the above measure for the different types of wear can be is of great interest. Taking into account discrete events and continuous forms of wear separately (see Eq. (2.3.7)) we can find [24, 85]

$$D = \begin{cases} \frac{\theta}{\theta_b} & \text{for simple continuous wear} \\ \sum_i \frac{\theta_i}{\theta_{bi}} & \text{for multilevel continuous wear (including creep)} \\ \frac{n}{n_b} & \text{for simple cyclic wear} \\ \sum_i \frac{n_i}{n_{bi}} & \text{for multilevel cyclic wear} \\ \sum_i \frac{\theta_i}{\theta_{bi}} + \sum_j \frac{n_j}{n_{bj}} & \text{for intermittent continuous and cyclic wear.} \end{cases} \tag{2.3.13}$$

Any combination of the above forms of wear can also occur. It can be seen that the damage development measure is equivalent, for example, to the Odquist-Katchanov law of creep for the continuous type of wear, equivalent to the Palmgren-Miner fatigue law for cyclic wear, and to Archard's law of friction. Thus, the damage measures introduced above are simple to interpret, and they have a good experimental base.

In terms of system behaviour, the damage evolution measure simply describes the technical condition in applications of mechanical and civil engineering. Such defined measures are hard to measure directly. They can sometimes be measured on samples under laboratory conditions. Hence, for technical application we need to use symptoms S (see Sects. 2.2.2 and

2.2.3) of the system condition which are measurable physical quantities covariable with the damage measure. This means the mapping $S = \phi(D)$, where $\phi(D)$ is the symptom operator.

Introduction of the local (designated by r) damage measure

$$D(r, \theta) := \frac{E_d(r, \theta, \Delta)}{E_{db}(r, \Delta)} \tag{2.3.14}$$

results in the global damage measure through integration over the volume $\mathcal{V}$:

$$D(\theta, \Delta) = \frac{1}{\mathcal{V}} \int_{\mathcal{V}} D(r, \theta, \Delta) dr. \tag{2.3.15}$$

2.3.3 Damage and Symptom Evolution

The system analysed here is considered as an energy transforming system designed for a pre-specified demand, as shown in Fig. 2.23. The input power N_i, as the derivative of energy with respect to the life time θ, is transformed into the usable power N_u, and part is transformed into the dissipated power $N_d(\theta, V)$. Part of this dissipated power is stored internally as $E_d(\theta, V)$, causing the damage to the system, and the remaining power $V(\theta)$ is dissipated externally. In accordance with these assumptions, the dissipative power flow can be expressed as

$$N'_d(\theta, V) = \frac{\partial E_d(\theta, V)}{\partial \theta} = N_d[\theta, V(\theta)] - V(\theta). \tag{2.3.16}$$

Additionally, the above powers depend on the logistic vector Δ reflecting the history of the system under consideration. It must be remembered that in practice only part of the externally dissipated power $V(\theta)$ can be observed in terms of vibration, noise, temperature etc. Therefore, the goal is also to substitute the quantity $V(\theta)$ by the measurable quantity, which is the symptom, by creating a model of damage evolution.

Some constitutive equations must be assumed in order to create the energy processor model. These should be related to practice, of course. The first rule is:

1. balancing the energy flow, that is $N_i = N_u + N_d$,
2. the more the system is damaged, the greater is the externally dissipated power:

$$dV(\theta) = \beta dE_d[\theta, V(\theta)], \tag{2.3.17}$$

   with β a proportionality coefficient, constant in a first approach,
3. the internally dissipated energy is finite and bounded by E_{db},

$$E_d(\theta, V) = \int_0^\theta [N_d(\xi, V) - V(\xi)] d\xi \le E_{db}, \tag{2.3.18}$$

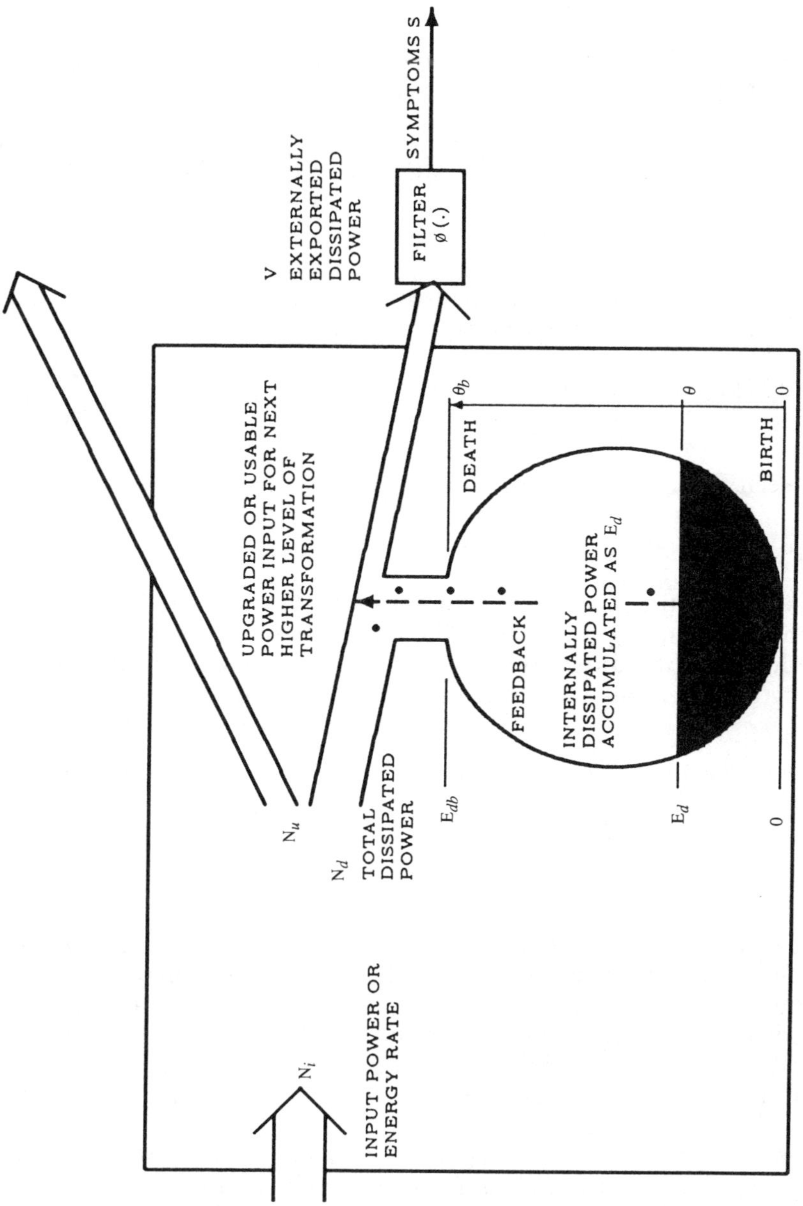

Fig. 2.23. Model of an operating system (open system) with limited dissipation capacity, destructive feedback and residual process observation

4. the (internal) structure of the system is assumed to remain unchanged within its lifetime, and this is expressed by a part of the total dissipated power in the form

$$\frac{\partial N_d[\theta, V(\theta)]}{\partial V(\theta)} = \alpha = \text{ const.} > 0, \rightsquigarrow V(\theta) = \alpha^{-1} N_d[\theta, V(\theta)]. \quad (2.3.19)$$

The increment of the internally dissipated energy $dE_d(\theta, V)$, taking into account Eqs.(2.3.16), (2.3.18), and (2.3.19), will turn into

$$dE_d(\theta, V) = \frac{\partial E_d(\theta, V)}{\partial V} dV + \frac{\partial E_d(\theta, V)}{\partial \theta} d\theta = (\alpha - 1)[\theta dV(\theta) + V(\theta) d\theta]. \tag{2.3.20}$$

Insertion of this equation into the constitutional Eq. (2.3.17) leads to the differential equation for the externally dissipated power,

$$\frac{dV(\theta)}{V(\theta)} = \frac{\beta(\alpha - 1)d\theta}{1 - \beta(\alpha - 1)\theta}. \tag{2.3.21}$$

It can be seen from Eq. (2.3.21) that with $V \to \infty$ it follows

$$\theta \to \theta_b = [\beta(\alpha - 1)]^{-1}. \tag{2.3.22}$$

With $V \to \infty$ the time θ_b is called the breakdown time (lifetime) of the system, and it is also the measure of the operability of the system.

The solution of the differential Eq. (2.3.21) is obtained with this notation, giving the internally and externally dissipated power,

$$\left.\begin{aligned} N_d(\theta) &= N_{d0}\left(1 - \tfrac{\theta}{\theta_b}\right)^{-1} = N_{d0}(1 - D)^{-1}, \; N_{d0} := N_d(\theta)|_{\theta=0}\,, \\ V(\theta) &= V_0\left(1 - \tfrac{\theta}{\theta_b}\right)^{-1} = V_0(1 - D)^{-1}, \; V_0 := V(\theta)|_{\theta=0}\,. \end{aligned}\right\} \tag{2.3.23}$$

With respect to the example (2.3.13) D, on one hand, is the dimensionless life time of the system, and on the other hand, it is the damage measure (2.3.12). Eqs.(2.3.23) show that the total dissipated power N_d and the power of the residual process are colinear. Both will tend towards infinity if the life time θ tends towards the breakdown time (lifetime) θ_b. However, there will not be an infinite power because the input power is finite and the balance $N_i = N_d + N_u$ must be kept.

With these results the next step can be taken toward the symptom evolution model. As is known from practice, only a filtered part of the external dissipated power can be observed in terms of vibration, noise, temperature etc. This fact may be acknowledged by the symptom operator $\phi(.)$ dependent on V:

$$\frac{S(\theta)}{S_0} = \phi\left(\frac{V(\theta)}{V_0}\right). \tag{2.3.24}$$

The symptom operator $\phi(.)$ can be found by theoretical studies with a mathematical model and by a proper diagnostic experiment, where $1-D =:$

ΔD is the dimensionless residual life of the system under consideration. According to the definition of the damage measure (2.3.12) and its physical meaning (2.3.13), the residual life can be written as

$$1 - D \equiv \Delta D = \begin{cases} 1 - \frac{\theta}{\theta_b} & \text{for continuous wear,} \\ 1 - \frac{n}{N_b} & \text{for cyclic wear.} \end{cases} \tag{2.3.25}$$

As discussed in [88, 86] this quantity may be also named the **damage capacity** of a system. The damage capacity of a new system is thus $\Delta D = 1$, and just before the system's breakdown its damage capacity is totally exhausted and approaches zero.

With this new interpretation of the argument of the symptom operator we will look once more for the Eq. (2.3.24) of the life symptom of the energy transforming system. If one identifies the type of symptom operator $\phi(.)$ in a diagnostic experiment or assumes it by prior knowledge, the inverse operator $\phi^{-1}(.)$ can be calculated. As the result one again obtains the damage capacity

$$\Delta D := 1 - D = \phi^{-1}\left(\frac{S_0}{S}\right). \tag{2.3.26}$$

This seems to be a very important step forward, which means that when one knows the damage evolution of the system expressed by its symptom life curve $S(\theta)$ or $S(D)$, its residual life or its damage capacity ΔD can be recalculated. As a generalization, considering the evolution of energy transforming systems, one can say that the foundation of symptom-based diagnostics is proven analytically.

From Eq. (2.3.26) it is seen how important the knowledge of the type of symptom operator $\phi(.)$ is in symptom-based diagnosis. Starting from Eq. (2.3.24), in a first approach $S(\theta)/S_0$ can be approximated by the first term of its power series as

$$\frac{S(\theta)}{S_0} = \phi\left(\frac{V(\theta)}{V_0}\right) \approx \left(\frac{V(\theta)}{V_0}\right)^{\frac{1}{\gamma}} + \dots\,, \quad \gamma(\phi) > 0. \tag{2.3.27}$$

This approximation gives the symptom life curve of Pareto type, which is more easily recognizable when Eq. (2.3.23) is put into (2.3.27):

$$S(\theta) = S_0\left(1 - \frac{\theta}{\theta_b}\right)^{-\frac{1}{\gamma}}. \tag{2.3.28}$$

This equation is the symptom life curve with asymptotic behaviour at the breakdown time. The same symptom life curve expressed in terms of the damage measure reads

$$S(D) = S_0(1 - D)^{-1/\gamma} = S_0(\Delta D)^{-1/\gamma}. \tag{2.3.29}$$

These curves are illustrated in Fig. 2.24. The asymptotic behaviour of the

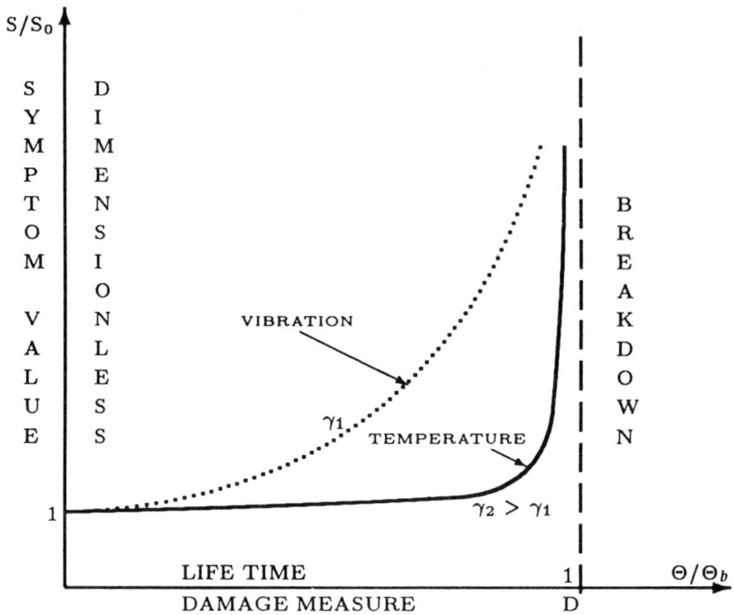

Fig. 2.24. Symptom evolution of the system condition dependent on the life time and damage: the symptom life curve with two different shape factors γ

symptom life curve can be noted at the breakdown time $\theta = \theta_b$ $(D = 1)$. It can also be seen that not every symptom dependent on γ may be sufficiently sensitive to reflect the technical condition of the system under investigation. For the Pareto symptom operator it can be stated in more detail that the smaller the value of γ the more sensitive is the given symptom. This is also illustrated in Fig. 2.24 when the temperature and vibration symptoms, measured for the same system in operation, are compared.

It can easily be seen from Eq. (2.3.29) that the damage capacity can be written as a function of the Pareto symptom operator:

$$\Delta D = \left(\frac{S_0}{S}\right)^{\gamma} . \tag{2.3.30}$$

Hence, for the Paretian life symptom the system damage capacity can be calculated very easily and expressed graphically; in the log-log scale it is a straight line inclined by the power γ. This fact is shown in Fig. 2.25 together with the residual life or damage capacity ΔD_e determination when the symptom value S_e has been observed or caculated. If there is a set of systems of the same type in operation, but of different life history, during their operation one can then observe a bunch of symptom life curves:

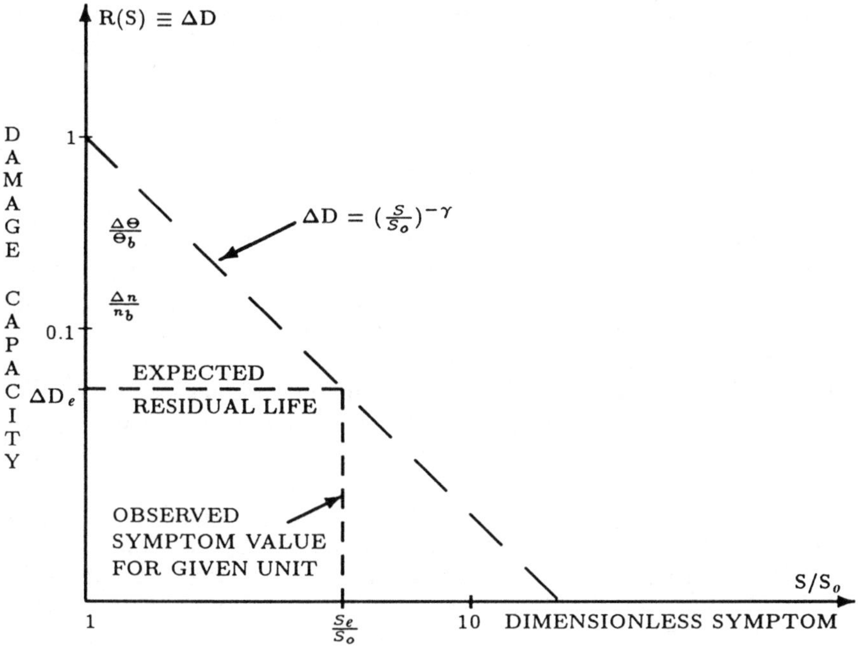

Fig. 2.25. Determination of system damage capacity for Paretian symptom

$$S(\theta, \Delta) = \frac{S_0(\Delta)}{[1 - \theta/\theta_b(\Delta)]^{1/\gamma}} = \frac{S_0(\Delta)}{[1 - D(\Delta)]^{1/\gamma}}, \quad \Delta = \Delta(Q, M, L, P, R). \tag{2.3.31}$$

When one looks for this symptom-life curve dependent on several parameters, the logistic vector Δ is included, or even dependent on the process. For example, for the load history L it can be concluded that assessment of the conditions of mechanical systems should be supported by statistics applied to suitably chosen symptoms during routine diagnostic practice (monitoring).

2.3.4
Mechanical Properties of Operating Systems in the Holistic Approach

If symptoms S evolve during the system's operation, how do the primary characteristics of mechanical systems, the mass, the stiffness, and the damping ratio behave? The essential characteristics of mechanical systems during their life-stages, expressed for an SDOF model, are the stiffness k, the inertia m, and the damping coefficient b (internal dissipation energy). The damage considered is cumulative, so it also contributes to the life time-dependent parameter modifications. Hence, following the damage measures proposed by Natke and Yao [33], the following equations describing the evolution of the model parameters [85]) can be obtained:

$$\left.\begin{array}{lcl} m(D) & = & m_0(1 \mp a_m D)^{\alpha_m} \\ k(D) & = & k_0(1 \mp a_k D)^{\alpha_k} \\ b(D) & = & b_0(1 \mp a_b D)^{\alpha_b} \end{array}\right\} \tag{2.3.32}$$

with exponents $\alpha_m, \alpha_k, \alpha_b \geq 0$, and damage intensity coefficients: $0 \leq a_m \leq 1$, $0 \leq a_k \leq 1$, $0 \leq a_b \leq 1$, which can be estimated by a suitable identification technique. D is the measure of cumulative damage according to Eq. (2.3.13), which in simple cases can be substituted by the dimensionless life time or dimensionless number of events.

Other models describing the system parameter changes during the life do exist [89][11]. The model presented above is very convenient for adjusting the type of modifications α_i, and the intensity or wear participation factor a_i. Now the question arises: can these characteristics serve as symptoms of the technical condition? According to the general definition of a symptom (2.3.24) and with the previous one, $S = \phi(D)$, the question can be answered positively. But these damage measures, when equated to the symptoms (2.3.32), can usually be measured only in laboratory conditions. However, various quantities can be measured indirectly as symptoms. It is easy to measure the deflection x due to a known force (static, dynamic; deterministic, random) [85]. It is shown in [85] that the static deflection as well as the dynamic response due to random white noise, taken as the symptom of damage, give the same result expressed in a dimensionless manner:

$$\frac{S_x(D)}{S_x(0)} = (1 - a_k D)^{-\alpha_k}, \tag{2.3.33}$$

if one acknowledges only the degradation of the stiffness $\alpha_k \neq 0$ in (2.3.32). If this is compared with the symptom-life curve (2.3.29), $\gamma_x = (\alpha_k)^{-1}$ can be calculated for the deflection as the symptom. Sometimes it is easier to measure the eigenperiod of the vibration $T(D)$. It can be shown that under the same assumptions of stiffness degradation the dimensionless symptom for the eigenperiod is

$$\frac{S_T(D)}{S_T(0)} = (1 - a_k D)^{-\frac{\alpha_k}{2}}, \tag{2.3.34}$$

and it provides a symptom exponent twice as great as for the deflection: $\gamma_T = (2/\alpha_k) > \gamma_x = 1/\alpha_k$. But additionally to the easiness of measuring an eigenperiod, this symptom has another important advantage [85].

[11] The authors are aware of the overdetermined definition in (2.3.32). However, it allows a more flexible and more sensitive adjustment. It is the simplest approximation, for example of $m_0 exp(\alpha_m D)$.

In [90] a measure of structural degradation for earthquake engineering investigations is introduced called *softening*. It is defined as

$$D_s := 1 - \frac{S_T(0)}{S_T(D)}. \tag{2.3.35}$$

It is calculated according to Eq. (2.3.34):

$$D_s = 1 - (1 - a_k D)^{\alpha_k/2}, \tag{2.3.36}$$

one obtains the measure of damage as defined in Eq. (2.3.12), and it follows simply that

$$D_s = D \quad \text{if} \quad a_k = 1, \ \ \alpha_k = 2. \tag{2.3.37}$$

This means that softening can be equated to the energy-based measure D of damage. In some cases the parameters a_k and α_k can thus be identified by the measured fundamental eigenperiod as a symptom of the vibrating system. Additionally, the accumulative damage D can be assessed directly.

As a generalization of this example and in accordance with the statement made in Sect. 2.2, in the diagnosis of conditions it is very important to choose an *appropriate symptom*, i.e. a symptom with maximum sensitivity with respect to the damage D and minimum sensitivity with respect to environmental noise.

It will be helpful to illustrate this holistic approach to system dynamics and its relation to diagnostics and system condition evolution. Very often, industrial systems act under severe environmental conditions like corrosion. Interactions of tall chimneys with wind, of platform legs with wind and water etc. are other examples. The latter two systems can be modelled by cantilevered beams with decreasing cross-sections due to corrosion, and with a fatigue process due to wind loading. A holistic model of such a system was created, and the fundamental eigenperiod of vibration was chosen as a symptom [88]. The result of the simulation is shown in Fig. 2.26, which illustrates the symptoms and condition evolutions of a steel chimney and a steel platform leg. As is seen, the normalized eigenperiod symptoms increase quite substantially up to ten times at the end of the system's life. The recalculation of the symptom dependent on life time to the damage capacity according to Eq. (2.3.26) is shown in Fig. 2.27. It is seen from the graph that the beginning is not a straight line, but the remaining part is straight (Paretian). Such a graph can be used for residual life assessment and for determining the limit value of life with respect to safety, as it is done there with $\Delta D_l = 0.1$ as the limit (of safety) value.

As a conclusion from this example, with the aid of a holistic mathematical model of the system under consideration and simulations of the dynamic behaviour, the symptom-based diagnosis can already be used at the design stage in order to assess the evolution of the life time condition. The result can be the locating of structural weak points where design improvements are necessary, and if the system exists, it will give the trans-

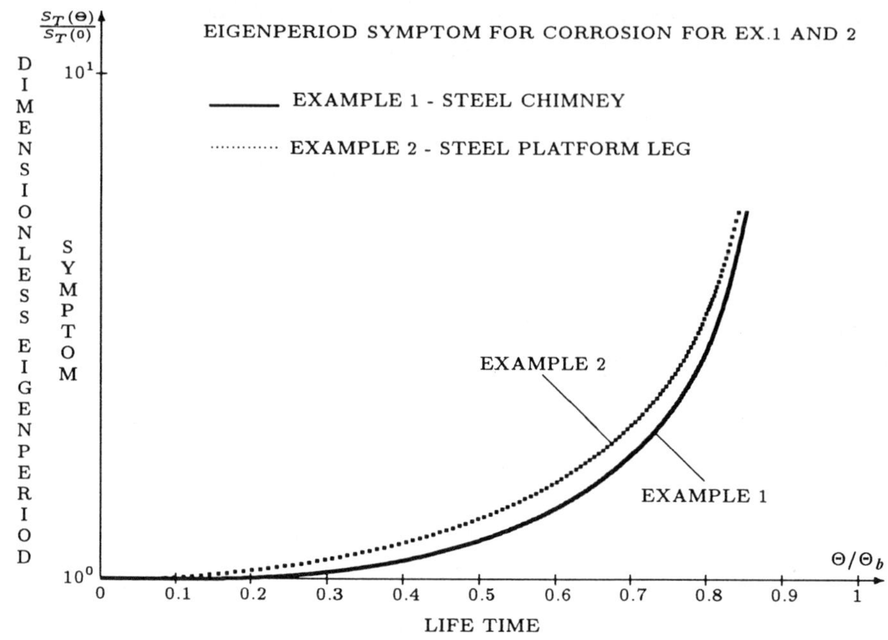

Fig. 2.26. The behaviour of eigenperiod symptoms for steel structures ageing due to corrosion

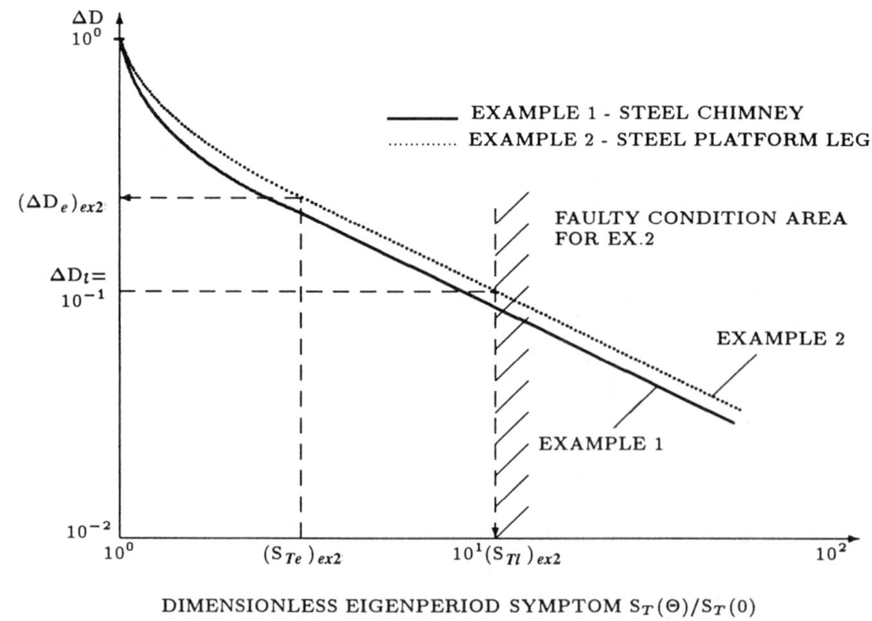

Fig. 2.27. The damage capacity ΔD for the eigenvibration periods and their use for symptom limit value determination, condition assessment, and trend prediction

ducer locations for system monitoring. Additionally, such simulations can provide in a first assessment of the system breakdown time θ_b.

2.3.5 Damage Capacity and Symptom Reliability of Operating Systems and their Diagnostic Use

As is mentioned in Sect. 2.3.3, if a set of systems of the same type is operating, these systems are generally of different origin with different histories designated by the logistic vector $\Delta(Q, M, L, P, R)$. Additionally, very often only a few statistical data are available. The appropriate symptom will be $S(\theta, \Delta)$. $N \gg 1$ systems of the same type are considered. Due to a lack of information Δ will be treated as a random variable. Therefore $S = S(\Delta)$ is also a random variable. Proper statistics for the symptom S are now required in order to assess the condition of a unit from the sample.

The best characteristic describing the operational properties of a set of units of the same type is the reliability. There are various definitions, and according to [91, 92] reliability is the characteristic of the ability of a component or a system to perform a specific function. For the purpose of performance assessment the life time-based reliability $R_\theta(\theta)$ is used and defined as the probability that the system will operate in specific conditions at a life time θ smaller than the breakdown time θ_b. The latter is, of course, a random variable,

$$R_\theta(\theta) = P(\theta_b - \theta > 0). \tag{2.3.38}$$

The relative residual life can be substituted by the damage measure increment, which is called the damage capacity ΔD here (see the previous section),

$$\Delta D := \frac{\Delta\theta}{\theta_b} := \frac{\theta_b - \theta}{\theta_b}. \tag{2.3.39}$$

The system reliability then follows as the probability of the system's residual life equal to the probability of the system's damage capacity which is greater than zero:

$$R_\theta(\theta) = P(\Delta D > 0). \tag{2.3.40}$$

As shown above, the internally observed dissipated energy is the measure of damage evolution. Additionally, this is reflected by the symptom $S(\theta, \Delta)$. Consequently, the system's reliability may be expressed by the symptom reliability based on the symptom life curve.

Corresponding to the breakdown time, the breakdown value of the symptom is defined as

$$S_b = S(\theta, \Delta)|_{\theta=\theta_b}. \tag{2.3.41}$$

For reasons of safety, if the systems are in continuous use the limit value of GOOD condition S_l is introduced, calculated from known properties of the random variable S_b by applying the statistical decision theory [24, 93].

In terms of the observed symptom $S(\theta, \Delta)$ at time θ the symptom reliability is defined as follows: the **symptom reliability** $R(S)$ is the probability that a unit classified as being in GOOD condition $(S < S_l)$ will be in operation; it is designated by the measured and required symptom value S [86]. Taking it initially as the conditional probability with S_l the symptom limit value, it follows

$$R(S) \equiv P(S_b > S | S < S_l) \equiv P_G(S_b > S). \tag{2.3.42}$$

As clarification of this definition, it is assumed that a set of units is measured at a calendar time θ_0, obtaining a set of symptom readings $\{S\}$, and that the restrictions given in the argument of definition (2.3.42) have been fulfilled. But even a short time later, at $\theta_0 + \delta\theta$, the symptom will be changed to $S + \delta S$, with $\delta\theta, \delta S > 0$, and a few units may approach the symptom breakdown value: $S + \delta S = S_b$. In order to predict how many units may fail, and using the definition (2.3.42), the total probability can be written as a conditional probability with the fact $S < S_b$:

$$\begin{aligned} & P_G(S + \delta S \leq S_b | S < S_b) \\ & = P_G(S + \delta S < S_b | S < S_b) + P_G(S + \delta S = S_b | S < S_b) \\ & = P_{GG}(S + \delta S) + P_{GF}(S + \delta S). \end{aligned} \tag{2.3.43}$$

Here $P_{GG}(.)$ and $P_{GF}(.)$ are the probabilities that the unit will will not fail and will fail, respectively, with the corresponding symptom value. The probability of failure can be assessed further [86]:

$$\left.\begin{aligned} P_{GF}(S + \delta S) &= P_G(S + \delta S \leq S_b | S < S_b) \\ = \frac{P_G(S < S_b \leq S + \delta S)}{P_G(S < S_b)} &= \frac{P_G(S) - P_G(S + \delta S)}{R(S)} = \\ &= \lambda(S)\delta S - \ldots \approx \lambda(S)\delta S, \end{aligned}\right\} \tag{2.3.44}$$

with

$$\lambda(S) := -\frac{d \ln R(S)}{dS}. \tag{2.3.45}$$

$\lambda(S)$ can physically interpreted as failure/fault intensity per unit symptom increment. The probability (2.3.44) is the assessment of failure probability $P_{GF}(.)$ for a unit being in GOOD condition with $S < S_l$.

Now the symptom reliability can be defined in more detail as below:

$$\left.\begin{aligned} R(S) &:= P_G(S) = P_{GG}(S) + P_{GF}(S), \\ p_G(S) &= p_{GG}(S) + p_{GF}(S), \\ R(S) &= \textstyle\int_S^\infty p_G(S_u) dS_u. \end{aligned}\right\} \tag{2.3.46}$$

It should be added that this defined symptom reliability is the generalization of the life-based reliability (Eq. (2.3.38)), and both expressions are equal only for $S_l = S_b$ [86].

Looking at the symptom reliability as calculated from the definition (2.3.46), we should remember that we know how the system deteriorates: it is in accordance with the theory described by Eq. (2.3.28). But with a set of systems in operation that have different histories, some individual pdfs $p_{GD}(D)$ of their damage measures will exist. The transformation of the probabilities can be performed with this relation. From probability theory the relationship between the pdf $p_G(S)$ and $p_{GD}(D)$ follows in accordance with [94],

$$p_G(S) = p_{GD}(D)\left|\frac{d(D)}{dS}\right|. \tag{2.3.47}$$

From (2.3.30) it follows that

$$D = 1 - \left(\frac{S_0}{S}\right)^{\gamma} = 1 - \left(\frac{S}{S_0}\right)^{-\gamma}. \tag{2.3.48}$$

Therefore,

$$\frac{dD}{dS} = \gamma\frac{S^{-\gamma-1}}{S_0^{-\gamma}} = \frac{\gamma}{S}\left(\frac{S_0}{S}\right)^{\gamma}, \quad S \geq S_0, \ \gamma > 0 \tag{2.3.49}$$

holds true.

The pdf $p_{GD}(D)$ is assumed to be generally symptom-dependent:

$$p_{GD}(D) = b_k\left(\frac{S(D)}{S_0}\right)^{k}, \quad k < \gamma\ , (b_k = 1 \text{ for } k = 0). \tag{2.3.50}$$

b_k is the normalization coefficient.

Using (2.3.46), one finds that

$$R(S) = \int_S^{\infty} p_G(x)dx = \int_S^{\infty} b_k\gamma\left(\frac{S_0}{x}\right)^{\gamma-k}\frac{dx}{x} = b_k\frac{\gamma}{\gamma-k}\left(\frac{S_0}{S}\right)^{\gamma-k}, \quad \gamma > k \geq 0. \tag{2.3.51}$$

As can be seen, independent of the method of observation of our set of systems, i.e. the value of the exponent in Eq. (2.3.50), the reliability distribution is of the Pareto type. If for simplicity $k = 0$ and $b_k = 1$ are set,

$$R(S) \equiv P(S < S_l | S_b - S > 0) = \left(\frac{S_0}{S}\right)^{\gamma}, \quad \gamma > 0, \tag{2.3.52}$$

finally follows with the same exponent γ as for the symptom life curve (see Eq. (2.3.28)).

What does this mean physically? For the Paretian life symptom Eq. (2.3.27) always gives a Pareto distribution of the symptom S, independent of the type of damage distribution (for any k in (2.3.50)) in a set of units $p_{GD}(D)$. Consequently, this behaviour is typical of evolutionary (hierarchical) processes independent of the type of system [95, 96]. If we now compare the right-hand side of the Paretian reliability (2.3.52) with the symptom life curve (2.3.28), it can easily be found that the following relationship holds true:

$$R(S) = \left(\frac{S_0}{S}\right)^{\gamma} = 1 - D \equiv \Delta D(S). \tag{2.3.53}$$

As is obvious, the symptom reliability is equal to the residual damage capacity, but now with a probabilistic meaning. This means that having determined the symptom reliability function from the diagnostic experiment, we can interpret it as the statistical equivalence of the residual damage capacity of a given unit in terms of the residual lifetime or residual number of cycles. Hence, the following relation of equivalence, following the equivalence of the damage measure (2.3.13), can be written and used:

$$R(S) = \left(\frac{S_0}{S}\right)^{\gamma} = \begin{cases} \Delta D & \text{residual damage capacity} \\ \Delta\theta/\theta_b & \text{residual life (dimensionless)} \\ \frac{\Delta n}{n_b} & \text{residual number of cycles (dimensionless).} \end{cases} \tag{2.3.54}$$

The diagnostic use of the above very important relation can be explained better in Fig. 2.28, where the symptom reliability function is shown as a result of the observation of a system set for a given value of S_e. This allows us to determine the expected damage capacity ΔD, or the equivalent residual lifetime $\Delta\theta/\theta_b$, or cycles $\Delta n_e/n_b$. A reliability graph of this kind is a straight line for the Pareto reliability and, if presented in double log-scale, it is very simple. It can be of great prognostic value, for example for the monitored machinery set in operation.

The theory of damage evolution, and the meaning elaborated here in detail for the Pareto life curve and distribution are also valid for other right-hand side skew long-tailed distributions, such as Weibull and Fréchet in a first approximation [97]. It is worthwhile to add here that both reliability approximations are complementary, i.e. if S has Weibull properties, then S^{-1} has Fréchet properties. It is noted that the Pareto pdf is the first (asymptotic) approximation of the Fréchet pdf. There is one more argument for using the Weibull distribution, since it can be applied in an advanced model of proportional hazards [98]. One can show that, dependent on the chosen symptom $S = \phi(V)$, (see Eq. (2.3.24)), the results shown in Table 2.2 below can be obtained.

Hence, following the results of Table 2.2, for application one can choose different symptom-life curves and symptom reliabilities, dependent on the observed tendency in a set of operating systems (OS). It may be useful here to recommend the use of the Fréchet description of the observed system life properties. The Pareto distribution is the asymptotic approximation of

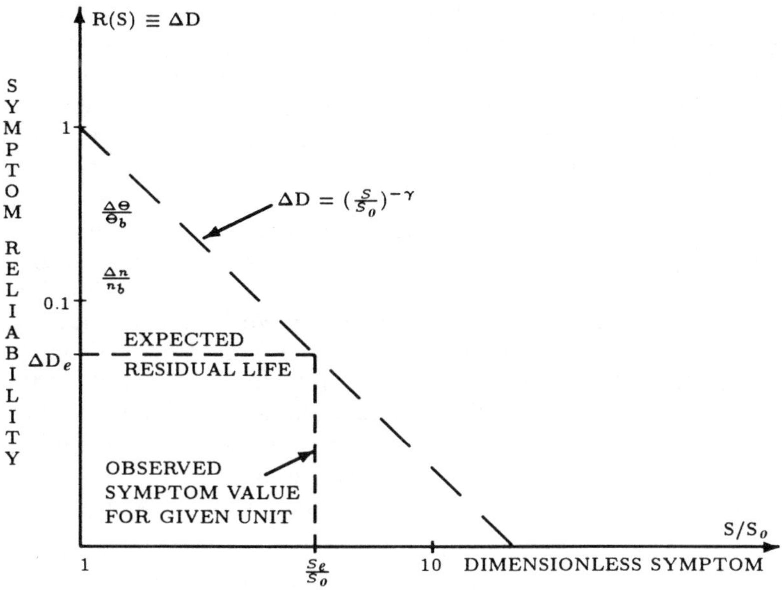

Fig. 2.28. Example of the prognostic use of the symptom reliability and/or damage capacity $R(S)$ for a set of systems as the equivalence of the residual lifetime

Table 2.2. Various symptom operators, life curves and symptom reliabilities for the diagnosis of systems in operation

Symptom operator $V > V_0,\ \gamma > 0$ $\Phi(\frac{V}{V_0}) \simeq$	Symptom life curve $S_0 > 0,\ \frac{S(D)}{S_0} =$	Symptom reliability, or Damage capacity $R(S) =$	Remarks, Symptom model
$\left(\ln \frac{V}{V_0}\right)^{1/\gamma}$	$[-\ln(1-D)]^{1/\gamma}$	$\exp[-(\frac{S}{S_0})^{\gamma}]$	**Weibull,** $S \geq S_0$
$[-\ln(1-\frac{V_0}{V})]^{-1/\gamma}$	$[-lnD]^{-1/\gamma}$	$1-\exp[-(\frac{S}{S_0})^{-\gamma}]$	**Fréchet,** $S \geq S_0$ (reciprocity of Weibull type symptom)
$(\frac{V}{V_0})^{1/\gamma}$	$(1-D)^{-1/\gamma}$	$(\frac{S}{S_0})^{-\gamma}$	**Pareto,** $S \geq S_0$ (asymptotics of Fréchet)
$1+\left(1-\frac{V_0}{V}\right)\cdot\frac{1}{\gamma}$	$1+\frac{1}{\gamma}D$	$1+\left(1-\frac{S}{S_0}\right)\cdot\gamma$	**Uniform,** $S \geq S_0$ $S \sim S_0$ (approximation of Pareto and exp(.))
$\exp\left(\frac{1}{\gamma}(1-\frac{V_0}{V}\right)$	$\exp(\frac{1}{\gamma}D)$	$1-\gamma\ln\frac{S}{S_0}$	**Exponential,** $S \geq S_0$

Fréchet, and the Fréchet symptom life curve has much greater dynamics and sensitivity. The Weibull distribution and corresponding life curve can be seen as complementary, because it can be shown that if S obeys a Fréchet distribution, then S^{-1} is Weibull distributed and vice versa. Having assessed (from the data base) the type of symptom reliability and its shape coefficient γ, diagnosis can be performed, i.e. the assessment of the present and the future condition of the system under question. This can be done independently, either in the symptom domain S or in the residual life domain ΔD.

In order to assess the machine condition in the symptom domain, we require some symptom limit values S_l, as already introduced, or standard values for comparison. In the damage measure domain the residual or remaining life will be assessed, as was already shown by the relationship (2.3.54). But in order to determine the symptom limit value S_l, or the alarm value S_a, the statistical decision theory [99] has to be applied: the Neyman-Pearson rule of risk assessment. If the availability of the machines is known (or the machine group), $0 < G \leq 1$[12], and if the allowable risk is specified, $0 < A \ll 1$, expressed by the false condition assessment, the limit values (breakdown and alarm) can be calculated as follows (see Sect. 5.1):

$$\left.\begin{aligned} G \cdot R(S_l) &= A, \\ G \cdot R(S_a) &= cA. \end{aligned}\right\} \tag{2.3.55}$$

As can be seen, the symptom reliability $R(S)$ is used here again, and only the allowable probability of unneeded repair A (false decision) should be specified, as well as the value of the coefficient c. Usually, dependent on the age and the maintainance quality of machines, A is chosen between 0.02 and 0.1, which corresponds to $2\% \div 10\%$, and for the alarm coefficient c values between 2 and 4. This means that unnecessary repairing of $(2 \div 10)\%$ of our machinery stock is taken into account in order to avoid the breakdown condition. This rule of symptom limit value calculation gives good results in the vibration condition monitoring of machines [24].

The residual life assessment will start with the relationship (2.3.54). Having measured the life symptom value S_e for the machine under consideration, one can write

$$R(S_e) = 1 - D_e \equiv \Delta D_e. \tag{2.3.56}$$

On the other hand (and from the above considerations), we can also set the safety limit of the remaining life, because it is shown that

$$R(S_l) = \frac{A}{G} \equiv \Delta D_l \quad \text{– limit remaining life.} \tag{2.3.57}$$

[12] For non-repairable systems it is $G = 1$.

Now the basic relations for the condition assessment of the system can be written

$$\left.\begin{array}{lll}\text{GOOD condition:} & & \\ & \text{If} & S_e \leq S_l \\ & \text{or} & \Delta D_e > \Delta D_l, \\ \text{FAULTY condition:} & & \\ & \text{If} & S_e \geq S_l \\ & \text{or} & \Delta D_e \leq \Delta D_l. \end{array}\right\} \tag{2.3.58}$$

This is the method of condition assessment, and in the case of the periodic monitoring of operating systems with the incremental life step $d\theta$, even a probabilistic approximation of the breakdown time can be calculated,

$$\theta_b = N_b d\theta = \frac{n_e}{1 - R_e} d\theta, \tag{2.3.59}$$

and the residual number of observations,

$$\Delta N_b = \frac{n_e R_e}{1 - R_e}, \tag{2.3.60}$$

where $R_e = R(S_e)$ and n_e is the successive number of given observations with the value S_e, when counting has started with units as *new*, i.e. for $\theta = 0, n = 0$.

Knowing this, we can even predict the next symptom value after the life time increment $d\theta$. This assessment of the remaining life, and a prediction of the corresponding symptom S_p can be performed. With $D_p = D_e + \delta D_p$ it follows:

$$\frac{S_p}{S_0} \equiv \frac{S(D_e + \delta D_p)}{S_0} = \phi[(1 - D_e - \delta D_p)^{-1}] = \phi\left[\left(1 - \frac{n_e + p}{N_b}\right)^{-1}\right], \tag{2.3.61}$$

where $\delta D_p = (pd\theta)/(N_b d\theta)$ is the dimensionless prediction step related to the number p of observations, if the condition monitoring is periodic.

For the symptom models of the Weibull or Fréchet type the symptom value can be predicted using the equation

$$S_{n_e+p} = S_{n_e}\left(1 + \frac{pN_b}{n_e \gamma \Delta N_b}\right), \quad p = 1, 2, \ldots \ll n_e \leq N_b. \tag{2.3.62}$$

The theory presented in brief here has great inference power and can be generalized far beyond the domain of the systems considered here [96]. It has been generalized lately into the theory of energy transforming systems (ETS), [100, 101], where new fractal-like abilities and behaviour have been revealed. This gives a physically based method for the reliability and diagnostic modelling of OS shown briefly below.

It will be noted that for non-repairable systems in civil and mechanical engineering the evaluation of system performance can also be carried out reliably by application of the crossing theory. The excellent book by T.T. Soong and M. Grigoriu [102] must be mentioned here.

2.3.6 Application of the Damage Evolution Theory

Operating systems are mostly of complex design, function, energy flow, etc. As is postulated in the general system theory (GST), this complexity is a fundamental reality of most systems [103]. From this point of view OS consist of many subsystems (assemblies, subassemblies, components) with their specific energy fluxes and dissipations. Hence the question arises: should every subsystem be treated as an energy transforming systems (ETS)? Considering this more deeply, one will come to the conclusion that every subsystem of an OS with a real or virtual boundary can be treated as ETS. Moreover, subsystems must sometimes be treated as ETS because their failure probability is high, substantially lowering the mission completion for the OS. Consequently, from the point of view of the energy flux, OS must be treated as an ensemble of ETS units organized hierarchically according to their design goal, specifications, and functional properties [101]. One should know that each ETS module has its own internal life $\theta_{(i)}$, and the breakdown time $\theta_{b(i)}$. For this reason, all the dynamic processes of energy transformation proceed multidimensionally. Fig. 2.29 is an attempt to explain this complex situation of the interdependency of subsystem lives and energy fluxes in the OS (more can be found in [101]). When one looks at this figure, several important conclusions can be noted. First, the self-similarity of energy transforming processes at each level of the hierarchy of the OS is seen. In other words, the self-similarity, or fractal-like energy transformation processes at every level of OS energy flux decomposition [96], can be observed. Secondly, at each level of the hierarchical ETS model one can establish an energy balance equation, which will take into account all the energy fluxes down from the given level of hierarchy. Thirdly, all the ETS outputs of a given hierarchical level contain all the required information (for example, damage evolution) from the lower levels of hierarchy. But the greater the distance the levels have, the harder the deciphering of their contents becomes.

Two cases from mechanical engineering and from civil engineering will be mentioned here as examples of the application of the ETS concept. Following this theory, several computer programs have been prepared to transform the symptom data base and to look for the most appropriate symptom reliability distribution, for example Weibull, Fréchet, etc.[104], and polynomial approximation, [105]. The program calculates some measure of goodness of fit, such as the coefficient of R and χ^2, the exponent γ of given distribution, the symptom limit value S_l , the residual life ΔD_e, and the residual observation number before the breakdown ΔN_b, (see

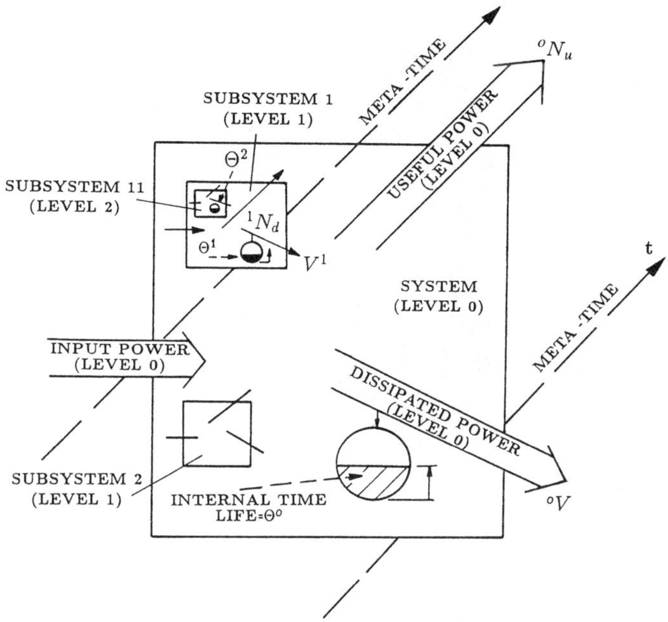

Fig. 2.29. Operating system as the hierarchical structure of ETS modules and the resulting interdependence of internal times immersed in the time flow of the meta-system

Eq. (2.3.60) and neighbouring ones). The series of these programs are programmed in 'MATLAB'.

As the reader remembers, the ETS theory permits one to understand the system damage capacity $\Delta D(S)$ in a deterministic and probabilistic way. The damage capacity can be understood deterministically for a given system which can be observed from its birth to its death. However, when observing a group of operating systems, the statistical concept of system reliability should be applied. Thus the computer results from the above-mentioned MATLAB programs can have a deterministic or probabilistic meaning.

Considering the deterministic meaning of $\Delta D(S)$ and taking respective symptom models (Weibull etc.), the life of the system in operation can be analyzed, for example, with the holistic model. It shows the modifications of the system properties and the operational characteristics during the damage accumulation caused by operational forces. Such an example is discussed in [88], where the life of a steel chimney under corrosion and fatigue caused by wind is investigated. Several symptoms are considered by simulation, the fundamental eigenperiod has been found as the best result for damage observation. $S_T(D)$ in Fig. 2.30 shows the application of the mentioned MATLAB programs for diagnosis of the residual life or damage capacity. From the figure one can see the simulated life of the chimney up to the symptom readings number 37, and the forecast of the symptom value for $p = 3$ steps. It can be seen that the residual life of

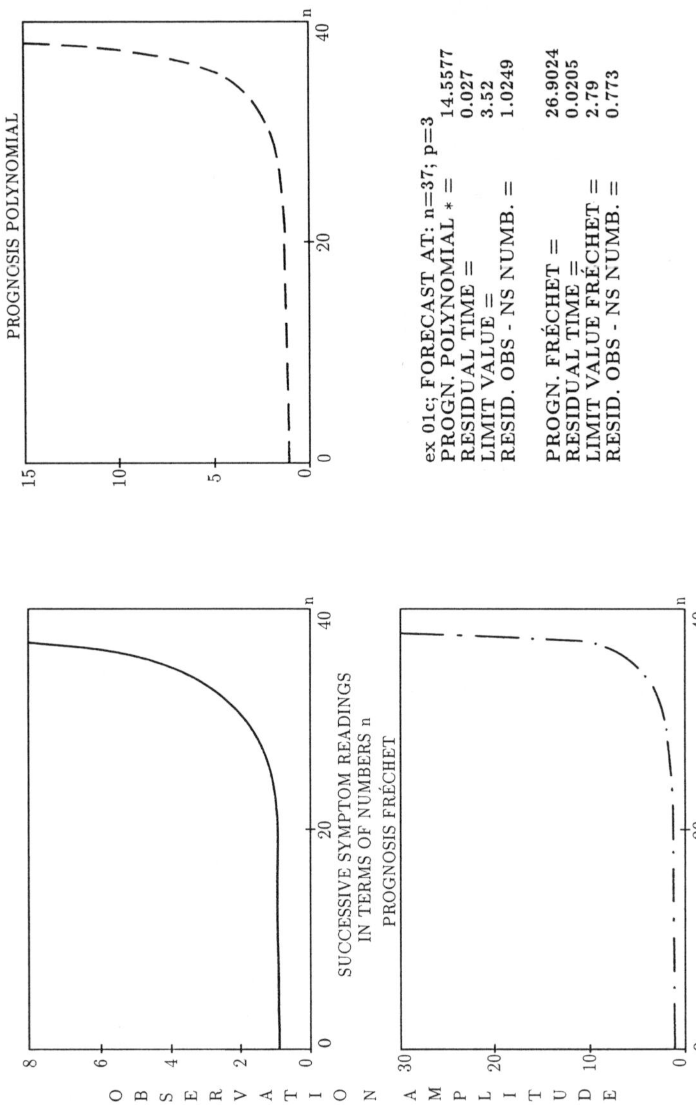

Fig. 2.30. Condition forecasting of the steel industrial chimney under the corrosion action by the computer program based on the ETS theory

the system is exhausted because the two different models of the symptom used for prediction give similar results (resid. obs. nos. 1.02 and 0.77).

If we look for the application of the ETS theory in the probabilistic case, when a group of operating units will be observed, Fig. 2.31 can now be interpreted. Here a group of $M = 15$ diesel engines were observed in terms of rms-amplitudes of acceleration each 10,000 km of service. These symptom readings are taken as the data base. The symptom reliability models were found with the best fit. Here it is seen that the Fréchet

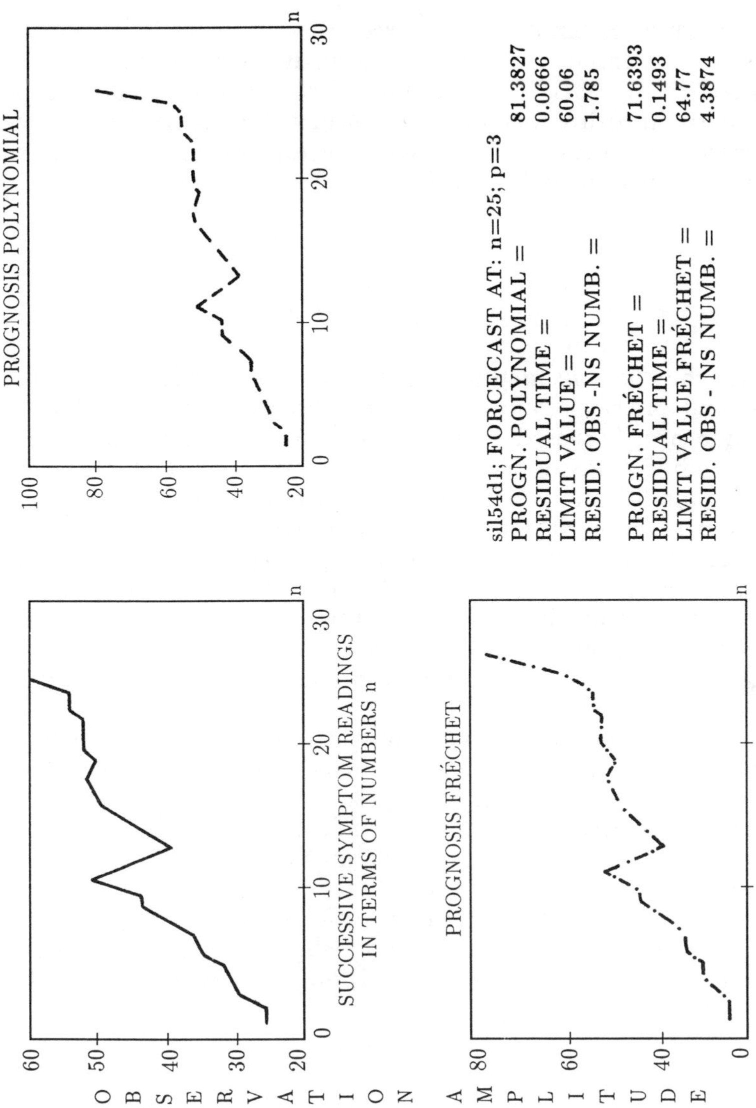

Fig. 2.31. An example of diesel engine diagnosis by the computer program cem8.m based on ETS theory

symptom distribution is most appropriate, and this is also true for the polynomial approximation. The upper left figure shows the observed behaviour of the engine number 'sil54d1' measured at point d1 with the symptom rms-vibration acceleration of the casing. The upper right figure gives us the forecast of the engine behaviour at the forecasting horizon $p = 3$ for the polynomial approximation, and the lower left the same forecast for the Fréchet distribution. The abbreviated message of the program concerning the symptom limit values, forecast symptom values, and the residual time and observation number are also shown for both the polynomial and Fréchet forecasts. It can be stated that the polynomial forecast is pessimistic here. In reality this was true, because the engine broke down shortly afterwards. More examples of accurate forecasting of the life of diesel engines are given in [97].

2.4 Summary

In this Chapter the *classical* methods of diagnosis are reviewed, and the mathematical and physical foundations are presented. The descriptions of damage/faults are discussed in relative detail. They are traced back to parameter modifications of the related mathematical model. However, model structure modifications with respect to changes of the number of DOF and the introduction of nonlinearities are discussed additionally.

Symptoms as sensitive quantities play a large role in the detection of faults, in decision-making and in the localization of faults. Known quantities are enumerated and discussed. In addition, emphasis is placed on subsystem modelling as an economic procedure for fault detection, localization and quantification as preparation for the state condition adjustment described in Chap. 4.

The last section contains a discussion of damage initiation and damage evolution models based on physical phenomena taking place during system operation. The main statement of this section is the introduction of various damage measures, and of the damage capacity and the symptom reliability. It is shown that an analytic relationship exists. The diagnostic use of the models is discussed. The energetic background is essential in damage investigations. Operating systems (including systems in service) require physically based holistic consideration which leads to energy transformations. The result is the energy transforming system theory used for system life prediction.

CHAPTER 3

Fault-Related Processes: Monitoring, Measurements, Processing of Signals

The process of a faulty system must be observed in order

- to detect the fault just in time, when re-adjustment or repair with minimum costs is possible,
- to perform the system measurements for model adjustment of the current state for fault detection, localization, finding the type, extent, evolution etc.,
- to assess the fault as a basis for decision-making on further actions (operational, repairing, etc.).

The equation of motion (1.3.1) written for the faulty process in the state space domain at the life instant θ and at time t, suppressing the θ-notation, has the form

$$\dot{\mathbf{x}}(t) = \mathbf{A}\mathbf{x}(t) + \mathbf{B}\mathbf{f}(t) + \mathbf{F}\mathbf{p}(t) + \mathbf{n}(t) \tag{3.0.1}$$

with the measuring (or output) equation

$$\mathbf{y}(t) = \mathbf{H}\mathbf{x}(t) + \mathbf{D}\mathbf{f}(t) + \mathbf{q}(t) + \mathbf{m}(t), \tag{3.0.2}$$

where

$\mathbf{x}(t)$	is the state vector, dots indicate differentiation with respect to time
$\mathbf{A}$	is the system (or state) matrix, quadratic of order n
$\mathbf{B}$	is the input (or distribution) matrix; it is an (n,m)-matrix
$\mathbf{f}(t)$	is the vector of external excitation with m components
$\mathbf{F}$	is the fault entry matrix
$\mathbf{p}(t)$	with its components represents external disturbances and input sensor faults
$\mathbf{n}(t)$, $\mathbf{m}(t)$	describe noise.

All the matrices and vectors are of proper dimensions so that they are composable.

H is the measurement (or output) matrix
D is the input-output matrix, which for our systems is generally **0**, except for improper transfer functions, where for the SDOF model the zero polynomial is of the same order as the characteristic polynomial (see [106])
q(t) describes the output sensor faults
y(t) is consequently the measured dynamic response with generally fewer components than **x**(t).

Equation (3.0.1) represents an adjusted model (to the life time considered), and therefore it is a verified and validated model, which means in other words that it is a trained model which already has experienced the learning phase. The system faults are included in the system matrix **A**. The system fault evolution between the time θ_i and the time θ_{i-1} can be described by

$$\Delta \mathbf{A}_i := \mathbf{A}_i - \mathbf{A}_{i-1}, \tag{3.0.3}$$

if the subscripts denote the corresponding life times. The comparison of suitable quantities (including reduction of noise) of the θ-dependent models then results in a trend description.

What can be stated with respect to Eqs.(3.0.1) and (3.0.2) is:

- *high sensitivity* with respect to faults as a local property, and *robustness* to unknown disturbances as a global property, are contradictory requirements which must be optimized
- the distinguishability of the various terms is important, and this also includes fault and error separation
- dynamic models as energy-equivalent models describe global system properties, while static models with many degrees of freedom enable us to represent local properties as, for example, stress maxima[1].

$\mathbf{p}(t)$ must first be excluded. External disturbances which excite the system can be assumed to be superimposed in $\mathbf{f}(t)$ if they are known (measured), otherwise they should be modelled and identified or assumed as random and incorporated in $\mathbf{n}(t)$. Input sensor faults will be excluded by a sensor check (possibly by calibration) before the measurement or by request between monitoring steps. The same method should be applied to exclude output sensor faults. The remaining noise has to be minimized.

The reader is reminded that the state space description (3.0.1) as a first order ODE can be advantageous. However, when writing the state matrix **A** one should not forget that its structure dependent on the matrices **M**, **C**, and **K** is known [5] for the state vector $\mathbf{x}(t) = (\mathbf{u}(t)^T, \dot{\mathbf{u}}(t)^T)^T$ as:

$$\mathbf{A} = \begin{pmatrix} \mathbf{0} & \mathbf{I} \\ -\mathbf{M}^{-1}\mathbf{K} & -\mathbf{M}^{-1}\mathbf{C} \end{pmatrix}. \tag{3.0.4}$$

[1] Therefore it is doubtful whether simultaneous fault detection and localization with the use of one model is optimum.

3.1 Monitoring

Monitoring is essential in the assessment of system condition evolution. Weak point analysis enables us to find locations and quantities which can serve as symptoms. For example, in rotating machinery, bearings or shaft-undercuts can be the weak points, so that one may consider these parts for the locations of transducers for observation. If the values of the measured symptoms exceed a pre-defined alert threshold, then the monitoring of symptoms as a result of the studies of possible faults can be performed, as described in Fig. 3.1. The following three decisions are needed in the process of system monitoring and assessment:

1. When comparing the inspection results with those coming from weak point analysis using the current mathematical models, we need to define the alert threshold and to dedect a possible exceedance.
2. From a detailed study of the system we need to decide whether the current mathematical model should be modified so that it will fit the current state (significance of modifications).
3. When comparing results of diagnosis/assessment with those of prediction/simulation using the modified models, we need to decide whether any corrective actions should be taken.

These decisions are mainly based on measured data.

Monitoring and diagnosis have to distinguish between

- noise and modifications of the system,
- the properties of various elements and/or subsystems,
- the performance of the units in different life times,
- the parameters of the model, and
- the possibly varying influence factors affecting the process observed.

The performance and the parameter monitoring are model-based, and are sometimes very sophisticated through thermodynamic modelling, as in the case of a jet engine [107]. Other types of condition monitoring (diagnostics) use either readings of tolerance units (fits) or of symptoms, or try to build some models later on.

3.1.1 The Goal of Condition Monitoring

The reader is reminded of what the diagnosis of the system is: it is the determination of the present and future technical condition together with the cause finding of observed significant symptom/state modifications and sometimes their assessment. Monitoring is the observation of generally a few particular quantities of the system in operation. The objectives are to determine substantial system modifications with the help of symptoms or state vector components before significant operational consequences (product quality, safety decrease etc.) appear. In addition, these measurements serve for model adjustment in order to perform the life time-dependent verified and validated mathematical model. This can be done

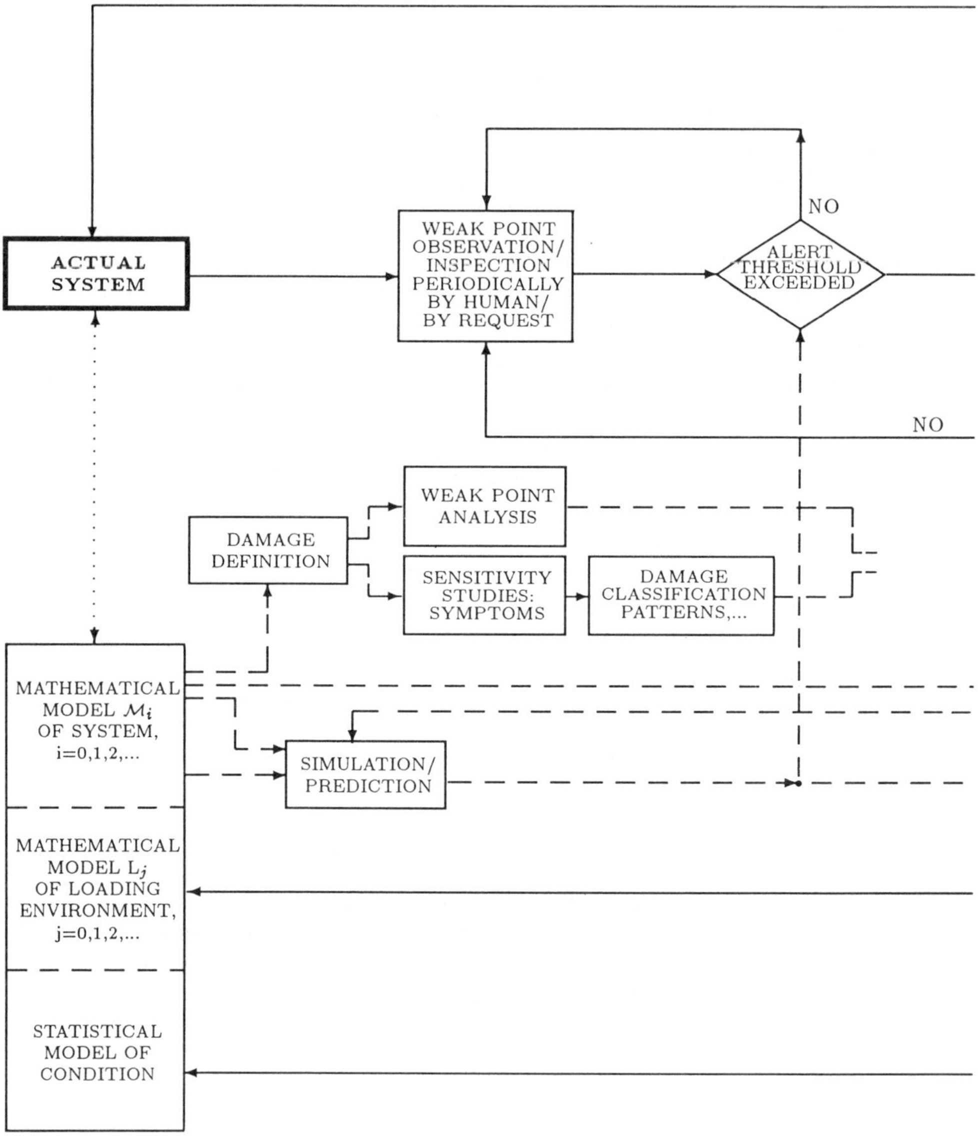

Fig. 3.1. Flow-chart of diagnosis including monitoring

in a two-stage process, as already mentioned: observing a few quantities within an alert system and switching to the real measurement of further quantities: monitoring on demand. Monitoring, of course, can be done periodically, where the required period is found computationally or by experience. The direct monitoring dependent on the task, and dependent on how critical the system behaves, will generally be done permanently.

Observability in the above sense is required. If the fault one is interested in can be expressed by a significant change in a particular modal

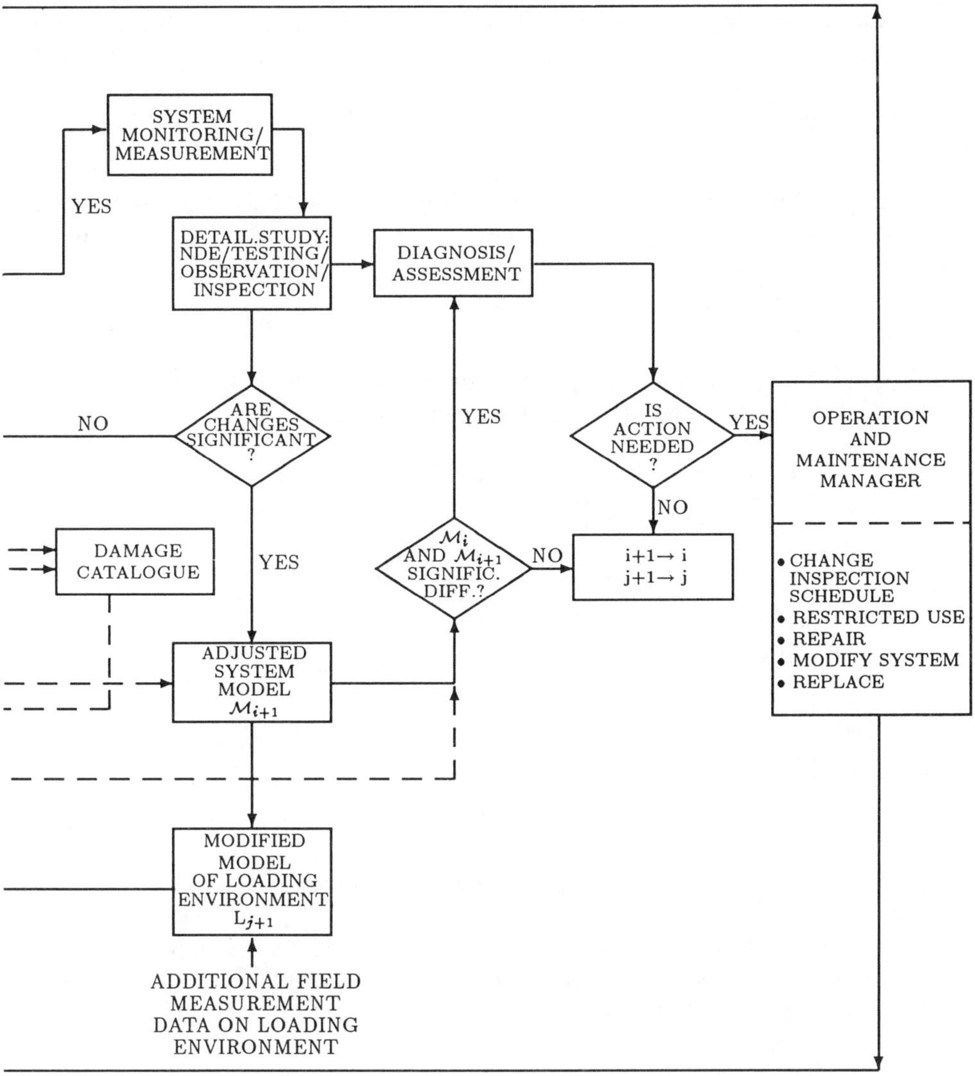

Fig. 3.1. (continued)

quantity or in a set of them (including dynamic responses decomposed in eigenvectors), then observability can be investigated by, for example, the Hautus criterion, the Kalman criterion [108], or by the Grammians [106]. In their spectral decompositions the state space Eq. (3.0.1) with the structure shown in Eq. (3.0.4) and the measuring Eq. (3.0.2) without fault and noise terms give the equations

$$\mathbf{A} = \mathfrak{Q}\Lambda\mathfrak{Q}^{-1}, \tag{3.1.1}$$

$$\mathbf{B} = \mathfrak{Q}\mathfrak{Q}^{-1}\begin{bmatrix} \mathbf{N} \\ \mathbf{0} \end{bmatrix}, \tag{3.1.2}$$

with **N** designating the locations of the non-zero forces, and with the modal matrix

$$\mathfrak{Q} = \begin{bmatrix} \mathbf{Q} \\ \mathbf{Q}\boldsymbol{\Lambda} \end{bmatrix}, \tag{3.1.3}$$

corresponding to the state vector $\mathbf{x}(t) = (\mathbf{u}^T(t), \dot{\mathbf{u}}^T(t))^T$. If the eigenvalues assembled in the diagonal matrix $\boldsymbol{\Lambda}$ are complex, then n conjugate complex values exist, which lead to the modal matrix

$$\mathbf{Q} = \left[\mathbf{Q}_0, \bar{\mathbf{Q}}_0\right], \tag{3.1.4}$$

designating with $\bar{(.)}$ the conjugate complex of (.). The frequency response matrix with initial conditions equal to zero follows as

$$\left.\begin{aligned} \mathbf{F}(j\omega) &= \mathbf{H}(j\omega\mathbf{I} - \mathbf{A})^{-1}\mathbf{B} \\ &= \mathbf{H}(j\omega\mathbf{I} - \mathfrak{Q}\boldsymbol{\Lambda}\mathfrak{Q})^{-1}\mathfrak{Q}\mathfrak{Q}^T \begin{bmatrix} \mathbf{N} \\ \mathbf{0} \end{bmatrix} \\ &= \mathbf{H}\mathfrak{Q}(j\omega\mathbf{I} - \boldsymbol{\Lambda})^{-1}\mathfrak{Q}^T \begin{bmatrix} \mathbf{N} \\ \mathbf{0} \end{bmatrix}. \end{aligned}\right\} \tag{3.1.5}$$

If only the displacements are measured,

$$\mathbf{H} = [\mathbf{H}_0, \mathbf{0}], \tag{3.1.6}$$

then it follows that

$$\mathbf{F}(j\omega) = \mathbf{H}_0\mathbf{Q}(j\omega\mathbf{I} - \boldsymbol{\Lambda})^{-1}\mathbf{Q}^T\mathbf{N}. \tag{3.1.7}$$

Complete observability is given if there is no vector $\mathbf{q}_i$, $i = 1, \ldots, n$, with $\mathbf{H}_0\mathbf{q}_i = \mathbf{0}$. Correspondingly, the system (expressed by the model) is completely controllable if there is no $\mathbf{q}_i$, $i = 1, \ldots, n$, with $\mathbf{q}_i^T\mathbf{N} = \mathbf{0}$. The correctness of these statements can be seen directly. Consequently, the observability of faults can be investigated with the above observability criterion.

3.1.2 Symptoms, Processes and their Characteristics

One problem to be solved is which quantity can serve as a symptom (see Sect. 2.2). Various chosen symptoms can have the same result, because neither the model is unique nor, consequently, symptoms for the diagnostic model. This dilemma can be solved by taking into consideration the OS with respect to energy flux and the hierarchy of ETS. The experience of the analyst can be supported by considering the density of energy flux in the OS. This is because the greater the energy density flux, the greater is the dissipation, i.e. damage at a given location. And at these locations, or adjacent, we should choose the discretization points for the dynamic model equal to measurement points using the prior knowledge from system analysis and from the past adjusted mathematical models. Additionally, theoretical analysis of weak points, of sensitivities, and of possible faults are decisive within these investigations.

The process is defined as the set of input and output quantities. Each system in operation, dependent on its design task, may have three principal inputs: feeding with energy and material, control changing the operational menu, and it is subjected to different environmental interactions. Sometimes for testing purposes we also need an artificial excitation instead of the operational forces in order to perform the measurements needed. When looking for outputs of OS we have only two outputs, the upgraded and dissipated energies, as was already shown in Fig. 2.23. This figure may also serve as the input/output illustration scheme for other OS-like structures, vehicles etc. when material feeding is excluded. The observation requirements and possibilities are now briefly outlined for the diagnostics of different types of OS.

Machines. In the case of a machine it is characteristic to have the feeding inputs, i.e. the energy input in terms of mechanical, electrical, chemical forms, and the material input for further processing into goods. Of course, the input for changing the production menu has to be controlled, and there is also the input of environmental disturbances. For the purpose of diagnosis a quantity is needed which has some relation to the damage intensity and location, or to the causes of it. This may be the process reflecting the current demand on the energy (or fuel) supply, as the energy efficiency depends on machine degradation advancement. For precision machinery and equipment which is sensitive to environmental vibrations, some quantities depicting the environmental forces or kinematic interactions should be observed. It may sometimes be necessary to observe these inputs by means of suitable time, frequency, and amplitude characteristics in order to correlate them afterwards with the symptom of condition, or other system damage characteristics. As far as the outputs of a machine are concerned, there are several possible methods of observation. Initially, the output of products is usually monitored by special transducers for control and quality assessment purposes. These processes are often used for diagnostic purposes, as deviation from optimal working process parameters and characteristics is directly connected with damage evolution (for more see, for example, Chap. 10 in [22], and also [109]). The dissipated energy output from the machine has a great diversity of physical processes, and almost all outputs can be used for diagnostic purposes. They are mostly of dynamic nature and can be mapped for any $\theta = const.$ into the three domains already mentioned: the time, frequency, and amplitude domains. This mapping serves for further processing and comparison with damage characteristics. In contemporary use in machine diagnostics it is mostly the vibroacoustic processes that are observed; wear debris and heat, the latter for thermography, are sometimes in use as well. As far as the nature of dynamic processes in machines is concerned, they are mostly periodical, and this is due to the rotational or reciprocal nature of machine operation, so their analysis is based on the Fourier series, and signal processing techniques are mostly applied. The smooth running of machines can be assessed by observing their structural velocities. If higher frequencies are of interest, then acceleration (owing to its multplication by ω^2) should be measured as the best quantity chosen.

Structures. As already mentioned, if material feeding is absent the same scheme of machine investigation can be taken for structural diagnostics. Energy feeding and control will be in use here only for very advanced types of structures (smart and active systems). Hence, most of the inputs come from dynamic environmental disturbances. Test forces are generally applied for identification and model updating. Many of the environmental dynamic interactions are of low frequency content (smaller than 2 Hz), i.e. wind and sea waves. Strong excitations, such as hurricanes and earthquakes, are causes of severe damage. Output quantities of the structures bear the information of the specific loads, and they also contain the residual process effects, which are dynamic in nature. If one excludes the check of thermal resistance here, and probably leakages as well, the same types of characteristics as in the case of machines can be used. If stresses are of interest, one should know that they can only be measured indirectly by strains. For simple structures it can be shown that the stress (σ) can be calculated by knowing the structural velocity ($\sigma \sim v$). In this context one should also remember the modal decomposition of the dynamic response, for example of σ.

Vehicles. Vehicles also have no material feeding (with the exception of spare parts etc.). Here the inputs are: energy (fuel), control, environmental interactions, and testing force inputs. In the case of motors and engines the current fuel consumption can be a symptom. The environmental interactions (forces and kinematics) play an important role, often being a cause of damage (a rough road, gusty and turbulent atmosphere, etc). For output quantities of vehicles the same residual processes as for machines can be considered. The desired product of the vehicle is the efficiency, the quality and safety of the transport of goods, people, etc. Different processes and measures for quality assessment will have different sources of diagnostic information. These may be the speed of travel, the vibrational and the acoustic comfort in terms of some process amplitude measures, etc. Of course, for the internal control of the vehicle engine, and the vehicle motion control, etc., many process parameters and characteristics are available, and at the same time they can be used for diagnostic purposes in some kinds of performance analysis [22] and process parameter diagnostics [109]. But these branches of diagnostics are outside the scope of our book.

In this paragraph the processes and signals of mechanical origin are considered in a broad sense. These are vibroacoustical processes taking place inside the system and in its environment. The frequency range of vibration starts close to zero and goes up to the ultrasonic of the magnitude of Megahertz (MHz). As is well-known, every quantity of a vibroacoustic field can be represented in the proper frequency range and with a physical dimension. For purposes of machine diagnosis we may use the vibration displacement amplitude in the range of 0 to 1 kHz, the vibration accelera-

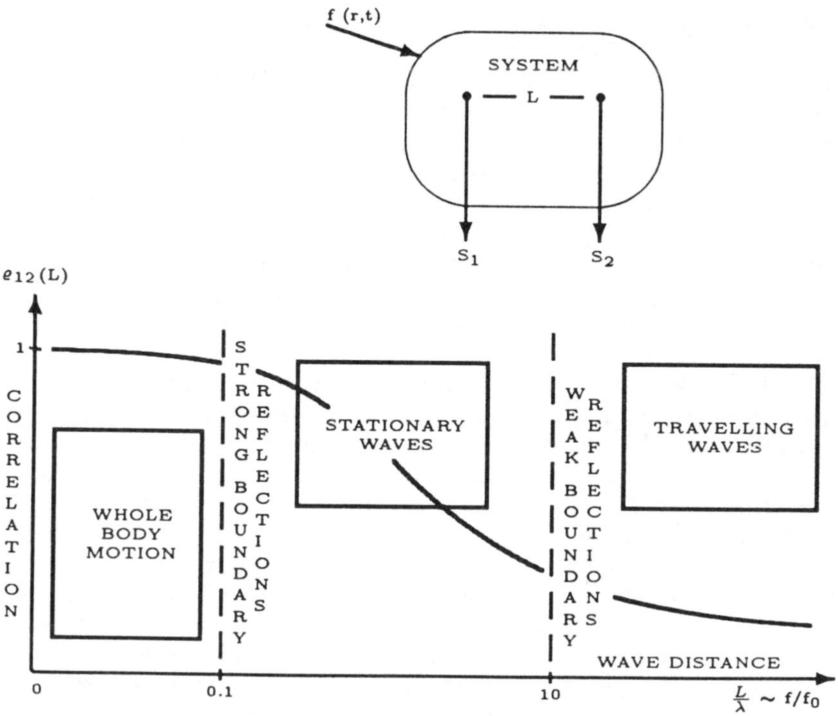

Fig. 3.2. Wave distance between two points on the structure $(L/\lambda)$and the prevailing type of motion with the correlation between signals S_1 and S_2

tion amplitude in the range of 1 - 10 kHz, either the so-called stress wave amplitude of the range 40 - 50 kHz, or even the acoustic emission in the material of frequency magnitude of 1 MHz. For structures the frequency ranges mentioned above are about 10^{-2} smaller. In some applications these quantities can serve directly as symptoms, and in other cases the quantities of the vibroacoustic field should be transformed into proper symptoms for detection and location purposes. The question arises of what field quantity we should measure dependent on the wave frequency content and on wavelengths. The qualitative relation of some vibrational quantities is shown in Fig. 3.2.

This figure was created by the analogy to acoustic reverberant fields in rooms [110]. The detailed shape of the graph and the exact values of the correlation coefficient depend on the type of the propagated wave (spherical, cylindrical, etc.) and the width of the related spectra. But the common feature is the spatial uncorrelation of two signals coming from the same source $f(r, t)$, with the increased frequency (f_0 the lowest natural frequency of the system). One should remember this when seeking damage detection by means of correlation or coherence functions. This correlation will decrease when additional sources of force come into action (impacts due to clearences, etc.) or if the wave distance increases due to wear (the crack in the load path, for example). The graph also presents the dominant type of structure motion. One can note that, when compared with the wavelength, for large distances between boundaries there will be only a boundary effect with adjacent points of the boundary. Inside the structure, dependent on the internal damping, travelling waves will prevail. For small structures (in comparison to the wave length) the whole body motion will prevail, and it will not be possible to distinguish between different signals, even those coming from different sources. Hence, for diagnostic purposes we should use the proper mixture of the range between standing and travelling waves. There is also a need to elucidate the detection distance of the fault or damage. According to Huygen's principle, each point of an incident wave can be treated as the source of a new wave. Hence, if damage is assumed as the wave source, the spatial incoherence of this wave should be taken into consideration. This will substantially depend on the distance of a few wave lengths (see Fig. 3.2). This property gives a clear indication that the distance between adjacent transducers should be $\Delta r_i = (r_i - r_{i-1}) < a\lambda$, where r is the spatial coordinate, a is a constant with a value between 1 - 3, dependent on the type of wave, the internal damping in the structure, etc.

To sum up the discussion on the mapping of the spatial properties of a structure: the medium and the high frequency ranges should be taken, and the transducers should be located within the range of a few wave lengths for overall diagnostic supervision and as close as possible to the evolving damage, provided that we have some prior knowledge from theoretical analysis or from experience with the type of system under investigation.

3.2 Measurements

Measurements are goal-orientated. Suitable quantities have to be chosen, and they have to fulfil some pre-assumptions. Because measurements are generally incomplete and uncertain, they have to be planned and carried out carefully. Incompleteness can concern the number of measuring points, the required frequency contents of the signals, and so on. Measurement errors can be deterministic and irregular. Deterministic measurement errors are the worst case, and they must be detected and removed, possibly computationally. Irregular measurement errors are generally modelled stochastically. They can be reduced in the estimates.

Test conditions should be optimized dependent on the goal and the system properties. Test optimization concerns the placement of sensors, their choice, and also the choice of test signals, exciters and their locations. Restrictions are given by the availability of the hardware, including financial and time conditions. The remaining conditions follow from the choices made before, and the system properties. They concern the frequency range, the type of pre-processing etc. Often the test signal cannot be chosen, and the operational conditions must serve for excitation. Then the most appropriate operation range has to be chosen which provides maximum sensitivity to and maximum information on the damage under consideration. The record length, the sampling etc. have to be chosen dependent on the assumptions to be made about the excitation (e.g. coloured noise). Some of these problems are discussed in the following from a simplified but practical point of view.

Mechanical, civil, aeronautical engineers and others are generally not experts in measuring techniques. Therefore they need support in performing measurements. One possibility is to use an expert system, for instance SAMBA[2] [111] as a system for applying measurements in civil engineering which can also be applied easily in other fields of engineering. Included in the knowledge-based system SAMBA is an assessment of systems which is restricted to particular civil engineering structures. However, due to its modular structure it is easy to modify the assessment component.

In this section only a few recommendations are given concerning environmental and boundary conditions, sensors[3] and the test/measuring set-up. A discussion of the (dynamic) properties of the various forcing signals can be found by the reader elsewhere [5, 7].

3.2.1 The Goal of Measurements

In the context of this book two main goals of measurements are emphasized: these are the monitoring and the measurements for model adjustment. No distinction will be made between monitoring devices and the sensors for the adjustment measurements. Both resulting sets of measured data will be used for the model adjustment.

Model adjustment takes system modifications into account. Therefore the theoretical studies of possible faults lead to dynamic variables/parameters sensitive to the parameters possibly changed (Fig. 3.1) and to be measured. Otherwise, if the information on model parameter changes is missing in the measurements, an adjustment cannot be performed. In other words, if parameter changes, for example, affect particular modal quantities (degrees of freedom), then the measured quantity (e.g. the dynamic response) must contain these degrees of freedom (in this example the normal mode deflection) in a non-negligible manner, and it must be measured. If, for example, the frequency range of the decisive sensor (with respect to the symptom considered) is insufficiently chosen, so that the corresponding eigenfrequency is outside the measuring range, then this information cannot be obtained from the measured signal. This means that the required frequency range has to be determined before the measurement takes place. A second requirement comes from the model itself: the measured signals should be used with a minimum of manipulations, employing suitable algorithms in order not to introduce additional

[2] Abbreviation for the German *System zur Anwendung der Meßtechnik im Bauwesen*

[3] Note: sometimes it is desirable to distinguish between sensors, pick-ups and tranducers. Sensor is the element of a measuring instrument to which a measurand is directly applied; sometimes the pick-up is meant. The conversion part that provides an output with a given relationship to the input quantity is called the measuring transducer. Pick-up is the total device. Here, generally, the word sensor is used.

errors. For instance, an interpolation of the signals of various measuring points should be avoided.

Observability is another problem to be checked in the context of model adjustment. Additionally, the problem connected with long-term measurements (stability) has to be solved by the proper choice of sensors, interval-wise calibration, and computational correction etc.

3.2.2 Environmental and Boundary Conditions

If the environmental conditions affect the object under investigation, then they are a part of the system and also have to be modelled. For example, the dynamic behaviour of a cable-stayed bridge depends on the temperature. Further environmental influence factors are: external disturbances like wind, vibrational and acoustic effects from outside the object, humidity, moisture content and geotechnical factors including changing groundwater levels. The SAMBA expert system mentioned above provides these effects by geodetic measurements. Very often these environmental effects are not included in the related mathematical model, while they are within the measured data. In so far they have to be recorded and stored and taken into account for comparative measurements: this means they must be considered parametrically. One consequence is that they should be kept constant during measurements.

Equation (3.0.2) describes the situation with measured external disturbances. In other words, these disturbances have to be detected and modelled, or they must be avoided. The worst case is when externally deterministic disturbances are present but one does not know about them. If environmental effects cannot be modelled as external disturbances because an interaction exists between the object considered and its environment, then subsystem modelling and synthesis have to be applied. The interaction forces are the interface forces between adjacent subsystems [36]. Examples of this are ground-structure, structure-structure, and fluid-structure interactions.

A very special but no less important feedback problem exists with the measuring system applied. For example, when one observes lightweight systems the weights of the sensors may have to be taken into account. Electromagnetic effects can come from the sensors, too, as well as from the vibration of not fixed (to the system to be observed) cables, induction from neighbouring installed electrical lines etc. In addition, the use of artificial test signals by electromagnetic exciters, for instance, can introduce some feedback (it is a structural-electric-magnetic interacting system), and therefore the force should be measured directly at the input location of the system.

The boundary conditions of the system are pre-determined by the existing system in operation. It should be noted that the effect of the boundary conditions is restricted to the lower eigenmodes (see the asymptotic behaviour of the modes, [30]), which means that for a distance greater than 1 wavelength the influence of the boundary condition on the displacement vanishes [26]. The consequence for the observation of faults near the boundary conditions by means of displacements is now obvious. The measurement of reaction forces may be suitable in that case [29]. If changed boundary conditions due to test conditions are present in the measurements, then these effects have to be corrected computationally.

If tests with a full-scale material model or with a scaled-down model have to be performed, then the boundary conditions have to be designed and considered in the physical interpretation of the test results and, if necessary, computationally corrected (see here the appropriate expositions in [5] and [7]). It should only be noted that one never should try to realize a rigid fixture which practically cannot be realized. It is much better to realize a defined elastic boundary condition and to change the results computationally for the required (e.g. rigid fixtures) boundary conditions.

3.2.3 Sensors: Properties, Calibration, Measuring Errors, Locations

It is important to choose appropriate sensors for the monitoring and measurements. The sensor has to be chosen dependent on the sensitive (with respect to the fault considered) quantity and on the goal. Again, the expert system SAMBA is referred to for help. Because it is impossible to give a complete introduction to measurement techniques only some statements are made on principles and only some practical details given.

Properties. The principle structure of a seismic sensor is shown in Fig. 3.3.

The first subsystem of the sensor, mounted on the system, is the sensoring part which moves with the system to be sensored. (It moves identically with the measuring point if a rigid connection is achieved in the frequency range of interest.) This structural part behaves as, and can be modelled as a dynamic SDOF or MDOF model. The movement of this structural model then has to be converted into an electrical quantity (w), when we restrict

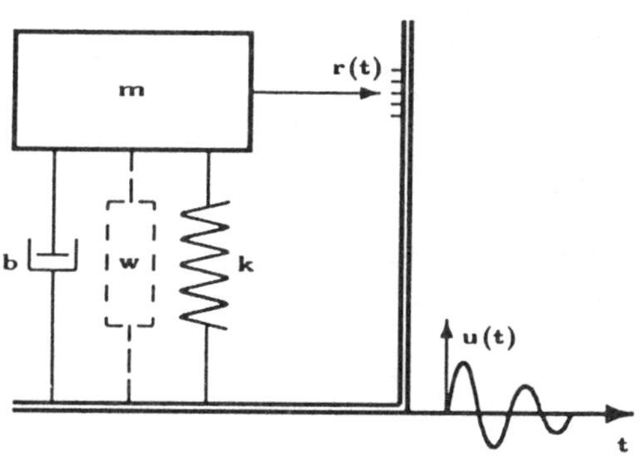

Fig. 3.3. Structure of a sensor in principle

ourselves to electrical measurements[4]. Consequently, we have to take into account the dynamic behaviour (input/output) of a structural system with its unit response function in the time domain, and its frequency response function in the frequency domain. Additionally, the properties and effects due to the conversion used have to be considered; this concerns the amplification, the cable shield etc.

If the user knows this, then he knows that a sensor has an eigenfrequency, a damping ratio, and that the dynamic response in the frequency domain has to be described by its real and imaginary part, or by its amplitude (magnitude) and the phase response function. Therefore, undistorted and non-delayed measured data can only be obtained in a small frequency range dependent on the construction of the sensor. If the fundamental frequency of the sensor is known, one should know that an SDOF model is a first approximation of a continuum (for example, modelled more accurately as an MDOF model), and therefore the first overtone of the MDOF model has to be considered next in order to find out the frequency range to be used without correction of the sensed structural dynamic response. In summary, one has to be aware that all the dynamic properties of the pick-up can theoretically occur and have to be taken into account. This, for example, concerns the transient dynamic responses, stochastic responses as well as the frequency dependency, which plays a role in calibration. Calibration for the static case is useful but not sufficient in dynamic measurements.

Because it is very informative, the sensor may be modelled as an SDOF model,

$$m\ddot{u}(t) + c\dot{u}(t) + ku(t) = p(t),$$

in the common denotation, and it should be investigated when it behaves as a displacement, velocity and acceleration sensor. With the relative coordinate $r(t)$ (see Fig. 3.4) and $p(t) = 0$, the equation of motion reads

$$m\ddot{r}(t) + c\dot{r}(t) + kr(t) = -m\ddot{u}(t).$$

If one emphasizes the inertia force in the pick-up (Fig. 3.4a) by construction, so that

$$1. \qquad |m\ddot{r}(t)| \gg |c\dot{r}(t) + kr(t)|$$

holds true, then

$$\rightsquigarrow \; m\ddot{r}(t) \doteq -m\ddot{u}(t), \rightsquigarrow \; r(t) \doteq -u(t):$$

the sensor works as a displacement sensor (seismometer). The spring only constrains the mass, while the damper restricts the amplitude.

If

$$2. \qquad |kr(t)| \gg |m\ddot{r}(t) + c\dot{r}(t)|,$$

---

[4] SAMBA is not restricted to the electrical measurements of mechanical quantities.

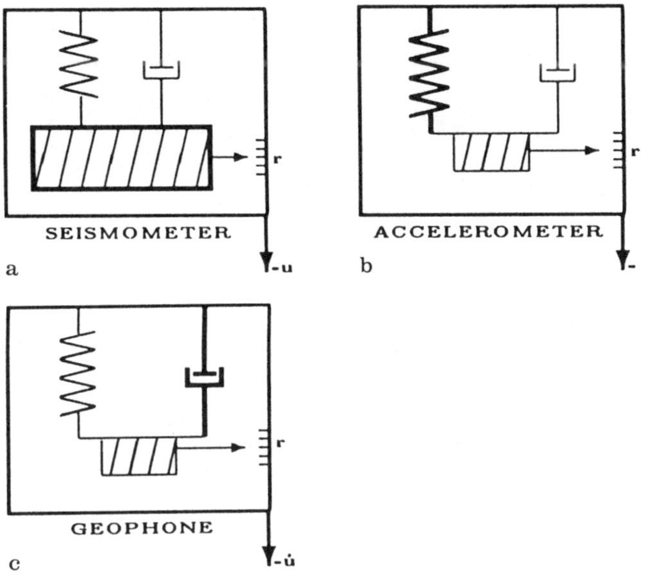

Fig. 3.4. Types of sensors in principle emphasizing different forces

(see Fig. 3.4b) it follows that

$$\rightsquigarrow \quad kr(t) \doteq -m\ddot{u}(t), \rightsquigarrow \quad r(t) \doteq -\frac{m}{k}\ddot{u}(t) = -\frac{1}{\omega_0^2}\ddot{u}(t).$$

The sensor observes accelerations,.

Finally, if the damping force is emphasized (Fig. 3.4c), then the sensor works as a velocity pick-up:

$$3. \qquad |c\dot{r}(t)| \gg |m\ddot{r}(t) + kr(t)|,$$

$$\rightsquigarrow \quad b\dot{r}(t) \doteq -m\ddot{u}(t), \rightsquigarrow \quad r(t) \doteq -\frac{m}{c}\dot{u}(t) = -\frac{1}{2\alpha\omega_0}\dot{u}(t).$$

When one compares the forces of these idealized sensors with the dominating forces in an SDOF model represented by the absolute values of the frequency response function, then the relation is obtained as represented in Fig. 3.5. For $\omega = 0$ and in its vicinity the restoring force approximately controls the equation of motion (accelerometer), in resonance the damping force dominates (velocity sensor), and for $\omega \gg \omega_0$ (resonance frequency) the inertia force controls the movement (displacement sensor).

In addition, the reader's attention is drawn to the pick-up shown in Fig. 3.6. This sensor works in the frequency range of 7 Hz to 100 Hz as a displacement pick-up, while the frequency range of 0 Hz to 3 Hz it functions as an accelerometer, as explained previously.

Finally, various mounting possibilities with their structural properties are mentioned: Fig. 3.7 follows [112]. The figure is self-explanatory.

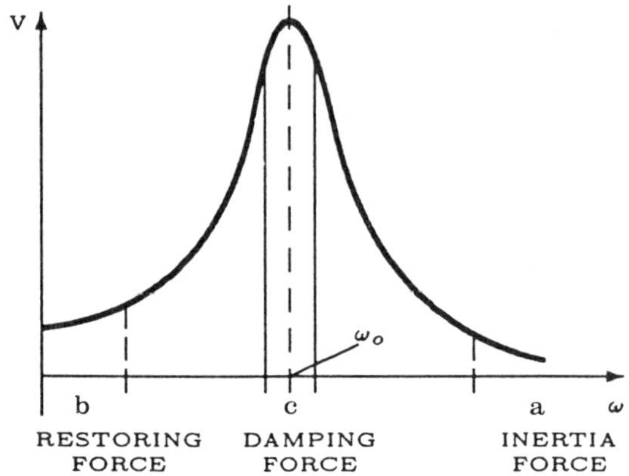

Fig. 3.5. Dominating forces of an SDOF model and the relation to the principle types of sensors shown by means of the absolute values of the frequency response functions

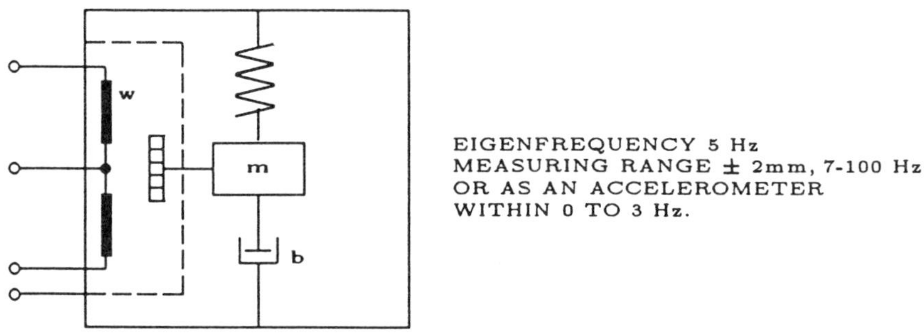

Fig. 3.6. A displacement sensor with a fundamental eigenfrequency of 5 Hz (differential bridge circuit)

Of the electric properties, dependent on the type of conversion etc., only the distortions by magnetic and electrical fields need to be mentioned. It must be repeated that the measuring cables should not be situated in the vicinity of power cables. Cable shields should also be used[5]. The fixing of cables has also been mentioned already, because their movement independent of the obje ct observed can induce additional signals.

With regard to the various measuring principles, the measuring chain, and their properties, the reader is referred to [113] to [115]. The classification of sensors following Ref. [113] should also be mentioned (Table 3.1).

Calibration. The environment affects the sensor and the total measuring chain, and consequently the measuring system output depends on these influence factors and on the measuring time. The calibration of sensors and

[5] Alternatively, optical fibres can be used, which also can serve as sensors.

Table 3.1. Classification of transducers

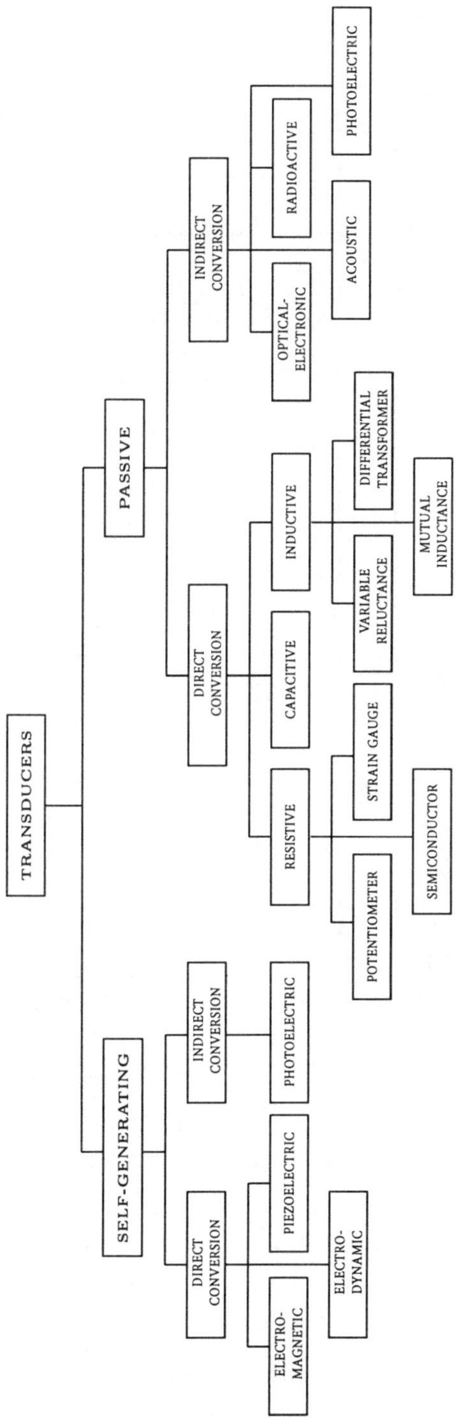

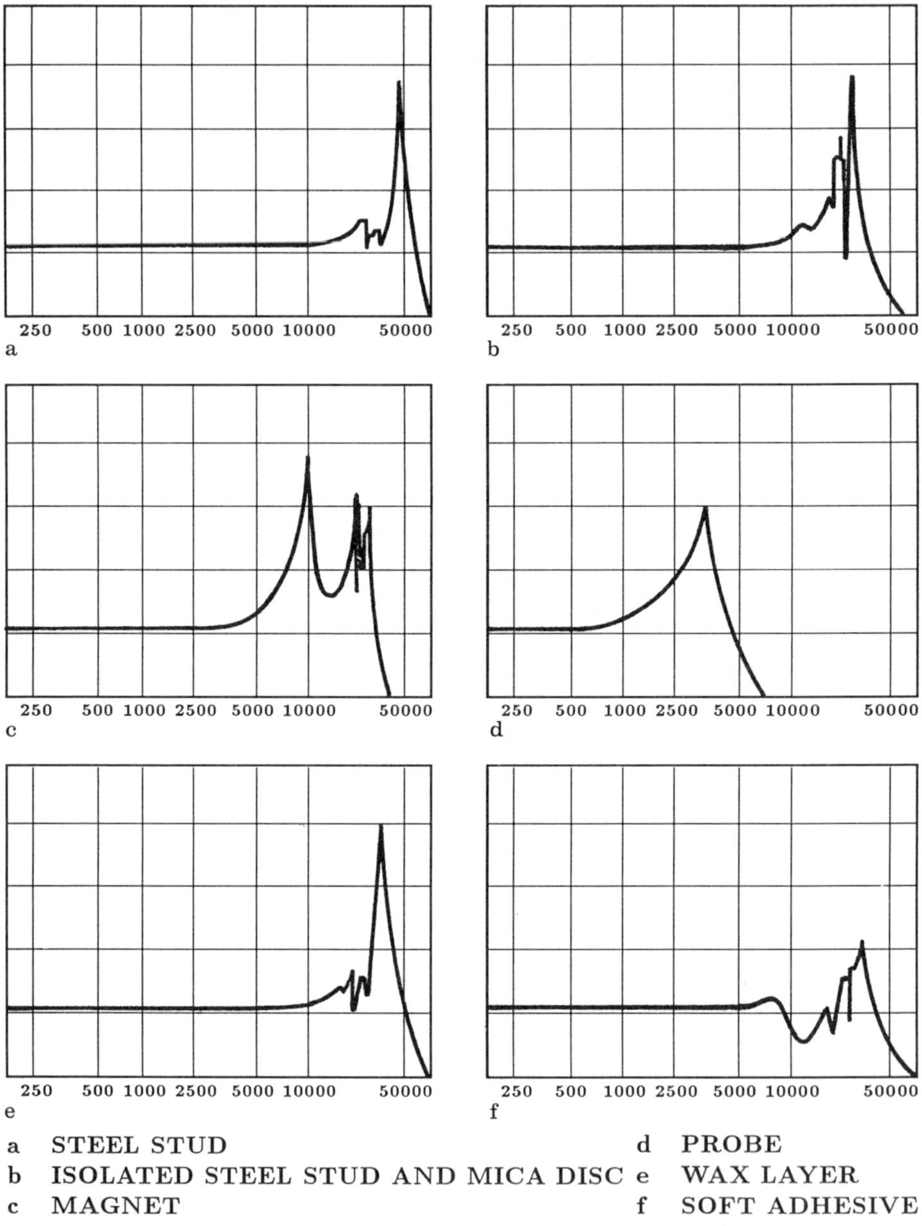

Fig. 3.7. Absolute values of frequency response functions dependent on different mounting types

the measuring chain and parts of it means determining the input/output relationship. Calibration concerns the frequency response function in one or several frequencies, the transfer factor (at the frequency 0), and the linearity. The total measuring chain as well as the sensor have to be calibrated. Calibration of the sensor includes determining its weight and the sensitivities with respect to environmental effects. Of course, the directional sensitivities of sensors have to be taken into account when fixing them for measurement and calibration. Various calibration procedures are distinguished. Calibration of the sensor and of the total measuring chain can be carried out:

- Statically: this means that the frequency response is measured for $\omega = 0$. This can be done for seismic sensors using the gravitational field of the earth. The sensor is balanced electrically and vertically in one direction, then it is rotated and measured in the opposite direction, again loaded with $1g$[6]. Pick-ups equipped with strain gauges can be loaded by known weights. Electrodynamic pick-ups can be calibrated using the reciprocity relationship, which means that a current is passed through the coil, producing a force. The force is measured by a dynamometer. This type of calibration can be done easily by the user. However, the frequency response function and the linearity cannot be determined.
- Direct dynamically: this method uses a known input by mechanical or electromagnetic vibrators. The dynamic response is then measured optically, acoustically etc. Piezoelectric accelerometers and force transducers are calibrated with this method, for example. The method is often used to calibrate the total measuring chain. Some transportable vibration exciters are only able to excite with a fixed amplitude and frequency. Possible transverse vibrations of the exciters are dangerous.
- Indirect dynamically: it is the same procedure as described above. The only difference is that the excitation is applied in the closest vicinity of the sensor. This method is applied for critical elements of a system and serves additionally for checking the operational integrity.
- Reciprocally: reciprocity calibration can only be applied for transducers with linear conversion elements working in two directions like electromagnetic and piezoelectric sensors. The electromagnetic velocity pick-up can be used as a velocity-sensing pick-up and as an exciter. The output of the exciter has to be measured and used for calibration. It is an absolute method using reference pick-ups.
- Optically: it is an absolute method using interferometers for reference pick-ups in the upper frequency range. It therefore supplements reciprocity calibration.
- By comparison: the output of the pick-up to be calibrated is compared with an already precisely calibrated reference pick-up. The calibration can be affected by the mounting of the two pick-ups. Relative movement between the pick-ups should be avoided.

The measuring chain can also be calibrated without the sensor. The following methods can be used:

- Calibration by substitution: a substitution is used instead of the transducer. This is generally a frequency generator with known output. The sensor has to be calibrated additionally.
- Insert calibration: a voltage source is inserted in series in the cable of the transducer. The source simulates the output of a self-generating transducer. The sensor must not produce an output signal during calibration.

[6] Gravitational acceleration

- Shunt calibration: the method is applicable only to variable-resistance transducers when it is not excited. A shunt resistance is connected across the "active" sensor resistance. The change in the bridge resistance is then measured.

It must be repeated that many effects can influence the measuring equipment, so that at least before and after the measurement (if possible with a calibrated sensor) the measuring chain should be calibrated (checked), otherwise the user does not know the real output of the system.

Measuring Errors. Figure 3.8 summarizes the possible errors in a test. The input is disturbed by the environment, and therefore the system is forced by $p(t)+n(t)$ instead by $p(t)$. Sometimes the loading is not by an external force but by a force with a feedback of the system: interaction. Additionally, the system is influenced by the environment and a possible feedback of the measuring system. Consequently, the system output is $z(t)$, as shown in Fig. 3.8. The measuring equipment itself is affected by the environment, and therefore the output $u(t)$ results as a superposition of $u_1(z)$ and $u_2(s_2)$. This consideration shows clearly how carefully and professionally tests have to be planned and conducted.

As already mentioned, measuring errors can be deterministic and irregular. It is much better to speak generally of measurement uncertainties, because one does not know the location, the quantity and the type of errors.

Deterministic errors can stem from influence factors, from the measuring equipment, the interaction with the forces, the test set-up etc. Examples of this are errors due to

- influence quantities (environment, user),
- nonlinearity,
- erroneous calibration,
- errors of adjustment,
- irreversible changes,
- reversible changes,
- drifts.

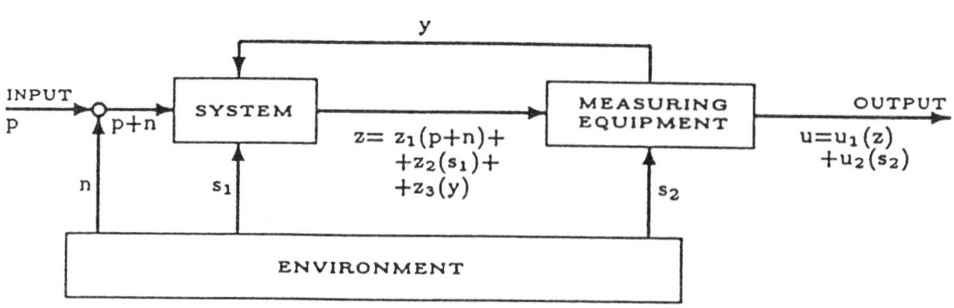

Fig. 3.8. Possible errors in a test

Deterministic errors should be avoided. If they are unavoidable, then they must be detected and corrected (at least computationally). Unknown but existing deterministic errors are the worst pitfall.

Irregular errors are

- reading off, sampling errors,
- noise due to
  - thermics,
  - the corpuscular nature of the material,
  - electricity.

Irregular errors are usually modelled stochastically. They should be minimized by applying estimators (see Chap. 4). Necessary numerical manipulations can amplifiy these errors. For example, the differentiation of measured signals will amplify the errors, and integration will smooth random errors with zero mean, but it will amplify the offset. The averaging of repeated measurements (without time lag due to sampling in series of various signal channels) is an easy method to apply. It reduces a normally distributed random error by $\sqrt{1/N}$, if N is the number of measurements. More detailed information is given by the t–distribution if a particular error is requested. Of course, confidence intervals can be requested and fulfilled [48].

Hints for reducing measurement errors concern

- the choice of a suitable measurand (see Sect. 3.1.2)
  - the velocity can be measured in order to obtain information on stresses, and to assess the smooth running of a machine
  - acceleration measurements should be made if higher frequencies are of interest, and if forces have to be investigated which cannot be measured directly
  - the choice of an appropriate location (see the next paragraph)
  - ...
- the choice of the right measuring method
  - absolute and relative measurements
  - contactable, contactless
  - active, passive
  - ...
- the choice of suitable instruments
  - maximum sensor sensitivity
  - minimum of relevant disturbance influences
  - consideration of dynamics, resolution, measuring interval
  - inaccuracy, linearity, frequency response
  - cable length, cable type, ground loops
  - type of mounting
  - additional instruments, such as pre-amplifier
  - calibration.

Locations. Prior information about the dynamic behaviour and possible faults determine the choice of measuring points. The results of system analysis and experience must be introduced if direct test optimization is impossible. The prior information concerns mainly

- stress peaks, the knowledge of maximum deflections etc. should serve for choosing measuring points (possible locations of faults)
- curvatures of modal shapes should be reconstructable through measured values (proportional to stresses)
- zero-crossings of modes should be known (for reconstruction and in order to know points for attachment, if necessary; these attachment forces will not interact with the system in this particular mode)
- measuring points should coincide with nodes (collocation points) of the mathematical model in order to avoid interpolation and possible additional errors.

Generally speaking, the discretization and modelling rules from analysis and simulation should be applied [9].

3.2.4 Test/Measurement Set-up

Absolute and relative measurements have to be distinguished. Relative measurements use a measuring rig, and the quantities are measured against it. This measuring rig must, of course, be structurally decoupled from the foundation of the test object. Mention has already been made of how to fix the sensors and cables, the latter in order to avoid motion relative to the test object. If resonance can occur, one should be aware of large displacements (reciprocally proportional to the modal damping ratios in superposition) and their consequences (space, measuring range of sensors etc.).

A test rig may be necessary if artificial test signals have to be applied. This must be decoupled from the object and a measuring rig that may possibly have been installed. The realization of pre-defined boundary conditions has already been discussed. If a free-free condition is to be realized, then an elastic suspension with a suspension eigenfrequency of less than 1/3 of the fundamental elastic eigenfrequency of the free-free system is required [5]. Additional exciters can sometimes be fastened as a pendulum with a counterweight for good action. Impulse hammers for quick investigations are very well known (for their restrictions see [5]).

3.3 Signal Pre-processing

The output signals of the sensors are generally analogue weak signals. They must be transmitted over some distances, they must be suitable for interconnected instruments, and they must fulfil some requirements for manipulation. Signal conditioning is therefore required to convert the signal in a form acceptable to the equipment to be handled. Signal transmission by cables (for telemetry transmission see elsewhere) has to consider cable properties. The weak signals have to be amplified to values suitable for the subsequent data acquisition and recording. Digital handling of the analogue data requires sampling which must fulfil particular demands. A quick-look evaluation seems to be advantageous, because often the measurement is not repeatable, or on-line evaluation for quick decisions is required. Finally, the signals often have to be filtered, and segmentation can help to detect and overcome signal trends or, more generally, instationarity.

3.3.1 Conditioning, Sampling, Pre-analysis

Interconnection. Cables for analogue signal transmission have special electrical properties: they possess an impedance which affects the signal. The output can be affected to a great extent, dependent on different cable lengths. The limitations of signal transmission over considerable distances depend on the interconnecting medium used, the nature of the monitoring equipment employed, and the context in which it is operated.

Voltage outputs generally can only be transmitted over relatively short distances if high precision is required, assuming that only a little current is allowed to flow. AC voltage outputs are less sensitive to induced noise. Current loops are preferable for transmitting a signal over a few hundred metres. The current is typically between 4 and 20 mA. Currents outside this range can therefore be used as fault indications.

Frequency modulation serves for transmitting a signal over long distances without loss of accuracy and with a low susceptibility to interference. It is a signal (carrier wave) with a time-dependent frequency $\omega(t)$, while amplitude modulation is a signal bearing the information by a carrier wave of constant frequency and amplitude.

Amplification. Amplification means increasing the output of a sensor to values suitable for further manipulation. One must be aware that not only the information in the signal will be amplified, but also the distortions contained in the signal.

Voltage amplifiers can consist of several stages. The voltage of the input signal is magnified to an identical waveform, but phase shifted (180°) with respect to the input signal. Circuits built from resistance, impedance or transformers (ac-amplifiers) are drift-free, but they cannot amplify low frequency input signals directly. In this case, carrier frequency amplifier systems are used. DC amplifiers can operate up to about 0 Hz, but they possess some unstable characteristics.

Amplifier characteristics can be modified using a feedback circuit through which a part (factor a) of the output voltage is re-applied to the input together with the original signal. This can easily be seen in the frequency domain, where feedback corresponds to a series of frequency response functions (F_1 and F_2) which are mathematically related multiplicatively, and the feedback then results in the total frequency response function (see Fig. 3.9)

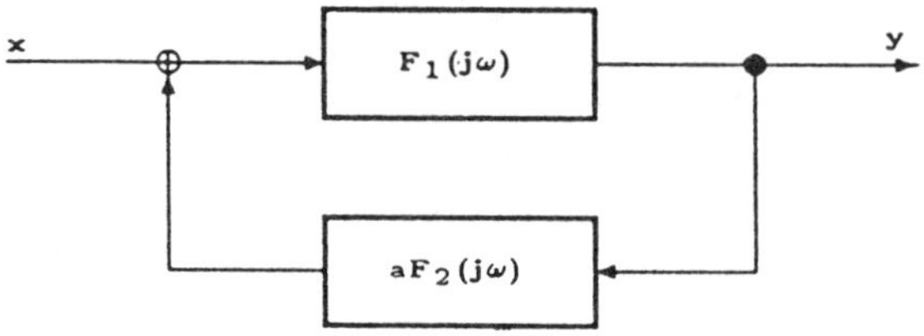

Fig. 3.9. Feedback circuit in order to modify the frequency response function of the primary device

$$F(j\omega) = \frac{F_1(j\omega)}{1 - aF_1(j\omega)F_2(j\omega)}.$$

Dependent on the feedback circuit frequency response function, the amplifier modification can be designed to the requirements. In this way it is possible to build filters, integrating as well as differentiating circuits, charge amplifiers etc. If, for example, the frequency response function of a sensor does not fulfil a phase shift-free transformation, then a connection in series without feedback and with its inverse frequency response function can help[7]:

$$F_2(j\omega) = F_1^{-1}(j\omega), \rightsquigarrow F(j\omega) \equiv 1.$$

However, it must be stated that its realization restricts this theoretical statement.

Charge amplifiers are used with piezoelectric sensors. The output is proportional to the charge of the transducer. In a professional design the output (dependent on the frequency contents) is minimum, dependent on the cable length, without affecting the sensitivity of the measuring system.

Impedance-transforming amplifiers are typical pre-amplifiers. They are used for capacitive sensors (in the context of bridge amplifiers). They decrease the impedance without amplifying the input. In order to minimize the connecting cable effects, which are not eliminated, the device should be mounted as close as possible to the sensor.

Quick-look. As already stated, the quick-look is a tool for checking and evaluating the currently measured signals. The peak and rms level graphics of every channel of a CAT-system are examples. One can take, for instance, the output signal of a measuring chain, various signal amplitudes, and spectra. The recording level indicator, recorder, oscilloscope, PCs inclusive an FFT analyzer etc. are the quick-look equipment.

Sampling. Digitized handling of the measured signals requires some precautions. During data acquisition the first step is the triggering of the

[7] As can be seen, in order to know the output one again has to know the transfer properties. Then the modification of the sensed signal or the input signal can be computed and, if necessary, corrected.

measuring equipment. Time series analysis also requires the phase information, which implies knowledge of the precise onset of the triggering point, especially in the case of a shock excitation, where it is necessary to start the measurement at such a time that all the required data are captured.

In many cases during data acquisition, the triggering point occurs well after the start of the signal, leading to a possible loss of useful data. On the other hand, if the onset of the triggering point is too early, noise is added to the signal, which gives rise to additional errors in the data acquisition process. In order to avoid these problems, the signal should be fully encompassed by time-windowing. This is achieved by triggering the measurement at a suitable point in time of the wave form which, put simply, is determined by the amplitude and slope of the incoming signal. The shape of the signal is thus important for determining the suitable triggering point. The criteria for determining the suitable triggering point for measuring as well for signal evaluation are discussed elsewhere [5, 48]. One should choose clear characteristics of the process. These considerations are also important when manipulating many channels in series [5].

Sampling itself should not be a problem if Shannon's sampling theorem is taken into account. It provides for the sampled signal to be reconstructed for all t (by the cardinal series, sometimes referred to as Whittaker's interpolation formula) in the frequency range considered without loss of information. It requires sampling with $\Delta t \equiv h = 1/(2f_g)$, if harmonics with frequencies up to the Nyquist frequency f_g should be detected in data sampled at intervals Δt. It should be noted that the harmonic with a frequency of exactly $1/(2\Delta t)$ cannot be properly detected, which means that the original signal can be reconstructed by the discrete series of the discrete Fourier transform (DFT) in the interval $[0, f_g)$. Taking into account the fact that measured signals generally contain frequencies higher than f_g, for example they are noise-corrupted, then these higher frequencies must be filtered out in order to avoid aliasing.

Pre-analysis. This is a part of signal analysis: windowing and simple evaluation, like peak and zero-crossing counting. The application of a (time) window is unconditional in measurements, because measured signals are finite length records: this means the application of a rectangular window

$$w_R(t) := \begin{cases} 1 \text{ for } -T < t < T, \\ 0 \text{ elsewhere.} \end{cases} \tag{3.3.1}$$

to the signal $x(t)$, which is thought of as being defined in the entire interval, $-\infty < t < \infty$:

$$x_T(t) := x(t) w_R(t). \tag{3.3.2}$$

Other windows serve for smoothing at the interval ends in order to avoid jumps of the time functions if periodical continuation is performed, as done in DFT. References can be found by the reader elsewhere, e.g. [5].

One should be aware that windowing (multiplication of functions in the time domain) results in a convolution integral when Fourier transformed:

$$\mathcal{F}\{x_T(t)\} = \int_{-\infty}^{\infty} X(j\omega')W(j\omega - j\omega')d\omega' \tag{3.3.3}$$

with capital letters designating the Fourier transforms of the corresponding time functions, the latter denoted by small letters. Only the Dirac function $\delta(\omega)$ (which corresponds to a window of an ∞-width) will give the exact transform, while Eq. (3.3.3) contains the error called leakage. A frequency (spectral) window is named a filter (see the next subsection). However, the reader is referred to the relevant references, for example [5, 46].

Statistics of narrow band processes are closely related to the above discussions, because the limit investigations of $w_R(t) \rightarrow \equiv 1$ results in $\frac{\sin \omega T/2}{\omega/2} \rightarrow \delta(\omega)$ when Fourier transformed.

Crossing analysis, distribution of peaks, frequency of maxima etc., also signal (time series) analysis, can be found by the reader in the well-known textbooks, for instance in [46, 5].

3.3.2 Filtering and Segmentation

Filtering. The aim of filtering is to suppress parts of the signal spectrum or to weight signals in the frequency domain. This is important so that the damage information in the related signal is not masked. Very generally, the signal $X(j\omega)$ is transformed into the signal $Y(j\omega)$ by

$$Y(j\omega) = W(j\omega)X(j\omega), \tag{3.3.4}$$

and then the filtered signal $y(t)$ is obtained by the inverse Fourier transform. $W(j\omega)$ is the **filter** function. As can be seen, it is a window in the frequency domain, because it is a multiplicative ω-dependent transformation: $W(j\omega)$ plays the role of a frequency response function. Consequently, each relationship (3.3.4) can be interpreted as a filter. A filter, a frequency window, corresponds to a convolution in the time domain. Therefore $W(j\omega)$ cannot be chosen arbitrarily when causality is to be maintained[8]. A filter affects the amplitude as well as the phase of a signal.

Filters are distinguished between

- active and passive filters dependent on the need for additional energy
- analogue and digital filters dependent on the type of signal processing
- their function (see Fig. 3.10):
  - high-pass
  - low-pass
  - band-pass

[8] Convolution in the time domain requires integration from $-\infty$, which can mean that the system already knows of a force before it starts.

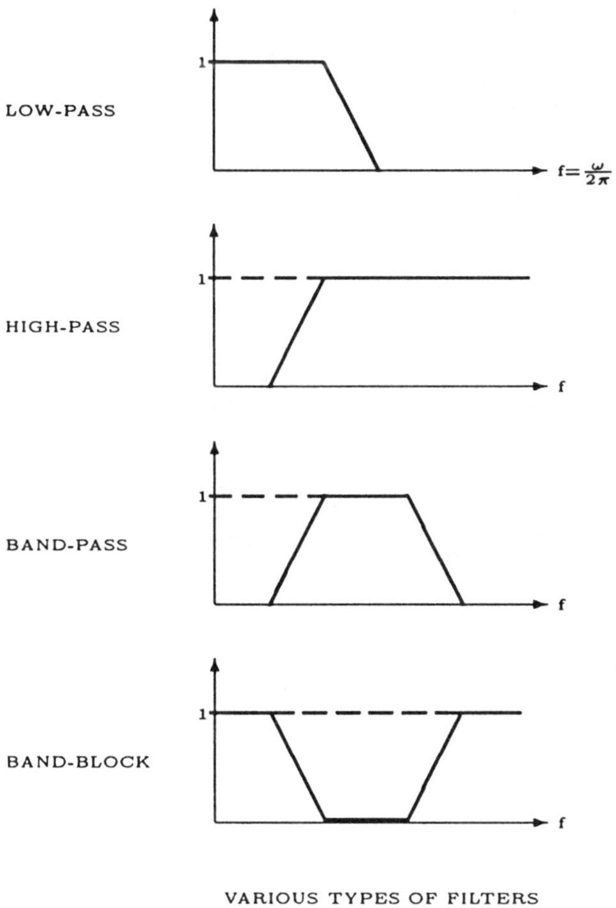

Fig. 3.10. Various types of filters

- band-rejection
- tracking

- their type of transfer property, for example Butterworth, which has a linear amplitude characteristic (constant amplitude transformation within the pass-band).

Some filter characteristics will now be explained with an analogue low-pass filter. Then follows a note on the digital filtering and an operator description.

An RC integrating circuit, as shown in Fig. 3.11 is characterized by its frequency response function

$$\frac{U_{out}}{U_{in}} = \frac{1}{1 + j\omega RC}. \tag{3.3.5}$$

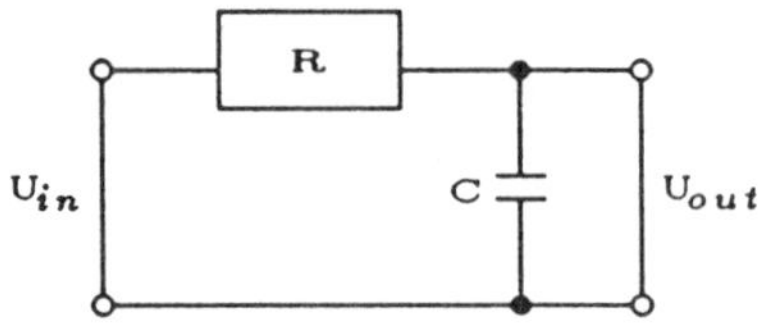

Fig. 3.11. Low-pass filter

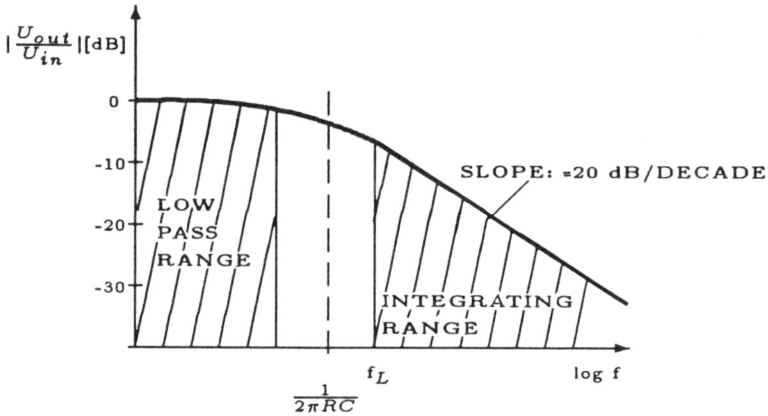

Fig. 3.12. Absolute values of the frequency response function of the RC circuit of Fig. 3.11

For $\omega RC \gg 1$ it follows

$$U_{out} \doteq \frac{U_{in}}{j\omega RC},$$

which means that it works as an integrator. If $\omega RC \ll 1$, then $U_{out} \doteq U_{in}$, and the output voltage equals the input voltage, while the circuit acts as a low-pass filter. If the time constant RC has large values, then only the static part of the input is unaltered. Figure 3.12 contains the absolute values of the frequency response function versus the frequency.

The **analogue realizations** of filters result in devices with non-ideal characteristics. Therefore one distinguishes between various realizations in practice: the Butterworth filter has a linear amplitude characteristic, while the Bessel filter has a linear phase characteristic; the latter will be applied if the shape of the signal is to be kept. Figure 3.13 represents a comparison of various filters with their main (disadvantageous compared with the ideal) characteristics. The Cauer low-pass filter is characterized by a steep amplitude decrease (it can be used as an anti-aliasing filter) but by an undulating property in the pass-band width and in the stop band.

Digital filtering has the advantage of reproducibility, while analogue filters are affected by the environmental conditions. Their time domain and equivalent frequency domain properties have to be considered when realized in a computer, which means with the consequences of digitization.

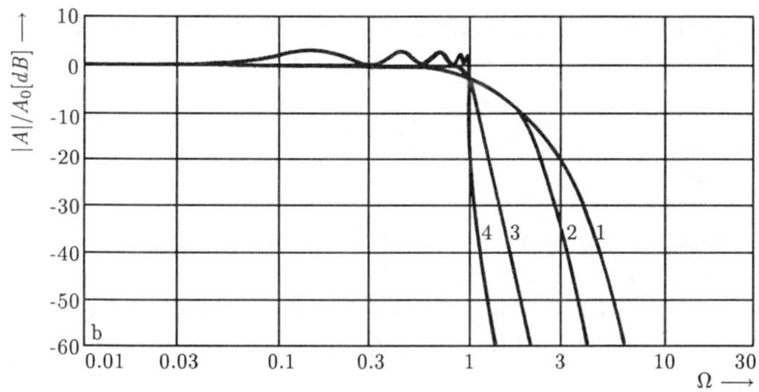

CURVE 1: LOW-PASS WITH CRITICAL DAMPING
CURVE 2: BESSEL LOW-PASS
CURVE 3: BUTTERWORTH LOW-PASS
CURVE 4: CHEBYSHEW LOW-PASS WITH 3dB WAVINESS

Fig. 3.13. Comparison of the magnitudes from various filters

A recursive digital filter generates the output time series y_n using previous output values as input values (feedback procedure):

$$y_n = cx_n + \sum_{k=1}^{m} g_k y_{n-k}. \tag{3.3.6}$$

g_k are discrete values of the filter impulse response function. Fourier transformed, Eq. (3.3.6) turns into

$$Y(j\omega) = cX(j\omega) + Y(j\omega) \sum_{k=1}^{m} g_k e^{-j\omega k\Delta t}. \tag{3.3.7}$$

Substitution of the exponential functions in the above equation by

$$z = e^{-j\omega\Delta t}$$

leads to the corresponding expression resulting from the z-transform. The related transfer function of the recursive filter follows as

$$W(z) = \frac{c}{1 - \sum_{k=1}^{m} g_k z^k}. \tag{3.3.8}$$

As can be seen, pre-determination of the location and type of the poles serves for the design of the filter.

The relationship between the RC low-pass filter and the recursive filter is performed as an example. The recursive filter may be defined as

$$y_n = (1-a)x_n + ay_{n-1}$$

with

$$a = e^{-\Delta t/RC}.$$

The frequency response function $W(j\omega)$ then follows with the approximation $a \doteq 1 - (\Delta t/RC)$ for $RC \gg \Delta t$ as

$$W(j\omega) \doteq \frac{1}{1 + j\omega RC},$$

which is the low-pass filter frequency response function (3.3.5).

In summary: if the reader is aware of the various transforms and the input and output relationships of (causal) systems in the discretized forms, then there is no problem in understanding and applying windowing and filtering.

Filtering can also be expressed by the truncation of series. Let us consider a linear equation

$$Af = g \ , \ A \ : \ F \rightarrow W, \ \ f \epsilon F, \ \ g \epsilon W, \tag{3.3.9}$$

where F and W are Hilbert domains. A is generally assumed to be a non-continuous invertible operator, which means that evaluating f for given A and g is an ill-posed problem, and as an algebraic operator it is ill-conditioned [72]. When we consider the singular value decomposition (SVD) of A, then a regularization of the generalized inverse operator A^+ of A is

$$T_\gamma g = \sum_{\sigma_n > 0} \sigma_n^{-1} F_\gamma(\sigma_n, g) < g, u_n > v_n \tag{3.3.10}$$

with σ_n the singular values and u_n, v_n the corresponding left and right vectors respectively of the SVD of A. $< .,. >$ denotes the inner product. The weighting F_γ is a real-valued function which acts on the spectrum of the generalized inverse operator A^+ like a filter. For example, the low-pass filter is defined as

$$F_\gamma(\sigma_n) = \begin{cases} 1 \text{ for } \sigma_n \geq \gamma, \\ 0 \text{ for } \sigma_n < \gamma \end{cases}$$

which truncates the sum with respect to a pre-given positive regularization parameter γ. Signals with high frequency noise are rescued from this noise by filtering.

Segmentation. It means splitting the signal into homogenous parts by means of an algorithm. The lengths of these parts are adapted to the local characteristics of the signal to be analyzed. The homogeneity of the segments can be defined in terms of a mean level or in terms of spectral characteristics. As can be imagined, segmentation is a tool for analyzing *non-homogenous* signals. These are signals which are instationary, or stationary with a (deterministic) trend etc. Therefore segmentation is suitably applied in recognition of modifications in order to obtain fewer false alarms and missed detections [116].

3.4
Signal Processing

As already mentioned, discussions of time series analysis can be found by the reader elsewhere [5, 48]. Signal processing contains the manipulation of the signals in order to obtain maximum information with minimum distortions. Here only methods are mentioned which increase the signal-to-noise ratio and give the condensation of information.

3.4.1
Increase of the Signal-to-noise Ratio

Synchronized averaging. The simplest method is the averaging of repeatedly measured signals with a defined beginning. The random distortions with zero mean will be reduced by $1/\sqrt{N}$ if N measurements are taken.

Correlation analysis. Correlation analysis in the time domain is well-known. It is the expectation applied to time shifted stationary signals:

$$\Phi_{xy}(\tau) = E\{x(t)y(t+\tau)\}. \tag{3.4.1}$$

If the ergodic theorem holds, then the estimates can be performed by time averaging over a sample of functions (with probability 1):

$$\Phi_{xy}(\tau) = R_{xy}(\tau) := \lim_{T\to\infty} \frac{1}{T}\int_0^T x(t)y(t+\tau)dt. \tag{3.4.2}$$

Noisy signals $\tilde{x}(t) = \overset{0}{x}(t) + n(t)$, $\tilde{y}(t) = \overset{0}{y}(t) + m(t)$, where the noise signals $n(t), m(t)$ are uncorrelated with each other and uncorrelated with the undisturbed signals $\overset{0}{x}(t), \overset{0}{y}(t)$, yield the relation

$$\begin{aligned}\Phi_{\tilde{x}\tilde{y}}(\tau) &= E\{\tilde{x}(t)\tilde{y}(t+\tau)\} = \Phi_{\tilde{x}\overset{0}{y}}(\tau) + \Phi_{\tilde{x}m}(\tau)\\ &= \Phi_{\overset{0}{x}\overset{0}{y}}(\tau) + \Phi_{\overset{0}{x}m}(\tau) + \Phi_{n\overset{0}{y}}(\tau) + \Phi_{nm}(\tau).\end{aligned}$$

When the uncorrelated signals are considered, all the expressions except the first one are equal to zero. For $\tilde{x}(t) = \tilde{y}(t)$ the latter term in the above equation is $\Phi_{nn}(\tau)$, which tends to zero with $\tau \to \infty$ due to the zero-mean of $n(t)$. In other words, if the measuring time T is large enough, then the mean-free stochastic disturbances vanish. Numeric handling is performed by applying the cyclic FFT algorithm.

Spectral analysis. In the frequency domain the Fourier analysis and the spectral analysis are well-known. While the Fourier analysis does not contain any expectation, the spectral analysis contains the expectation by application of the Wiener-Khintchine transformation, that is when the

Fourier transform of the correlation function is considered. However, if estimations of the spectral functions are performed numerically, that is discretely, then the result is deterministic and built up by the corresponding periodograms (finite discrete Fourier transforms). Here the expectation operator has to be included additionally and approximately by smoothing (filter) or other operations or by averaging [48, 5].

Cepstrum analysis. Another method should be mentioned: that is the cepstrum analysis. It is based on the natural logarithm of the spectral density functions. The multiplicative input/output relationship of systems is then turned into an additive relation. The cepstrum analysis is invariant against translation, rotation and scale changes by the procedures of the Fourier transform - logarithm procedure - (inverse) Fourier transform. These are interesting properties for synergetic pattern recognition [117].

3.4.2 Information Condensation

Some types of information condensation have already been mentioned (see Sect. 3.3.1). Here deterministic as well as stochastic methods (e.g histogram) have to be enumerated. The SVD (Eq. (3.3.10)) is thus another method of information condensation, because it permits the evaluation of the singular values and the corresponding vectors (thus in the frequency domain).

Process models like AR, MA, and ARMA[9] models also serve for information condensation. They will be discussed in Chap. 4. Non-parametric methods perform the estimation of the impulse response functions and the frequency response functions [4, 5]. Correlation analysis and spectral analysis have already been mentioned. The latter serves for estimating resonances of the system and their time-dependent shifts. Experimental modal analysis goes beyond this. It uses a model-based evaluation and results in various eigenquantities. Further methods exist, but they are not itemized here.

3.4.3 Inverse Filtering

Since the measured signals are bandlimited, the effects of bandlimitation on the input and output data must be removed prior to identification. A simple (suboptimal) method is to use the inverse of the mathematical model (the identified one or the one approximated by the prior mathematical model), which will be excited with the measured signal in order to obtain an estimate of the deconvolved signal. This approach is applied with the input residual WLS method in order to remove the bias approximately. The data are merely filtered with the inverse (identified or prior) mathematical model (whitening filter, see Fig. 4.1). This can be performed

[9] AR-MA stands for Auto Regressive - Moving Average.

in the image domain by applying the inverse frequency response function (Fourier transform) or inverse transfer function (Laplace transform) of the mathematical model: $\hat{\mathbf{P}}(j\omega) = \mathbf{F}^{-1}(j\omega)\mathbf{U}(j\omega)$, or $\hat{\mathbf{P}}(s) = \mathbf{H}^{-1}(s)\mathbf{U}(s)$.

Deconvolution as a non-parametric inverse problem is generally ill-posed [71, 72], and therefore needs to be numerically solved with measured data and special procedures (regularization). One approach uses mean-squared error optimum deconvolution. It is based on correlation analysis, takes the impulse response function of the mathematical model, and applies the corresponding auto-variance and cross-variance matrices [5, 118]. The correlation functions formally fulfil the same convolution integral as for deterministic signals. With the application of numerical integration the estimates, characterized with a hat, lead to the linear algebraic system of equations

$$\hat{\mathbf{\Phi}}_{pu} = \hat{\mathbf{\Phi}}_{pp}\tilde{\boldsymbol{\gamma}} \tag{3.4.3}$$

in

$$\tilde{\boldsymbol{\gamma}} := (\tilde{\gamma}_0, \tilde{\gamma}_1, \ldots, \tilde{\gamma}_{N-1}), \;\; \tilde{\gamma}_i = \tilde{g}_i h, \tag{3.4.4}$$

with $g_i := g(t_i)$ and h the time step. For $m = N - 1$, N the number of unknowns, the matrix $\hat{\mathbf{\Phi}}_{pp}$ of the auto-correlation functions of the input is regular and therefore invertible. However, the resulting vector of the aproximated unit impulse function values can be very erroneous. One should therefore consider more equations than unknowns in order to apply the WLS approach, thus leading to estimates instead of approximations:

$$\hat{\boldsymbol{\gamma}} = (\hat{\mathbf{\Phi}}_{pp}^T \mathbf{G} \hat{\mathbf{\Phi}}_{pp})^{-1} \hat{\mathbf{\Phi}}_{pp}^T \mathbf{G} \hat{\mathbf{\Phi}}_{pu}, \tag{3.4.5}$$

$\mathbf{G}$ is the suitably chosen weighting matrix.

3.5 Summary

The detection of faults and their locations in operating systems depends on many stages of reasoning and consideration. It depends on the particular monitoring and the type of diagnostics applied. Monitoring is an important stage of symptom-based diagnostics. Hence we have briefly elucidated the process and the possible links within the system and the damage or fault, and the signals which carry the fault-related information.

Measurements are the only possible way of obtaining the required data of the real process and of the existing system. Because these data are incomplete and uncertain, they are additionally affected by the measuring equipment used, and their manipulation requires the application of special precautions and carefully chosen procedures. From this point of view the measurements are discussed in some detail in relation to the topic. The fact that the engineer involved is generally not an expert in measuring techniques is also taken into account.

The information contained in the signals and time series can be extracted and condensed. Many procedures exist dependent on the goal to be achieved. However, all the methods use measured data, and when inverse problems have to be solved, these problems are generally ill-posed or ill-conditioned. This requires special procedures in order to obtain a unique and stable solution.

It is also noted here that ultrasonic testing can be combined with intensity investigations, and with a Volterra series expansion in order to handle nonlinear effects. Additionally, smart systems, which are structures with built-in measuring capabilities by, for example, multi-functional sensors combined with control devices (e.g. piezo-shells, special composite materials), can be applied. In active systems these are feedback-systems with control devices; the control forces can be used for fault detection (e.g. applying triangularization) and localization. These topics are under research.

CHAPTER 4

Model-Supported Diagnostics Via Parameter Estimation

Monitoring and measurements with some signal processing serve for the early detection of system state deviations from the normal (non-faulty, undamaged) state. This task will generally be done on-line while the system is in operation. The measured process signals within the system's lifetime contain some information about the system's state at the time of measurement. As already mentioned, the measured signals are noise corrupted and incomplete (e.g. for system description), and in the worst case superimposed by systematic errors. Both monitoring, and how the *process* could be identified, are discussed in Sect. 3.1. Monitoring can be performed model-supported, but due to the demands made by it, parameter estimation is generally excluded. A process description may therefore be available, but there is no system description at present.

Within model-supported diagnostics the best available knowledge base is the validated model of the system under consideration. The adjustment and validation of the mathematical model are done, of course, by the use of measured data. Here diagnosis means early fault detection in its broader sense, localization, type of fault, finding out the causes of faults, and safety assessment and cures, as well as prediction of the state condition. This can be done with the use of the adjusted and validated mathematical models $\mathcal{M}_i, i = 1, \ldots, N_1$ of the existing system. Because the necessary computations with the models mentioned are often time-consuming, this is generally done off-line. This innovatory process has the advantages of

- further use of the measurements already existing
- providing the substitution of missing measurements
- systematic adjustment of the mathematical model at every life time and therefore at every system condition (life time-dependent) using prior knowledge
- establishing verified and validated mathematical models
- simulation and prediction of past and future forces (finding causes and cures): safety assessment, and
- trend forecasts.

Two learning processes are involved here. The first one concerns the system and process in the normal state ($\rightsquigarrow$ the validated reference model as one part of the knowledge base), and the second one is a self-learning process with respect to the life time properties of the system and the resulting process by model adjustment at every required state.

As already mentioned, the problems to be solved are the feature extraction of possible system modifications from the measured signals and the suppression of the dominating normal conditions (this includes, of course, the use of known [estimated] normal conditions) in the available information. The approach discussed here is the application of system identification, and in particular of the adjustment (correction, updating, reconciliation, calibration) of mathematical models on the basis of measured data. The correction can be done in an inverse formulation and in a forward strategy by the application of the Bayesian approach and multi-hypothesis testing. The forward identification method is applied, for example, in [31] and will not be discussed here. The inverse formulation with the parameter estimation is restricted to the frequency domain[1], and here emphasis will be placed on the unconstrained weighted least squares (WLS) with some penalty terms, although other estimators are also discussed.

4.1 The General Procedure

The concept of model-supported diagnostics is discussed in Sect. 1.2, shown in Fig. 1.1, and extended in Fig. 3.1. Here the mathematical models are assumed as spatially discretized, because in practice the systems seldom allow continuous modelling. The initial mathematical model serves for sensitivity studies ($\rightsquigarrow$ essential parts of the mathematical model, e.g. with respect to load paths and resulting stiffnesses) and for performing a catalogue of possible faults based on pregiven causes and the resulting consequences for safety requirements. When the system has been constructed and the first tests can be made, the initial mathematical model will be verified and validated. This model, which is a model with known confidence and sufficiently small errors, is used for updating the prior sensitivity analysis, and the previous theoretical studies of possible faults can be improved. These initial fault investigations are very important for the localization, diagnosis and assessment of system modifications during the system's life time.

Observation of the system (monitored data and additional measurements, periodically or on request) yields data dependent on the life time θ_i for an adjusted (with respect to the current state at time θ_i) mathematical model. Parameter deviations between the adjusted model and the reference model – when submodels are used see Sect. 2.2.6, Eqs. (2.2.24)–(2.2.27), checked by a pre-defined quality criterion - provide information on the faulty process and the system modifications. Of course, the significance of the detected modifications has to be investigated (Chap. 5).

The use of the adjusted mathematical models at previous times and the current time permits the user to go into details, which means locating faults within the mathematical model and assessing them, providing that the mathematical models describe the quantities to be considered sufficiently accurately and in sufficient detail. In general, a dynamic model with a few

[1] The main experience of the senior author is with regard to the frequency domain.

degrees of freedom will not describe stress concentrations if these are the quantities taken for decision (Chap. 5) [111].

4.2 Model Adjustment – Preparations

As already mentioned, model adjustment is a part of system identification [5]. The prior model has to be corrected with the current measured data in order to obtain a verified and validated mathematical model which represents the current state. If the model structure is adequate to the system behaviour, then model adjustment is reduced to parameter estimation. In the following it is assumed that model structure identification is performed (for detection and possible model structure deviations see Sects. 2.1.4 and 2.2.3), so that the problem is reduced to parameter estimation. Estimation methods will be applied due to the measurement errors modelled randomly and due to the incompleteness of measurements: the use of samples instead of the ensemble. Systematic errors must be avoided, otherwise they have to be known in order for them to be corrected computationally. The ill-posedness of the inverse problems [72] should be taken into consideration; in consequence, the Bayesian approach is emphasized, since it allows the inclusion of prior information.

The first step in parametric identification is the choice of the model formulation (type) and its parametrization ($\leadsto$ the prior mathematical model). The second step concerns the measurement. The measured data have to fulfil special requirements for model adjustment. Then the residuals performed between the measured/identified quantities and the corresponding quantities of the model to be adjusted have to be considered. The last step is the estimation of the parameters required for damage detection etc.

4.2.1 Models

Possible structural model descriptions are summarized in Table 4.1 for linear systems.

The equations of motion as well as their solutions in the various domains can be used. The parameter matrices can also serve directly as a theoretical basis with their properties decomposed by the related eigenvectors, and then the equations of motion are taken as constraints. Here, in general, the equations of motion in the form of Eq.(1.3.1) or in the image domain, as in Eq.(2.1.29), will be taken as the model. The assumption of the (equivalent) viscous damping seems to be somewhat arbitrary. However, due to the lack of information and knowledge this is an easy path to take, and it will be generally accepted. The assumption of hysteretic damping (complex stiffness matrix [5, 6]) seems to be easy, too, but this damping model is restricted to harmonic excitation, and is therefore too restrictive for application to operating systems. The model structure of the

Table 4.1. Systematics of mathematical models

DOMAIN	TIME DOMAIN , IMAGE DOMAIN	
EQUATIONS OF MOTION	PARAMETRIC ↗	CLASSICAL AS 2nd ORDER ODE IEQ; FREE, FORCED MOTION
	PARAMETRIC ↘	STATE SPACE 1st ORDER ODE; FREE, FORCED MOTION
	NON-PARAMETRIC:	SEE SOLUTIONS
SOLUTIONS	PARAMETRIC:	MODAL COORDINATES
	NON-PARAMETRIC:	MATRICES OF IMPULSE RESPONSE FUNCTIONS; FREQ. RESPONSE, TRANSFER FUNCTION MATRICES
PARAMETER MATRICES	PARAMETRIC ↗	SPECTRAL DECOMPOSITIONS
	PARAMETRIC ↘	DIRECT
APPROXIMATIONS	DETERMINISTIC:	DIFFERENCES, NUM. INTEGRATION
	STOCHASTIC:	DIFFERENCES (COVARIANCE EQUIVALENT)
ENERGIES, WORKS		} MODAL, NON-MODAL

band-limited mathematical model will then be determined by the number of DOF given by the initial model and the frequency range of interest. If a fault introduces additional DOF, then the mathematical model must be changed (see Sect. 2.1.4), mainly by re-modelling [111]. In this context it will be noted that for the choice of the model formulation its aim is essential within diagnosis, including the quantities used for monitoring. The model sensitivities as discussed in Sect. 2.1.3 have to be taken into account.

The properties of the equations are important: the posedeness of the formulation. For example, determination of the stiffness matrix $\mathbf{K}$ or the corresponding influence matrix $\mathbf{G} = \mathbf{K}^{-1}$, assuming that the inverse exists, yields the equations

$$\mathbf{K} = \mathbf{P}_0\mathbf{U}_0^{-1}, \tag{4.2.1}$$

and

$$\mathbf{G} = \mathbf{U}_0\mathbf{P}_0^{-1}, \tag{4.2.2}$$

with the use of linearly independent force vectors assembled in the matrix $\mathbf{P}_0$ and with the resulting response matrix $\mathbf{U}_0$. If the test of the system is force-controlled, that means the forces to be applied are chosen, then they can be chosen such that the force matrix $\mathbf{P}_0$ is well-conditioned, while the response $\mathbf{U}_0$ contains the system properties which cannot be influenced. Therefore the dynamic response matrix $\mathbf{U}_0$ may be ill-conditioned. Consequently, Eq.(4.2.2) is well-conditioned and recommended as a basis. In the other case, when the test is diplacement-controlled, the reverse statement holds true, and the formulation (4.2.1) can be chosen well-conditioned. Consequently, the causality of the problem has to be taken into account. The resulting properties (posedeness of the problem) can require a different formulation, especially when measured data with measuring errors have to be considered ($\rightsquigarrow$ regularization methods: modification of the mathematical operator, see Sect. 4.4.3).

If the locations of parameter modifications due to faults are known, then it is possible, for example, to model these faults within the state space with a fault entry matrix and a corresponding vector with components representing the system faults (see Eq.(3.01)). Here another fault model will be applied. Subsystem modelling, as mentioned in Sect. 2.1.3 with adjustment factors as introduced in Eqs.(2.2.24)–(2.2.26) for the matrices of the submodels, is recommended as an economic and effective procedure. It can also be used for detecting the locations of modifications, if necessary by submodel halving (submodel refinement). This subject has already been mentioned in Sect. 2.2.6.

Sensor and calibration faults as systematic errors are tackled within monitoring and measurements (Chap. 3). In consequence, the irregular measurement errors, which are modelled stochastically, require the application of estimators (Sect. 4.2.3) for their reduction. In this context it will only be mentioned that the Fourier transform of measured signals which are corrupted by uncorrelated (to the signal) noise will not influence this uncorrelation, while the Laplace transform will result in correlated noise in the image domain when discrete transforms are applied.

4.2.2 Tests

Although measuring techniques and signal processing have already been discussed in Chap. 3, some remarks are repeated and some additional ones are made here.

Testing includes excitation and measurement. Natural excitation (a system in its environment and in operation/service) is often insufficient with respect to the amplitude level and the frequency content. The goodness of the test results also depends on the experimenter and on the available external test conditions, such as the test crew, equipment, time available etc.

The basic requirements for measurement technology are:

- repeatability of the measured data (same equipment, same team)[2],
- comparability of the data[3] (different equipment, same team),
- reproducibility of the measurements (different equipment, another team).

The different types of tests are necessary in order to detect different types of errors:

- If the irregular measuring errors are modelled stochastically, then repeated tests result in measured data whose measuring errors are supposed to be minimized by applying estimators. Systematic errors are not recognizable.
- The second type of testing should detect systematic errors coming from the test equipment if the same procedure is applied.
- The third test type serves to find out the subjective influence of the test teams[4]. The statistical errors are generally larger than under repeatability conditions, because systematic errors can exist "between the different labs".

Technical properties and requirements can be found by the reader in Chap. 3.

Within identification one has to require

- observability,
- controllability,
- identifiability.

These items are discussed, for example, in [4]. The reader can find the necessary and sufficient conditions for mechanical systems in [106, 108].

The experimenter, the environment, the equipment, various possibilities of error, and the type of model with its properties are mentioned next. We have to take into account the measurement uncertainties and the fact that only samples are available. Different estimators applied to the same mathematical model, but written in different formulations, will give different estimated values of the same theoretically defined parameters. Consequently, the model goal is decisive within diagnosis.

The measurands have to be defined first. In the context of TACAM the first step is structure identification, and the second step, if the structure is sufficiently approximated, is parameter estimation. In addition to the requirements already mentioned, parameter estimation demands the consideration of

- the range of model validity, and, if a test is possible and necessary,
- an optimum test design so that the resulting model is usable (has sufficiently small errors).

---

[2] The principle of non-distinguishability ($\rightsquigarrow$ concurrency theory) and weak and strong causality (chaos theory) are not discussed here.

[3] This is not a technical term.

[4] Possibly dependent on the day of the week: Monday work after the weekend, or Friday work looking forward to the weekend.

The first requirement concerns the force amplitudes, frequency range, and initial conditions for nonlinear systems. The optimum test design determines the sample size, the measuring points and excitation points to be chosen, and if work is being done in the frequency domain it determines the frequency spacings. The test design, if required and possible, can be based on the information matrix for formulating the optimality criterion, as is done, for example, in the paper [119]. Practical constraints often prevent optimization, and then prior knowledge and experience have to substitute it.

4.2.3 Residuals

The generation of residuals is called transformation into the parity domain. Residuals are mentioned in Sect. 2.2.2. There the input, output and generalized residuals are shown (Figs. 2.9–2.11), and Fig. 2.12 contains an overview. As there are differences between dynamic quantities of the mathematical model and the corresponding measured quantities, residuals serve for the detection of model deviations and, under particular assumptions, for system modifications, as already discussed in Sect. 2.2.6. Local residuals can be used additionally for the localization of modifications (Sect. 2.2.6). It is repeated that residuals are performed for a state condition at life time θ_i, where the prior mathematical model for the state condition θ_{i-1} is the basis for the parametrized model to be adjusted. The parameters then estimated will describe the modification between the state at time θ_i and θ_{i-1}. As a reference time, of course, a life time other than θ_{i-1} can be chosen. In the following the distinction in notation with respect to the life time coordinate will be suppressed.

In general: with the dynamic quantity $\boldsymbol{\alpha}_r \in C^n$ of the mathematical model and the corresponding measurements $\boldsymbol{\alpha}_r^m$ the residual vector is

$$\mathbf{v}_r = \boldsymbol{\alpha}_r - \boldsymbol{\alpha}_r^m \in C^n, \quad r = 1(1)N. \tag{4.2.3}$$

The dynamic quantity $\boldsymbol{\alpha}_r$ is a function of the parameters $\mathbf{a}$ to be estimated, $\boldsymbol{\alpha}_r(\mathbf{a})$, $\mathbf{a} \in R^J$, $J = I + R + S$ is the sum of the number of parameters to be adjusted in accordance with Eqs.(2.2.24)–(2.2.26). Of course, the measurements $\boldsymbol{\alpha}_r^m$ can also be estimates, for example, estimated eigenfrequencies.

The properties of the residuals are that:

1. they contain the randomly modelled measurement error which, at most, should be additive and uncorrelated, and
2. their sensitivity with respect to the parameters to be estimated is important, including the localization properties of the estimates.

Dependence on measurements. If only the output is additive noise corrupted with zero mean, designated by $\mathbf{r}(\mathbf{t})$, or in the frequency domain by $\mathbf{R}(j\omega) = \mathcal{F}\{\mathbf{r}(t)\}$, and ω is the frequency, then the output residual is

$$\mathbf{v}_{Or} := \mathbf{U}(j\omega_r, \mathbf{a}) - \mathbf{U}^m(j\omega_r) \tag{4.2.4}$$

with

$$\mathbf{U}(j\omega_r, \overset{o}{\mathbf{a}}) - \mathbf{U}^m(j\omega_r) = \mathbf{R}(j\omega_r), \tag{4.2.5}$$

and fixed values ω_r. A reminder: it is

$$\mathbf{U}(j\omega) = \mathcal{F}\{\mathbf{u}(t)\},$$

and it holds true that

$$\mathbf{U}^m(j\omega_r) = \overset{o}{\mathbf{U}}(j\omega_r, \overset{o}{\mathbf{a}}) + \mathbf{R}(j\omega_r).$$

$\overset{o}{\mathbf{a}}$ is the 'true' value of $\mathbf{a}$. The residual (4.2.4) is equal to the random error with the expectation equal to zero at $\mathbf{a} = \overset{o}{\mathbf{a}}$. The resulting WLS estimator is linear with respect to the measurement noise, but nonlinear with respect to the parameters to be estimated, because the parameter matrices depend on the vector $\mathbf{a}^T = (\mathbf{a}_M^T, \mathbf{a}_B^T, \mathbf{a}_K^T)^T$ included in the dynamic stiffness matrix which has to be inverted:

$$\mathbf{U}(j\omega, \mathbf{a}) = [-\omega^2\mathbf{M}(\mathbf{a}_M) + j\omega\mathbf{B}(\mathbf{a}_B) + \mathbf{K}(\mathbf{a}_K)]^{-1}\mathbf{P}(j\omega). \tag{4.2.6}$$

Another reminder: the parameters to be estimated are dependent on the chosen submodels describing the defined subsystems designated by their indices. This spatial property will relate to the physical subspace.

The input residual is defined with the excitation vectors as

$$\begin{aligned} \mathbf{v}_{Ir} : &= \mathbf{P}(j\omega_r, \mathbf{a}) - \mathbf{P}^m(j\omega_r) \qquad (4.2.7) \\ &= [-\omega_r^2\mathbf{M}(\mathbf{a}_M) + j\omega_r\mathbf{B}(\mathbf{a}_B) + \mathbf{K}(\mathbf{a}_K)]\mathbf{U}^m(j\omega_r) - \mathbf{P}^m(j\omega_r). \end{aligned}$$

As can be seen, this residual vector is linear in the parameters to be estimated, but the random disturbances are weighted by the dynamic stiffness matrix. At the true parameters $\overset{o}{\mathbf{a}}$ it follows that

$$\mathbf{v}_{Ir}(\overset{o}{\mathbf{a}}) = \mathbf{S}(j\omega_r, \overset{o}{\mathbf{a}})\mathbf{R}(j\omega_r) \tag{4.2.8}$$

due to

$$\mathbf{S}(j\omega_r, \overset{o}{\mathbf{a}})[\overset{o}{\mathbf{U}}(j\omega_r) + \mathbf{R}(j\omega_r)] - \overset{o}{\mathbf{P}}(j\omega_r) \text{ with } \mathbf{P}^m = \overset{o}{\mathbf{P}},$$

because only the output is corrupted by additive noise. Obviously, the residual (4.2.8) results in correlated random errors even if $\mathbf{R}(j\omega)$ represents white noise. It is well known [4] that for least squares estimation using a generalized model, the condition for unbiased estimates corresponds to the condition of "white residuals". The method of generalized least squares [4] applied to generalized models can therefore be seen as an attempt to overcome the bias problem by introducing "whitening filters" [120, 121]. However, only the dynamic stiffness matrix is known for the

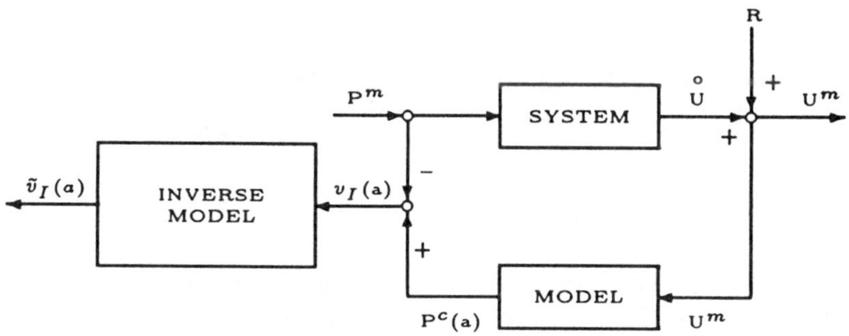

Fig. 4.1. Inverse filtering of the input residual

prior mathematical model, $\mathbf{a} = \mathbf{e} := \{1\}$, and in order to avoid iterative adjustment, it follows that

$$\tilde{\mathbf{v}}_{Ir}(\overset{o}{\mathbf{a}}) := (-\omega_r^2\mathbf{M} + j\omega_r\mathbf{B} + \mathbf{K})^{-1}\mathbf{v}_{Ir}(\overset{o}{\mathbf{a}}) \doteq \mathbf{R}(j\omega_r). \tag{4.2.9}$$

This filtering procedure is shown in Fig. 4.1. The input residuals very clearly show the necessity for statistical weighting. The resulting loss function (objective function) of the WLS now contains the inverse covariance matrix of $\mathbf{R}$ as a weighting [5].

The input is generally also noise-corrupted, and then the input residuals directly contain this additional noise. However, when a test is performed with artificial excitation, this input noise plays a minor role within the input residual method, because the input is controllable within certain limits.

If eigenquantities of the damped system are identified, $\hat{\lambda}_{Br}$, $\hat{\mathbf{u}}_{Br}$ for the rth eigenvalue and eigenvector respectively, the modal equation error is

$$\mathbf{v}_{Mr} := [\hat{\lambda}_{Br}^2\mathbf{M}(\mathbf{a}_M) + \hat{\lambda}_{Br}\mathbf{B}(\mathbf{a}_B) + \mathbf{K}(\mathbf{a}_K)]\hat{\mathbf{u}}_{Br}. \tag{4.2.10}$$

The residuals of the undamped system correspond to Eq.(4.2.10). In addition to (4.2.10), the orthonormalization can be taken into consideration, and the residuals read [122]

$$\mathbf{v}_{MNr} := \left\{ \begin{matrix} \mathbf{v}_{Mr}(\mathbf{a}) \\ 2\hat{\lambda}_{Br}\hat{\mathbf{u}}_{Br}^T\mathbf{M}(\mathbf{a}_M)\hat{\mathbf{u}}_{Br} + \hat{\mathbf{u}}_{Br}^T\mathbf{B}(\mathbf{a}_B)\hat{\mathbf{u}}_{Br} - 1 \end{matrix} \right\}. \tag{4.2.11}$$

Additionally, identified generalized masses can be taken into account [5, 123]. When compared with (4.2.10) the extended residual (4.2.11) significantly influences the measurement error propagation. If one assumes that the estimates are erroneous with the variances $\sigma_{\lambda r}^2$ and σ_{ur}^2 and if one neglects the correlation between the errors of the eigenvectors, linearization leads to

$$E\{\mathbf{v}_{MNr}\mathbf{v}_{MNr}^T\} \doteq \mathbf{G}_{MNr}^{-1} = (\mathbf{U}_{MNr}^m\mathbf{U}_{MNr}^{mT})^{-1} \tag{4.2.12}$$

with the definition

$$\mathbf{U}_{MNr}^{m} := \begin{bmatrix} \hat{\lambda}_{Br}^2 \mathbf{M}(\mathbf{e}) + \hat{\lambda}_{Br}\mathbf{B}(\mathbf{e}) + \mathbf{K}(\mathbf{e}) & [2\hat{\lambda}_{Br}\mathbf{M}(\mathbf{e}) + \mathbf{B}(\mathbf{e})]\hat{\mathbf{u}}_{Br} \\ 2\hat{\mathbf{u}}_{Br}^T[2\hat{\lambda}_{Br}\mathbf{M}(\mathbf{e}) + \mathbf{B}(\mathbf{e})] & 2\hat{\mathbf{u}}_{Br}^T\mathbf{M}(\mathbf{e})\hat{\mathbf{u}}_{Br} \end{bmatrix} \cdot \begin{bmatrix} \sigma_{ur}\mathbf{I} & \mathbf{0} \\ \mathbf{0} & \sigma_{\lambda r} \end{bmatrix}. \tag{4.2.13}$$

As can be seen, in the above definition (4.2.13) the upper left-hand matrix corresponds to the residual $\mathbf{v}_{Mr}$ without the orthonormalization, which is ill-conditioned, because it becomes singular for the eigenvalues of the initial model.

A further generalization can be achieved by taking into account the complete orthogonalization properties [5] instead of only the one equation (without the null matrices).

If only the eigenvalues are identified, then the equation error residuals can also be taken. However, the present eigenvectors now depend on the parameters $\mathbf{a}$ to be estimated; therefore the eigenvalue problem has to be additionally included in the computation, and the resulting procedure is nonlinear in the parameters. In special cases, however, the eigenvectors of the initial model can be taken (consequently, the iterative procedure can be avoided), because the sensitivity of the eigenvectors with respect to the parameters a_j, $j = 1(1)J$, is of second order compared with that of the eigenvalues [5]. – With regard to further properties, for example the formal similarities of the residuals mentioned so far, see [124].

Finally, the partial residuals (see Sect. 2.2.2) will be considered instead of the equation errors. They are extensively discussed in [5]. For example, if only eigenvalues are measured, then the residual vector is written as

$$\mathbf{v}_0 = \{\omega_{0r}^2(\mathbf{a}) - \hat{\omega}_{0r}^2\}, \tag{4.2.14}$$

which can be divided by the initial known values in order to obtain dimensionless quantities.

If the measured, identified quantities are used directly in the residuals, then weighting with the inverse covariance matrix of the corresponding residuals is sufficient to minimize the stochastic measuring errors (Markov estimation [5]). The unbiased estimator will give a lower bound of the variances and covariances of the estimates by the inverse Hessian matrix taken at the estimates $\mathbf{a} = \hat{\mathbf{a}}$ (Rao-Cramér inequality [69]). The Hessian matrix is defined as:

$$\left[\frac{\partial^2 \mathcal{J}(\mathbf{a})}{\partial a_\rho \partial a_\sigma}\right], \quad \mathcal{J}(\mathbf{a}) \text{ the loss function.} \tag{4.2.15}$$

If the residuals are built up from derived quantities, then filtering similar to that used for the input residuals for parameters $\mathbf{a}$ at $\mathbf{e}$ has to be applied in order to reduce the bias of the estimates.

The correlation of the measurement errors has already been discussed. The residual sensitivity with respect to the measurement errors is another property to be considered. Residuals which are performed by estimates, such as identified eigenquantities which already contain an error reduction due to the included estimation applied to measured values, depend on the corresponding standard deviations instead on measurement errors. In contrast, those residuals which are built up with input and output quantities contain the measurement errors (sometimes amplified, see above). Consequently, it is necessary to work with a measurement error as small as possible, which means that prior error reduction is recommended by using repeated measurements and taking their averages. In the case of normally distributed random errors, this results in a noise reduction by a factor $1/\sqrt{N}$, if N repeated measurements are used.

Sensitivity with respect to the parameters. The parameter sensitivity of the residuals,

$$\frac{\partial \mathbf{v}_r(\mathbf{a})}{\partial a_j} = \frac{\partial \boldsymbol{\alpha}_r(\mathbf{a})}{\partial a_j}, \quad j = 1(1)J, \tag{4.2.16}$$

is part of the estimation procedures (see the next subsection). Knowledge of them is useful with respect to the choice of the parameters to be estimated in connection with the choice of the submodels (Sect. 4.3.1). Dynamic quantities $\boldsymbol{\alpha}_r(\mathbf{a})$ which are insensitive with respect to the parameters to be estimated do not contain sufficient information to permit accurate estimation of these parameters: in this formulation this is an ill-posed problem. These sensitivities are discussed in detail in [5].

A few examples will demonstrate some interesting properties. With regard to the eigenvalues of the model of the undamped system, Eq.(2.1.38), it follows Eq.(2.1.40) with respect to the adjustment factors a_{Ki} of the subsystem modelling (2.1.37):

$$\frac{\partial \lambda_{0r}}{\partial a_{Ki}} = \hat{\mathbf{u}}_{0r}^T \mathbf{K}_i \hat{\mathbf{u}}_{0r} =: k_{gri}, \tag{4.2.17}$$

which, when linearized, means

$$\Delta\lambda_{0r} \doteq k_{gri} \Delta a_{Ki}.$$

Here the generalized stiffness k_{gri} with respect to the rth eigenvector of the unmodified (initial) model determines the rth eigenvalue sensitivity for small changes in a_{Ki}. This means that the ith submodel stiffness matrix is decisive for this sensitivity. It is emphasized that the sensitivity (4.2.17) is independent of the eigenvector changes due to a_{Ki}.

The residual sensitivities of the matrix eigenvalue problem of the associated undamped system with respect to the inertia parameters $a_{M\sigma}$ and the stiffness paramters $a_{K\iota}$ (see Eqs. (2.2.24), (2.2.25)),

$$\left.\begin{array}{lllll} \mathbf{M}^c & := & \mathbf{M}(\mathbf{a}_M) & = & \sum_{\sigma=1}^{S} a_{M\sigma}\mathbf{M}_\sigma \\ \mathbf{K}^c & := & \mathbf{K}(\mathbf{a}_K) & = & \sum_{\iota=1}^{I} a_{K\iota}\mathbf{K}_\iota, \text{ or} \\ \mathbf{G}^c & := & \mathbf{G}(\mathbf{a}_G) & = & \sum_{\iota=1}^{I} a_{G\iota}\mathbf{G}_\iota, \end{array}\right\} \tag{4.2.18}$$

follow in accordance with[5] (4.2.11) as

$$\mathbf{v}_{MNr} = \left\{ \begin{array}{c} [-\hat{\lambda}_{0r}\mathbf{M}(\mathbf{a}_M) + \mathbf{K}(\mathbf{a}_K)]\hat{\mathbf{u}}_{0r} \\ \hat{\mathbf{u}}_{0r}^T\mathbf{M}(\mathbf{a}_M)\hat{\mathbf{u}}_{0r} - 1 \end{array} \right\} : \tag{4.2.19}$$

$$\frac{\partial \mathbf{v}_{MNr}}{\partial a_{M\sigma}} = \left\{ \begin{array}{c} -\hat{\lambda}_{0r}\mathbf{M}_\sigma\hat{\mathbf{u}}_{0r} \\ \hat{m}_{gr\sigma} \end{array} \right\}, \quad \frac{\partial \mathbf{v}_{MNr}}{\partial a_{K\iota}} = \left\{ \begin{array}{c} \mathbf{K}_\iota\hat{\mathbf{u}}_{0r} \\ 0 \end{array} \right\}, \tag{4.2.20}$$

and with the generalized mass with respect to the σth submodel matrix

$$\hat{m}_{gr\sigma} := \hat{\mathbf{u}}_{0r}^T\mathbf{M}_\sigma\hat{\mathbf{u}}_{0r}. \tag{4.2.21}$$

The role of the submodel parameters with respect to the identified eigenvectors of the total system is obvious, and can be easily assessed.

The input residual sensitivities in the Laplace domain with respect to all the adjustment parameters are (see Eq. (4.2.7))

$$\left. \begin{array}{rcl} \frac{\partial \mathbf{v}_{Ir}}{\partial a_{M\sigma}} & = & s_r^2\mathbf{M}_\sigma\mathbf{U}_r^m \\ \frac{\partial \mathbf{v}_{Ir}}{\partial a_{B\rho}} & = & s_r\mathbf{B}_\rho\mathbf{U}_r^m \\ \frac{\partial \mathbf{v}_{Ir}}{\partial a_{K\iota}} & = & \mathbf{K}_\iota\mathbf{U}_r^m \end{array} \right\} \tag{4.2.22}$$

with $\mathbf{U}_r^m := \mathbf{U}^m(j\omega_r)$. They are independent of the adjustment factors.

For the output residuals it follows with the transfer function matrix

$$\mathbf{H}_r(\mathbf{a}) := \mathbf{H}(j\omega_r, \mathbf{a}) := [s_r^2\mathbf{M}(\mathbf{a}_M) + s_r\mathbf{B}(\mathbf{a}_B) + \mathbf{K}(\mathbf{a}_K)]^{-1} : \tag{4.2.23}$$

$$\left. \begin{array}{rcl} \frac{\partial \mathbf{v}_{Or}}{\partial a_{M\sigma}} & = & -s_r^2\mathbf{H}_r(\mathbf{a})\mathbf{M}_\sigma\mathbf{U}_r(\mathbf{a}) \\ \frac{\partial \mathbf{v}_{Or}}{\partial a_{B\rho}} & = & -s_r\mathbf{H}_r(\mathbf{a})\mathbf{B}_\rho\mathbf{U}_r(\mathbf{a}) \\ \frac{\partial \mathbf{v}_{Or}}{\partial a_{K\iota}} & = & -\mathbf{H}_r(\mathbf{a})\mathbf{K}_\iota\mathbf{U}_r(\mathbf{a}). \end{array} \right\} \tag{4.2.24}$$

The sensitivities of the output residuals are dependent on the parameters to be estimated, which makes the assessment more difficult.

The sensitivities are summarized in [5, 124, 125]. It is again noted that the parameters a_j as adjustment factors of the submodels are related directly to the chosen subsystems. Consequently, the relationships between the estimates and the possibly faulty/damaged subsystems are given. The corresponding sensitivities of the dynamic quantities and the defined submodels are performed by the above derivatives with respect to the parameters.

4.2.4 Estimators

Basic knowledge of statistics and stochastics is required [94, 12, 13]. Fig. 4.2.4 gives a classification of estimation procedures [126, 4, 5]. Be-

[5] $\hat{\lambda}_{Br}$ is substituted by $\hat{\lambda}_{0r} = \omega_{0r}^2$.

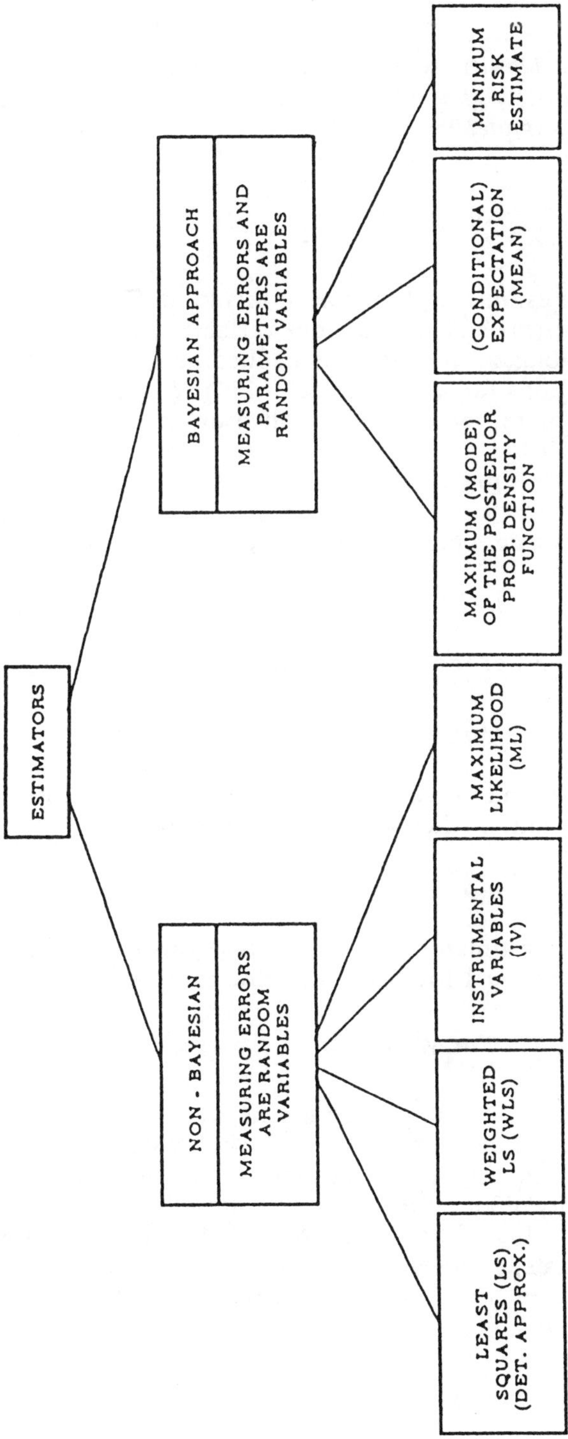

Fig. 4.2. Classification of estimators

fore the procedures are discussed, however, some terms regarding their properties will be defined.

Some definitions of estimator properties. The sample elements may be given vectorially as $\mathbf{x}$. $\{\hat{\alpha}_N(\mathbf{x})\}$ may be a set of estimation functions (estimates).

Example: The mean

$$\hat{\alpha}_N(\mathbf{x}) := \bar{x}_N = \frac{1}{N}\sum_{i=1}^{N} x_i$$

is an estimation function of this kind. □

$\hat{\alpha}_N(\mathbf{x})$ is also a random variable, which again depends on N random variables $\mathbf{x}$. $\hat{a}\epsilon\{\hat{\alpha}_N(\mathbf{x})\}$ is defined as an element of the set of estimation functions (an estimate), for which N is given.

The estimate $\hat{\alpha}_N(\mathbf{x})$ is called unbiased, if

$$E\{\hat{\alpha}_N(\mathbf{x})\} = \overset{o}{a} \tag{4.2.25}$$

holds true; otherwise it is called biased with the bias

$$b = E\{\hat{\alpha}_N(\mathbf{x})\} - \overset{o}{a} \neq 0. \tag{4.2.26}$$

The bias of the single estimate $\hat{a}$ is

$$b(\hat{a}) := \hat{a} - \overset{o}{a}\,. \tag{4.2.27}$$

If the random variable is $\hat{a}$, and $E\{\overset{o}{a}\} = \overset{o}{a}$, then (4.2.26) holds true for the expectation when (4.2.27) is considered,

$$b(\overset{o}{a}) = E\{\hat{\alpha}_N(\mathbf{x}) - \overset{o}{a}\} = E\{\hat{a} - \overset{o}{a}\} = E\{b(\hat{a})\}. \tag{4.2.28}$$

An estimator or estimate is consistent if the estimate converges with probability 1 for $N \to \infty$ to the true value,

$$\lim_{N\to\infty} W\{|\hat{\alpha}_N(\mathbf{x}) - \overset{o}{a}\,| \geq \varepsilon\} = 0 \tag{4.2.29}$$

for every $\varepsilon > 0$. It is sufficient for the consistency of an estimate that its standard deviation with increasing N converges to zero (the proof follows from the Chebychev inequality). This makes no statement concerning the quality of an estimate due to a finite N. Consequently, consistent estimates can be biased for a finite N, and an unbiased estimate does not need to be consistent. However, a consistent estimate is asymptotically unbiased. An estimate is consistent in the squared average if, in addition, the variance is zero,

$$\lim_{N\to\infty} E\{[\hat{\alpha}_N(\mathbf{x}) - \overset{o}{a}]^2\} = 0. \tag{4.2.30}$$

The measure for the relative efficiency of estimates $\hat{a}_1, \hat{a}_2$ is:

$$\eta := \frac{E\{(\hat{a}_1 - \overset{o}{a})^2\}}{E\{(\hat{a}_2 - \overset{o}{a})^2\}} = \frac{\sigma^2(\hat{a}_1)}{\sigma^2(\hat{a}_2)}, \tag{4.2.31}$$

which means that an estimate $\hat{a}_2$ is more efficient than $\hat{a}_1$ if its variance is smaller ($\eta > 1$) than the variance of $\hat{a}_1$.

An estimator is called asymptotically efficient if it contains the smallest variance of all unbiased estimates with the same information content[6].

Deterministic approximation. The least squares (LS) approach of an overdetermined system of equations is well known,

$$\mathbf{Aa} = \mathbf{f}, \;\; \mathbf{A} \in R^{N\times J}, \;\; \mathbf{a} \in R^J, \;\; \mathbf{f} \in R^N, \tag{4.2.32}$$

and this can be understood as a deterministic method. Its solution will be obtained with the Gaussian transformation matrix, and its inverse matrix as a pseudo-inverse:

$$\left.\begin{array}{rcl} \mathbf{A}^T\mathbf{Aa} & = & \mathbf{A}^T\mathbf{f}, \\ \mathbf{a} & = & (\mathbf{A}^T\mathbf{A})^{-1}\mathbf{A}^T\mathbf{f}. \end{array}\right\} \tag{4.2.33}$$

It can be shown that the problem (4.2.33) is often ill-conditioned [71, 72], and that the LS solution is unique and of minimum energy (norm), which makes sense in mechanics. The solution of an underdetermined system of equations can be found in the references quoted above, and elsewhere.

The question arises of how the system of Eq. (4.2.32) will be obtained in the context of residuals. In the frequency domain the residuals are defined in (4.2.3), and are assembled in the total residual vector

$$\mathbf{v} = \mathbf{v}(\mathbf{a}) = \{\mathbf{v}_r(\mathbf{a})\} \in C^{\nu N}, \quad \nu \in \mathcal{N}. \tag{4.2.34}$$

Example: If N eigensolutions $\hat{\lambda}_{0r}, \hat{\mathbf{u}}_{0r}$ are estimated, then the set of available measurements consists of $N + Nn = (n+1)N$ values, that means $\nu = n+1$. □

When the quadratic loss function

$$\mathcal{J}_1 := \mathbf{v}^T(\mathbf{a})\mathbf{v}(\mathbf{a}) \rightarrow \min(\mathbf{a}), \tag{4.2.35}$$

is defined, and a real vector $\mathbf{v}$ is assumed, the necessary condition for the extremum is

$$\left.\frac{\partial \mathcal{J}_1(\mathbf{a})}{\partial a_j}\right|_{\mathbf{a}=\hat{\mathbf{a}}} = 2\left.\left(\frac{\partial \mathbf{v}^T(\mathbf{a})}{\partial a_j}\mathbf{v}(\mathbf{a})\right)\right|_{\mathbf{a}=\hat{\mathbf{a}}} = 0, \;\; j = 1(1)J. \tag{4.2.36}$$

[6] The information content is reciprocally proportional to its variance.

After introduction of the sensitivity matrix (Jacobi-matrix)

$$D\mathbf{v} := \left(\frac{\partial \mathbf{v}}{\partial a_1}, \ldots, \frac{\partial \mathbf{v}}{\partial a_J} \right) \tag{4.2.37}$$

of order $(\nu N, J)$, the necessary condition reads

$$(D\mathbf{v}^T \mathbf{v})|_{\mathbf{a}=\hat{\mathbf{a}}} = \mathbf{0}. \tag{4.2.38}$$

These are J equations to compute J estimates $\hat{\mathbf{a}}$. If the residual vector $\mathbf{v}$ is linear in the parameters to be estimated,

$$\mathbf{v} \stackrel{!}{=} D\mathbf{v}\, \mathbf{a} - \mathbf{b}, \tag{4.2.39}$$

then (4.2.38) leads to a linear system of equations with the solution

$$\hat{\mathbf{a}} = (D\mathbf{v}^T D\mathbf{v})^{-1} D\mathbf{v}^T \mathbf{b}. \tag{4.2.40}$$

It can be seen in a comparison of Eqs. (4.2.32)–(4.2.33) with the above LS equations that $\mathbf{A} \equiv D\mathbf{v}$ and $\mathbf{f} \equiv \mathbf{b}$.

The sufficient condition yields the Hessian matrix (4.2.15). The Gaussian transformation matrix $(D\mathbf{v}^T D\mathbf{v})$ must have maximum rank J and, additionally,

$$J \leq \nu N \tag{4.2.41}$$

must be fulfilled. The equality sign gives J equations for the J unknowns.

In the case of $\mathbf{v}$ nonlinear in the $\mathbf{a}$, an iterative procedure has to be applied.

As also can be seen, no statistics are involved in this procedure, with the inverse Gaussian transformation matrix interpreted as a pseudo-inverse.

Maximum-likelihood (ML) estimation. The starting point is the conditional pdf. If the measurements $U_1^m, U_2^m, \ldots, U_N^m$ for estimating the parameter vector $\mathbf{a}$ are statistically independent (see footnote 7), then it holds true

$$p_{\mathbf{U}^m}(\mathbf{x}) = p_{U_1^m,\ldots,U_N^m}(x_1, \ldots, x_N) = p_{U_1^m}(x_1) p_{U_2^m}(x_2) \ldots p_{U_N^m}(x_N). \tag{4.2.42}$$

The additive measuring error

$$\mathbf{R}(j\omega) = \mathbf{v}(\overset{o}{\mathbf{a}}) = \mathbf{U}(j\omega, \overset{o}{\mathbf{a}}) - \mathbf{U}(j\omega)^m \tag{4.2.43}$$

has the joint pdf $p_{\mathbf{R}}(\mathbf{x})$, which is identical to the conditional joint pdf of the residuals $\mathbf{v}$ for $\mathbf{a} = \overset{o}{\mathbf{a}}$:

$$p_{\mathbf{R}}(\mathbf{x}) = p_{\mathbf{v}}(\mathbf{x}_{\mathbf{v}} |\, \overset{o}{\mathbf{a}}). \tag{4.2.44}$$

With these pdfs and for the measured values $\mathbf{U}^m$, as the estimation function for the parameter vector $\mathbf{a}$, one obtains the likelihood function

$$L(\mathbf{U}^m, \mathbf{a}) := p_\mathbf{v}[\mathbf{v}(\mathbf{a})] \equiv p_\mathbf{v}[\mathbf{U}^m - \mathbf{U}(\mathbf{a})]. \tag{4.2.45}$$

The estimation of $\mathbf{a}$ consists of maximizing the likelihood function:

$$\max_{\mathbf{a}} L(\mathbf{U}^m, \mathbf{a}) \equiv p_\mathbf{v}[\mathbf{v}(\hat{\mathbf{a}})]. \tag{4.2.46}$$

Note: The maximum likelihood estimator provides for the pdf $p_\mathbf{R}(\mathbf{x})$ to have a unique maximum. The normal distribution with zero mean value possesses this maximum with $E\{\mathbf{R}\} \equiv E\{\mathbf{v}(\overset{o}{\mathbf{a}})\} = \mathbf{0}$.

Because $L(\mathbf{U}^m, \mathbf{a})$ is non-negative, the maximum locations of $L(\mathbf{U}^m, \mathbf{a})$ and $\ln L(\mathbf{U}^m, \mathbf{a})$ are equal. The advantage of the *ln* operation is the resulting summation. The estimated vector $\hat{\mathbf{a}}$ follows from

$$\left.\frac{\partial L(\mathbf{U}^m, \mathbf{a})}{\partial \mathbf{a}}\right|_{\mathbf{a}=\hat{\mathbf{a}}} = \mathbf{0}. \tag{4.2.47}$$

If the measurement errors are normally distributed, it follows

$$p_\mathbf{R}(\mathbf{x}) \equiv p_\mathbf{v}[\mathbf{v}(\overset{o}{\mathbf{a}})] = \frac{1}{(2\pi)^{N/2}(det\mathbf{C})^{1/2}} exp\left[-\frac{1}{2}\mathbf{v}^T(\overset{o}{\mathbf{a}})\mathbf{C}^{-1}\mathbf{v}(\overset{o}{\mathbf{a}})\right], \tag{4.2.48}$$

with

$$\mathbf{C} = \mathbf{C}_{RR} = \mathbf{C}_{\mathbf{v}(\overset{o}{\mathbf{a}})\mathbf{v}(\overset{o}{\mathbf{a}})} = E\{\mathbf{v}\mathbf{v}^T|_{\mathbf{a}=\overset{o}{\mathbf{a}}}\},$$

and finally

$$\ln L(\mathbf{a}) = -\frac{1}{2}\ln[(2\pi)^N \det \mathbf{C}] - \frac{1}{2}\mathbf{v}^T(\mathbf{a})\mathbf{C}^{-1}\mathbf{v}(\mathbf{a}). \tag{4.2.49}$$

As can be seen, the result is, apart from a constant, a quadratic loss function in $\mathbf{v}(\mathbf{a})$ weighted by the inverse covariance matrix of the measuring errors. This estimator is the WLS estimator, and because of the special weighting it is called the Markov estimation.

It can be shown that the maximum likelihood estimation gives consistent and asymptotically efficient estimates. However, this estimator provides that the pdf of the measuring errors, and therefore that of the measured values, is known.

Bayes estimation. Within the Bayesian approach all quantities are assumed to be statistical variables which are a priori unknown or uncertain. Therefore, the vector $\mathbf{a}$ is a random vector. Consequently, the parameters a_j are related to prior pdfs $p_\mathbf{a}(\mathbf{x}_a)$, which describe the uncertainties of the parameters. A uniform pdf which, for example, is defined over the total parameter range, is "non-informative" in this context. The Bayesian theorem [12] states that the a posteriori pdf of the parameters, which describes the uncertainty of the parameters as random variables after an input/output measurement, is related via the likelihood function with the a priori pdf of the parameters as follows:

$$p_{\mathbf{a}}(\mathbf{x}_a|\mathbf{U}^m) \propto L(\mathbf{U}^m, \mathbf{x}_a)p_{\mathbf{a}}(\mathbf{x}_a). \tag{4.2.50}$$

The a posteriori pdf of the parameters, of course, is conditioned by the measurements[7]. The estimates $\hat{\mathbf{a}}$ of the a posteriori most probable parameter values follow from

$$\begin{aligned} p_{\mathbf{a}}(\hat{\mathbf{a}}|\mathbf{U}^m) &:= \max_{\mathbf{x}_a} p_{\mathbf{a}}(\mathbf{x}_a|\mathbf{U}^m) = \max_{\mathbf{x}_a} L(\mathbf{U}^m, \mathbf{x}_a)p_{\mathbf{a}}(\mathbf{x}_a) \\ &= L(\mathbf{U}^m, \hat{\mathbf{a}})p(\hat{\mathbf{a}}). \end{aligned} \tag{4.2.51}$$

Assuming all the prior distributions as normal (with respect to the parameters and the measuring errors), e.g.

$$p_{\mathbf{a}}(\mathbf{x}_a) \propto exp[-\frac{1}{2}(\mathbf{x}_a - \mathbf{e})^T \mathbf{C}_{aa}^{-1}(\mathbf{x}_a - \mathbf{e})],$$

where $\mathbf{e}$ is the prior most probable value of $\mathbf{a}$ and $\mathbf{C}_{aa}$ is the covariance matrix of $\mathbf{a}$, the estimation of the a posteriori most probable values is identical with minimizing the quadratic loss function

$$\mathcal{J}_B(\mathbf{a}) = \mathbf{v}^T(\mathbf{a})\mathbf{C}_{RR}^{-1}\mathbf{v}(\mathbf{a}) + (\mathbf{a} - \mathbf{e})^T \mathbf{C}_{aa}^{-1}(\mathbf{a} - \mathbf{e}). \tag{4.2.52}$$

This particular result can be interpreted formally as the extended WLS (EWLS), and will now be written as:

$$\mathcal{J}_3(\mathbf{a}) := \mathbf{v}^*(\mathbf{a})\mathbf{G}_v\mathbf{v}(\mathbf{a}) + (\mathbf{a} - \mathbf{e})^T \mathbf{G}_e(\mathbf{a} - \mathbf{e}). \tag{4.2.53}$$

Here it is assumed that $\mathbf{v}$ could be complex, and $\{\ldots\}^*$ designates the conjugate complex and transposed vector $\{\ldots\}$. $\mathbf{G}_v$ and $\mathbf{G}_e$ are composable

[7] The probability distributions of the random variables x, y are called statistically independent if $P_{xy}(u,v) = P_x(u)P_y(v)$. This is with the probabilities W : $W[x < u, y < v] = W[x < u]W[y < v]$, which can be written as $W[x < u] = \frac{W[x<u,y<v]}{W[y<v]}$: the probability of the occurrence of $x < u$ is true only when those cases where $y < v$ occur are considered.

The fact that $x < u$ occurs only if $y < v$ occurs can be expressed as conditional probability:

$$P(x|y) = \frac{P_{xy}(u,v)}{P_y(v)}.$$

The Bayesian theorem now states

$$P(x|y) = \frac{P(y|x)P_x(u)}{P_y(v)}.$$

weighting matrices, which describe the "confidence" in the measurements and in the prior knowledge $\mathbf{e}$, respectively: they are the inverse covariance matrices $\mathbf{C}_{vv}^{-1}, \mathbf{C}_{ee}^{-1}$. It should already be noted here that the variances of the measuring errors can generally be estimated (approximated). However, if the prior values $\mathbf{e}$ are those from the prior mathematical model, then their covariances or variances are unknown and cannot be assessed ($\leadsto$ regularization, see next paragraph).

The extended WLS (EWLS) and their simplifications. The advantage of the EWLS is that the solution can be controlled by the penalty term[8]. This does not mean an arbitrary solution, but it means including additional information (see Sect. 4.4.3). In order to minimize Eq.(4.2.53) the necessary and sufficient conditions that $\hat{\mathbf{a}}$ is a (local) minimum of $\mathcal{J}_3$ are:

$$\left.\frac{\partial \mathcal{J}_3(\mathbf{a})}{\partial \mathbf{a}}\right|_{\mathbf{a}=\hat{\mathbf{a}}} = \Re[D\mathbf{v}^* \mathbf{G}_v \mathbf{v}(\mathbf{a})]|_{\mathbf{a}=\hat{\mathbf{a}}} + \mathbf{G}_e(\hat{\mathbf{a}} - \mathbf{e}) = \mathbf{0}, \tag{4.2.54}$$

$$\left.\frac{\partial^2 \mathcal{J}_3(\mathbf{a})}{\partial \mathbf{a} \partial \mathbf{a}^T}\right|_{\mathbf{a}=\hat{\mathbf{a}}} = \left.\frac{\partial \mathcal{J}_2(\mathbf{a})}{\partial \mathbf{a} \partial \mathbf{a}^T}\right|_{\mathbf{a}=\hat{\mathbf{a}}} + \mathbf{G}_e, \tag{4.2.55}$$

positive (semi-)definite, with

$$\mathcal{J}_2(\mathbf{a}) = \mathbf{v}^*(\mathbf{a}) \mathbf{G}_v \mathbf{v}(\mathbf{a})$$

the loss function of the WLS, which means taking (4.2.53) with $\mathbf{G}_e = \mathbf{0}$. Eq.(4.2.54) together with Eq.(4.2.39) with real values turns into

$$\hat{\mathbf{a}} = (D\mathbf{v}^T \mathbf{G}_v D\mathbf{v} + \mathbf{G}_e)^{-1}(D\mathbf{v}^T \mathbf{G}_v \mathbf{b} + \mathbf{G}_e \mathbf{e}). \tag{4.2.56}$$

As can be seen, the solution is influenced by the weighting $\mathbf{G}_e$ in the inverse matrix and the weighted prior knowledge $\mathbf{G}_e\mathbf{e}$. The Hessian matrix (4.2.55) of the EWLS is the superposition of the Hessian matrix of the WLS and the weighting matrix $\mathbf{G}_e$; this sum must be positive (semi-)definite. The estimate $\hat{\mathbf{a}}$ is the a posteriori most probable parameter vector.

The EWLS (4.2.53) differs from the WLS formulation by the additional penalty term within the loss function carrying the prior knowledge from, for example, the prior mathematical model or prior parameter estimates. The advantages of this extension are manifold:

- it includes additional information
- missing measurements can be substituted by theoretical knowledge, for example of the prior mathematical model
- the distance between $\hat{\mathbf{a}}$ and the prior knowledge $\mathbf{e}$ can be controlled by the choice of $\mathbf{G}_e$
- the loss function can be made convex by the additional term
- the additional term can be used for regularization: this means modifying the operator of the often ill-conditioned system of equations to be solved, so that the modified system of equations is now well-conditioned

[8] Now without the stochastic background!

- it can be used to construct an always convergent iteration procedure for solving the resultant system of equations nonlinearly in the parameters to be estimated[9].

The disadvantage is also obvious: it is the lack of knowledge of $\mathbf{G}_e$, and therefore the risk of strongly biased estimates if more guesswork is substituted in place of information. This weighting matrix can also be determined mathematically, which means without direct physical input. One of the possibilities is to apply cross-validation [128, 72], which means estimating the weighting together with the parameters (see Sect. 4.4 on algorithms).

With $\mathbf{G}_e \to \mathbf{0}$ the EWLS turns into the WLS, which means that one trusts only the measurements. Reciprocally, with $\mathbf{G}_v \to \mathbf{0}$ one trusts only the prior information, $\hat{\mathbf{a}} = \mathbf{e}$, and the information of the measurements is suppressed.

Additionally, the LS method Eq.(4.2.35) is obtained with $\mathbf{G}_e = \mathbf{0}$ and $\mathbf{G}_v = \mathbf{I}$. However, with statistical weighting (inverse covariance matrices) the results are Markov estimates (with minimum variances for unbiased estimates).

It should be noted that the various LS approaches can also be formulated while taking constraints into account. For example, the minimization of the loss function (for instance defined as a norm) may be subject to some additional information, such as positive definiteness, symmetry of matrices etc. The mathematical formulation can be found elsewhere [129]. It is discussed as a part of regularization in Sect. 4.4.3.

Instrumental variables. With the residual vector (4.2.39) linear in the parameters and a matrix $\mathbf{W}$ chosen such that $\mathbf{W}^T D\mathbf{v}$ is positive definite, the orthogonality method is applied,

$$\mathbf{W}^T\mathbf{v} = \mathbf{W}^T(D\mathbf{v}\mathbf{a} - \mathbf{b}) \tag{4.2.57}$$

with the demand

$$E\{\mathbf{W}^T\mathbf{v}|_{\mathbf{a}=\hat{\mathbf{a}}}\} = \mathbf{0}. \tag{4.2.58}$$

The matrix $\mathbf{W}$ is called the matrix of instrumental variables, and the method is the instrumental variables method (IV). It follows the estimate

$$\hat{\mathbf{a}} = (\mathbf{W}^T D\mathbf{v})^{-1}\mathbf{W}^T\mathbf{b}. \tag{4.2.59}$$

This estimate is unbiased with $E\{\mathbf{v}|_{\mathbf{a}=\overset{o}{\mathbf{a}}}\} = \mathbf{0}$,

$$E\{\hat{\mathbf{a}}\} = E\{(\mathbf{W}^T D\mathbf{v})^{-1}\mathbf{W}^T(D\mathbf{v}\,\overset{o}{\mathbf{a}} + \mathbf{v}|_{\mathbf{a}=\overset{o}{\mathbf{a}}})\} = \overset{o}{\mathbf{a}} + E\{(\mathbf{W}^T D\mathbf{v})^{-1}\mathbf{W}^T\mathbf{v}|_{\mathbf{a}=\overset{o}{\mathbf{a}}}\}, \tag{4.2.60}$$

if $\mathbf{W}^T$ and $\mathbf{v}|_{\mathbf{a}=\overset{o}{\mathbf{a}}}$ are uncorrelated (then the E operator can be applied to the single factors).

[9] The perturbation method by varying $\mathbf{G}_e$ systematically so that the previous solution lies within the convergence region of the new (perturbed) system [127].

This is the great advantage of the IV compared with the LS methods. Because $\overset{o}{\mathbf{a}}$ is unknown, it is difficult to find an optimum IV matrix. In practice, **W** is often determined recursively [130].

Note: Kalman filtering ([131] to [133]) is well-known for state and parameter estimation. Kalman filtering is a recursive estimation procedure using sequential measurement data sets. The prior knowledge of the parameters (expressed by their actual covariance matrix) is improved step by step by taking the prior parameter estimates and new data for the subsequent parameter estimation. This can be understood as a kind of regularization (see Sect. 4.4.3).

Discussion of various estimators. In addition to the classification given in Fig. 4.2.4, a hierarchical review is presented in Fig. 4.3.

The relations between various estimators have already been discussed: the Bayesian approach leads, on the one hand, with normally distributed random variables (parameters as well as measuring errors) to the EWLS, and on the other hand, with $p_{\mathbf{a}}(\mathbf{x}_a)$ =const. to the Maximum-likelihood

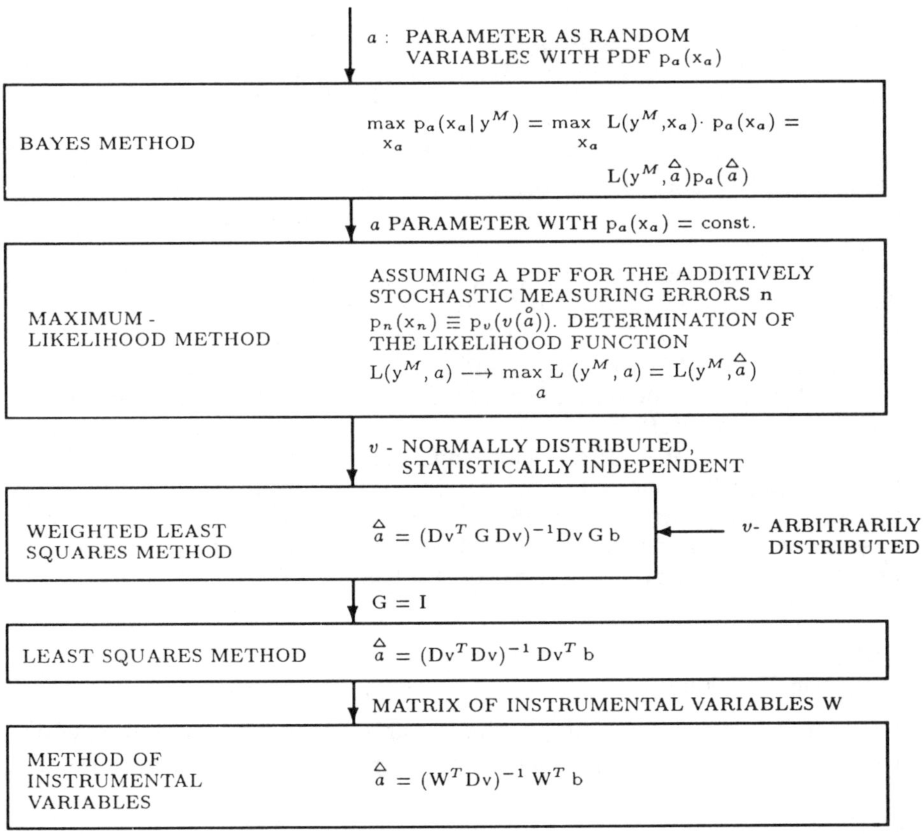

Fig. 4.3. Hierarchical classification of estimators

method for additive noise. The latter turns into the WLS for Gaussian pdfs and statistically independent measured values. Because in practice it is very seldom that any statistical knowledge of the measurements and their errors will be available, the assumption of normal distributions is in common use; consequently, the EWLS will be recommended. As a reminder: the disadvantages are the generally biased estimates and the lack of knowledge of the confidence in the prior knowledge. However, if the weighting $\mathbf{G}_v$ is chosen statistically based (filtering!), then the bias due to the correlation of the measuring errors can be drastically reduced, and the penalty term in Eq.(4.2.53) is very advantageous, but with, of course, additional expenditure (which has still to be described, see next section). The simplifications of the EWLS have already been mentioned: WLS, LS. If in particular cases the bias of the estimates from any LS method cannot be accepted, then the IV method can be applied. It is an easy method to apply; however, the determination of the instrumental variables can be difficult.

4.3 Model Adjustment – Methods

With the preparations made previously we are now able to formulate the model adjustment methods in order to perform a validated mathematical model at every life time θ required, assuming that the required measurements are available. Since fewer components of the dynamic response vector are measured than DOF of the bandlimited model exist, this case will be discussed in Sect. 4.5 in order not to interrupt the train of thought. Decisions on the significance of parameter modifications are discussed in Chap. 5.

The methods are formulated and discussed for the WLS. This means that the application of the EWLS, which is recommended, follows automatically with Eqs. (4.2.54) to (4.2.56). The application of the IV can also be performed easily by choosing the residuals and the parametrization.

In addition to the requirements of verification and validation (see Sect. 1.3.1) to be fulfilled from the adjusted model, within the localization problem the procedure of halving is already mentioned (Sect. 2.2.6). The stopping rule concerning the halving of the submodels is governed by the required accuracy with respect to the fault localization. The inaccuracy of the estimates will be judged by the estimates of the corresponding variances and covariances (see Sects. 4.6.1 and 5.2).

4.3.1 Subsystem Modelling

The first step is the modelling of the possibly occurring local faults, which are restricted to those causing parameter modifications[10]. Consequently,

[10] Other modifications, for example model structure modifications, must be handled as described in Sect. 2.1.4.

each element of the parameter matrices should be adjusted. However, because only a restricted number of measurements are generally available, and in order to reduce the expenditure, subsystem modelling (2.2.24)-(2.2.27) is introduced:

$$\left.\begin{array}{llllll} \mathbf{M}^c & := & \mathbf{M}(\mathbf{a}_M) & = & \sum_{\sigma=1}^{S} a_{M\sigma}\mathbf{M}_\sigma \\ \mathbf{B}^c & := & \mathbf{B}(\mathbf{a}_B) & = & \sum_{\rho=1}^{R} a_{B\rho}\mathbf{B}_\rho \\ \mathbf{K}^c & := & \mathbf{K}(\mathbf{a}_K) & = & \sum_{\iota=1}^{I} a_{K\iota}\mathbf{K}_\iota, \text{ or} \\ \mathbf{G}^c & := & \mathbf{G}(\mathbf{a}_G) & = & \sum_{\iota=1}^{I} a_{G\iota}\mathbf{G}_\iota. \end{array}\right\} \tag{4.3.1}$$

Estimates of the parameters a_j will then give estimates of the physical parameter matrices (4.3.1).

The motivation is summarized as follows:

1. It permits a reduction of the number of parameters to be estimated drastically; therefore
2. it is a kind of regularization (introduction of a coarser parameter topology) because, assuming a given set of measurements, one can perform an overdetermined system of equations instead of working with an underdetermined one; a smaller number of parameters can be estimated more accurately than a larger one,
3. the expenditure will be reduced, and
4. it meets current requirements concerning the handling of large systems ($\rightsquigarrow$ submodels, object-orientated modelling and programming).

The problems to be solved are the choice of the submodels and how to locate parameter modifications (fault detection and localization). The choice of the submodels is determined by

1. knowledge of uncertainties of the initial model, for example if
   - slender wings are modelled as Euler-Bernoulli beams, then the torsional stiffnesses result in values that are too small due to the simplifications. The flexibility matrix can be partitioned into

     $$\mathbf{G} = \begin{bmatrix} \mathbf{G}^{zz} & \mathbf{G}^{z\alpha} \\ \mathbf{G}^{\alpha z} & \mathbf{G}^{\alpha\alpha} \end{bmatrix}$$

     with respect to the bending z and the torsion α. Now each submatrix, extended by null matrices in order to build the matrices $\mathbf{G}_\iota$ in Eq.(4.3.1), can be taken for correction.
   - Often the coupling between subsystems is difficult to model, and then the submodelling can take this into account by introducing coupling elements (e.g. spring constants) to be estimated and additional factors with respect to the submodel matrices,
2. by prior knowledge from system analysis (see Fig. 3.1), including
   - physically realized subsystems,
   - sensitivity analysis,
   - studies of possible fault effects (selected regions),
3. direct comparison of computed quantities with the corresponding measured quantities, for example taking eigenvectors and the residuals per-

formed by the dyadic products, bearing in mind the spectral decompositions of the parameter matrices, which then also give the locations of the deviations in addition to the size of the deviations.

4. If nothing is known by modelling and simulation (in contrast to the studies requested to be made before, see Fig. 3.1!) before adjustment, then more or less arbitrary assumptions must be made. Dependent on the results with the assumptions made, recalculations will lead to the goal.

It is again emphasized that prior knowledge of the expected fault locations, and consequently of the parameter modifications, can be essential for the success of the model-based diagnosis (estimation procedure including the interpretation of the estimates). If the chosen parameter topology (choice of the a_j) is too coarse for the localization of significant parameter modifications, then sequential halving of the submodels concerned will lead to the result [26]. In order to reduce the expenditure, the submodels not influenced by the modification due to faults within the adjustment can be assembled in one submodel (see $\mathbf{M}', \mathbf{K}'$ in Eq.(4.3.30)). In consequence, if necessary, each matrix element can be estimated, and in this case the submodel matrices consist of only one non-zero element.

The number J of parameters to be chosen, and therefore the choice of submodels, must be smaller than the number of equations for the parameters in order to approximate the expectation:

$$\nu N > J. \tag{4.3.2}$$

In [68] it is shown that the submodels represented by the corresponding summand matrices in Eq.(4.3.1) cannot be chosen arbitrarily; they have to fulfil special requirements which are discussed in detail in [68]. Parametrization with submodels has not yet been solved generally.

If the adjustment factors are estimated by LS, WLS or ML, the involved Hessian matrices have to be positive definite in order to yield unique estimates for the factors as dimensionless parameters. Analytical investigations show [68] that for this reason

$$\left.\begin{aligned} \text{rank}\,[cs\mathbf{M}_1, cs\mathbf{M}_2, \ldots, cs\mathbf{M}_S] &= S, \\ \text{rank}\,[cs\mathbf{B}_1, cs\mathbf{B}_2, \ldots, cs\mathbf{B}_R] &= R, \\ \text{rank}\,[cs\mathbf{K}_1, cs\mathbf{K}_2, \ldots, cs\mathbf{K}_I] &= I \end{aligned}\right\} \tag{4.3.3}$$

are necessary conditions. This means that the *column strings* of all the summand matrices of the mass, damping and stiffness matrices repectively must be linearly independent. ($cs\mathbf{A}$: "the column sequenced vector structure of the elements of $\mathbf{A}$" [134]). This problem will be mentioned further in Sect. 4.5, accompanied by expenditure minimization using order-reduced models, which is based on [68] and [135].

Another difficulty exists. The incompleteness of the measurements can concern the frequency content. Let us assume that the prior mathematical model has a bandlimitation of $(0, \omega_b]$. The measurements may be restricted to $(0, \omega_1]$ with $\omega_1 < \omega_b$ and, for example, the estimated eigenquantities are restricted to this interval. Then one can generally not expect that the adjustment procedure will correct the mathematical model outside the frequency interval $(0, \omega_1]$. However, ω_1 must be chosen so large that the system modifications occur in the interval up to ω_1. Very often the modifications result in dynamics of higher frequencies [8, 26]. As can be imagined, the choice of the submodels (spectral decomposition!) together with the frequency information content of the measurements must fulfil special requirements. - If the parameters chosen are highly correlated (consequently a priori), then their estimates can be very erroneous. Highly correlated estimates will follow due to insufficient measurements. Or both causes yield correlated estimates (see Sect. 4.6.1).

4.3.2 Adjustment Based on Eigenvalues

It is relatively easy to measure accurately the eigenfrequencies of the associated undamped system, $\hat{\omega}_{0r}, r = 1(1)N$. These are scalars with a limited spatial information content (the spatial information is given only by the ordering of the eigenfrequencies $\leadsto$ DOF, see Sect. 2.2.6). The partial residual as well as the equation error can be used; both resulting procedures require computation of the adjusted eigenvectors. These methods are therefore adaptive with respect to the eigenvectors.

Partial residuals. First of all, only the stiffness matrix will be adjusted. The residual vector is defined as

$$\mathbf{v}_1^T := (\lambda_{01}(\mathbf{a}_K) - \hat{\lambda}_{01}, \ldots, \lambda_{0N}(\mathbf{a}_K) - \hat{\lambda}_{0N}). \tag{4.3.4}$$

The functional matrix (4.2.37) follows to

$$D\mathbf{v}_1 = \left[\frac{\partial \mathbf{v}_1}{\partial a_{K1}}, \ldots, \frac{\partial \mathbf{v}_1}{\partial a_{KI}}\right]. \tag{4.3.5}$$

It is an (N, I) matrix (with $\nu = 1$). The partial derivatives are given by Eq. (2.1.40) or by (see Eq. (4.3.11))

$$\frac{\partial \lambda_{0r}(\mathbf{a}_K)}{\partial a_{Kr}} =: k_{gri} \tag{4.3.6}$$

and with the parametrization

$$\mathbf{K}^c = \mathbf{K}(\mathbf{a}_K) = \sum_{\iota=1}^{I} a_{K\iota}\mathbf{K}_\iota. \tag{4.3.7}$$

As can be seen, the eigenvectors $\hat{\mathbf{u}}_{0r}^c = \hat{\mathbf{u}}_{0r}(\mathbf{a}_K)$ depend on the parameters to be estimated; consequently, the generalized stiffnesses (4.3.6) with respect to the stiffness parameter matrices of the submodels $\mathbf{K}_i$ also depend on the parameters: the eigenvectors also have to be estimated (theoretically).

The components of (4.3.4) follow to

$$\nu_{1r} = \sum_{i=1}^{I} a_{Ki}\hat{\mathbf{u}}_{0r}^T(\mathbf{a}_K)\mathbf{K}_i\hat{\mathbf{u}}_{0r}(\mathbf{a}_K) - \hat{\lambda}_{0r} = \sum_{i=1}^{I} a_{Ki}k_{gri}(\mathbf{a}_K) - \hat{\lambda}_{0r}. \tag{4.3.8}$$

With reference to (4.2.39) one obtains a pseudo-linear equation

$$\mathbf{v}_1 = D\mathbf{v}_1(\mathbf{a}_K)\mathbf{a}_K - \mathbf{b}_1 \tag{4.3.9}$$

with

$$D\mathbf{v}_1 = \begin{bmatrix} k_{g11}(\mathbf{a}_K) & \dots & k_{g1I}(\mathbf{a}_K) \\ \dots\dots\dots\dots & \dots & \dots\dots\dots\dots \\ K_{gN1}(\mathbf{a}_K) & \dots & k_{gNI}(\mathbf{a}_K) \end{bmatrix}, \quad \mathbf{b}_1^T := (\hat{\lambda}_{01}, \dots, \hat{\lambda}_{0N}). \tag{4.3.10}$$

Application of the WLS or EWLS leads to a system of equations nonlinear in the parameters. Therefore an iterative algorithm has to be applied; this includes the use of the eigenvectors as solutions to the associated matrix eigenvalue problem

$$[\lambda_{0r}(\mathbf{a}_K)\mathbf{M} + \mathbf{K}(\mathbf{a}_K)]\hat{\mathbf{u}}_{0r}(\mathbf{a}_K) = \mathbf{0} \tag{4.3.11}$$

for each r and for every iteration step with the resulting iterated parameter vector. This procedure automatically implies the one-to-one relationship of the computed eigenquantities to the identified eigenfrequency in each iteration step. Generally, this does not have to be an easy task; sometimes it is impossible to perform it without additional information [5]. It should be noted that the proof of the one-to-one relationship of the eigenquantities can be done using the generalized orthogonaltiy properties, which means with the iteration index k,

$$\hat{\mathbf{u}}_{0r}(\mathbf{a}_K^{(k+1)})\mathbf{M}\hat{\mathbf{u}}_{0r}^{(k)}(\mathbf{a}_K) = \delta_r^{k+1,k}. \tag{4.3.12}$$

The Kronecker δ must be approximately equal to one with the right relationship, otherwise it will be approximately zero. The procedure is summarized in Table 4.2.

As already mentioned, the iteration can be avoided in those cases where the eigenvector modifications due to the faults are small. Then the procedure can be performed with the eigenvectors of the prior model, and it is linear in the parameters; the solution is then given by Eq. (4.2.56) and its simplifications (see Table 4.2).

When the spectral decompositions of the stiffness matrix and the flexibility matrix were being discussed, it was stated (see Sect. 2.1.3) that the stiffness matrix adjustment is suitable for higher frequency content modifications and, vice versa, for the flexibility adjustment. Since, for example, cracks mostly introduce high frequency distortions, the stiffness formulation is emphasized. However, flexibility adjustment follows the same procedure in principle if inverse eigenvalues are taken:

$$\mathbf{v}_2(\mathbf{a}_G)^T = (\frac{1}{\lambda_{01}(\mathbf{a}_G)} - \frac{1}{\hat{\lambda}_{01}}, \dots, \frac{1}{\lambda_{0N}(\mathbf{a}_G)} - \frac{1}{\hat{\lambda}_{0N}}). \tag{4.3.13}$$

Table 4.2. Partial residuals: adjustment of the stiffness matrix by the use of measured eigenvalues (eigenfrequencies)

DESCRIPTION	FORMULA	NOTE (EQ. NO.)
PARAMETRIZATION	$\mathbf{K}^c = \mathbf{K}(\mathbf{a}_K) = \sum_{\iota=1}^{I} a_{K\iota}\, \mathbf{K}_\iota$	(4.3.7)
MEASUREMENTS	$\hat{\lambda}_{0r}, \quad r = 1(1)\, N$	
RESIDUAL VECTOR	$\mathbf{v}_1^T(\mathbf{a}_K) = (\lambda_{01}(\mathbf{a}_K) - \hat{\lambda}_{01}, \ldots, \lambda_{0N}(\mathbf{a}_K) - \hat{\lambda}_{0N})$	(4.3.4)
MATRIX EIGENVALUE PROBLEM	$(-\lambda_{0r}^{K(k)}\, \mathbf{M} + \mathbf{K}^{c(k)})\, \hat{\mathbf{u}}_{0r}^{K(k)} = \mathbf{0}$, k-ITERATION INDEX DUE TO $\mathbf{a}_K^{(k)}$ WITH $\mathbf{a}_K^{(0)} = \mathbf{e}$, $\mathbf{K}^{c(k)} = \mathbf{K}(\mathbf{a}_K^{(k)})$ etc. NORMALIZATION: $\hat{\mathbf{u}}_{0r}^{c(k)T}\, \mathbf{M}\hat{\mathbf{u}}_{0r}^{c(k)} = 1$	
ITERATION PROCEDURE	SEE EQ. (4.2.54) OR:	
	$\mathbf{a}_K^{(k+1)} = (\mathbf{D}\mathbf{v}_1^{(k)T}\mathbf{G}_v\mathbf{D}\mathbf{v}_1^{(k)} + \mathbf{G}_e)^{-1} \cdot (\mathbf{D}\mathbf{v}_1^{(k)T}\mathbf{G}_v\mathbf{b}_1 + \mathbf{G}_e\mathbf{e})$, $k = 0, 1, \ldots, \quad \mathbf{a}_K^{(k+1)} \longrightarrow \hat{\mathbf{a}}_K$ WITH $\mathbf{D}\mathbf{v}_1^{(k)} = \begin{bmatrix} k_{g11}(\mathbf{a}_K^{(k)}), \ldots, k_{g1I}(\mathbf{a}_K^{(k)}) \\ \cdots\cdots\cdots\cdots \\ k_{gN1}(\mathbf{a}_K^{k}), \ldots, k_{gNI}(\mathbf{a}_K^{(k)}) \end{bmatrix}$,	(4.2.56)
	$\mathbf{b}_1^T = (\hat{\lambda}_{01}, \ldots, \hat{\lambda}_{0N})$, $\mathbf{k}_{gr\iota}^{(k)} := \hat{\mathbf{u}}_{0r}^{c(k)T}\, \mathbf{K}_\iota \hat{\mathbf{u}}_{0r}^{c(k)}$	(4.3.10)
SIMPLIFICATION	WITH $\hat{\mathbf{u}}_{0r} \doteq \hat{\mathbf{u}}_{0r}^{(k)}$: $\quad \lambda_{0r} = \hat{\mathbf{u}}_{0r}^T\, \mathbf{K}\hat{\mathbf{u}}_{0r}$ AND (4.2.56) LINEAR IN $\hat{\mathbf{a}}$ (INDEPENDENT OF k).	EMPHASIZING DOF WITH HIGH EIGENFREQUENCIES WITHOUT SIMPLIFICATION: ADAPTIVE PROCEDURE WITH RESPECT TO THE EIGENVECTORS, BECAUSE THEY WILL ALSO BE ADJUSTED.

The parameter sensitivity is

$$\frac{\partial}{\partial a_{G\iota}}\left(\frac{1}{\lambda_{0r}(\mathbf{a}_G)}\right) = \hat{\mathbf{u}}_{0r}^T(\mathbf{a}_G)\mathbf{M}\mathbf{G}_\iota\mathbf{M}\hat{\mathbf{u}}_{0r}(\mathbf{a}_G) =: g_{gr\iota}(\mathbf{a}_G), \tag{4.3.14}$$

and equal to the generalized flexibility with respect to the submodel flexibility matrix $\mathbf{G}_\iota$ in the r-th eigenquantities. The sensitivity matrix $D\mathbf{v}_2$ and the vector $\mathbf{b}_2$ are built up in a similar way to the stiffness formulation (Table 4.3).

If only the inertia matrix $\mathbf{M}$ is to be adjusted,

$$\mathbf{M}(\mathbf{a}_M) = \sum_{\sigma=1}^{S} a_{M\sigma}\mathbf{M}_\sigma, \tag{4.3.15}$$

we introduce the reciprocal eigenvalue

$$\kappa_{0r} := \frac{1}{\lambda_{0r}}, \tag{4.3.16}$$

and write the matrix eigenvalue problem in the form

$$[-\kappa_{0r}(\mathbf{a}_M)\mathbf{K} + \mathbf{M}(\mathbf{a}_M)]\hat{\mathbf{u}}_{0r}(\mathbf{a}_M) = \mathbf{0}. \tag{4.3.17}$$

Then it can be seen that the same formalism is valid as in the case of the stiffness adjustment, if

- the residuals $\mathbf{v}_3$ with regard to the reciprocal eigenvalues are performed (see Table 4.4)
- the generalized stiffnesses are substituted by the generalized masses

  $$m_{gr\sigma}^N(\mathbf{a}_M) := \hat{\mathbf{u}}_{0r}^{NT}(\mathbf{a}_M)\mathbf{M}_\sigma\hat{\mathbf{u}}_{0r}^N(\mathbf{a}_M)$$

- the normalization is changed into

  $$\hat{\mathbf{u}}_{0r}^{NT}\mathbf{K}\hat{\mathbf{u}}_{0r}^N = 1, \ \hat{\mathbf{u}}_{0r}^N := \frac{\hat{\mathbf{u}}_{0r}}{|\sqrt{\lambda_{0r}}|},$$

  in accordance with the matrix eigenvalue problem (4.3.17).

Of course, the resulting adjustment procedure (see Table 4.4) is a nonlinear one.

If both matrices, the inertia matrix as well as the stiffness or flexibility matrix, are to be adjusted, then the procedures discussed can be applied group-wise iteratively. For example, in the first iteration step one starts with the adjustment of the inertia matrix, taking the stiffness or flexibility adjustment factors equal to 1. Then the next iteration step concerns the stiffness matrix with fixed inertia matrix adjustment parameters etc. The advantage of this procedure is that the same routine can be applied for both iterations, and merely a re-arrangement of the data has to be made.

Table 4.3. Partial residuals: adjustment of the flexibility matrix by the use of measured eigenvalues (eigenfrequencies)

DESCRIPTION	FORMULA	NOTE (EQ. NO.)
PARAMETRIZATION	$\mathbf{G}^c = \mathbf{G}(\mathbf{a}_G) = \sum_{\iota=1}^{I} a_{G\iota}\mathbf{G}_\iota$	
MEASUREMENTS	$\hat{\lambda}_{0r} \neq 0, \quad r = 1(1)N$	
RESIDUAL VECTOR	$\mathbf{v}_2^T(\mathbf{a}_G) = (\frac{1}{\lambda_{01}(\mathbf{a}_G)} - \frac{1}{\hat{\lambda}_{01}}, \ldots, \frac{1}{\lambda_{0N}(\mathbf{a}_G)} - \frac{1}{\hat{\lambda}_{0N}})$	(4.3.13)
MATRIX EIGENVALUE	$\kappa_{0r} := \frac{1}{\lambda_{0r}},$	
PROBLEM	$(-\mathbf{G}^{c(k)}\mathbf{M} + \kappa_{0r}^{(k)}\mathbf{I})\, \hat{\mathbf{u}}_{0r}^{c(k)} = \mathbf{0}$ k-ITERATION INDEX DUE TO $\mathbf{a}_G^{(k)}$ WITH $\mathbf{a}_G^{(0)} = \mathbf{e},\ \mathbf{G}^{c(k)} = \mathbf{G}(\mathbf{a}_G^{(k)})$ etc. NORMALIZATION: $\hat{\mathbf{u}}_{0r}^{c(k)T}\mathbf{M}\hat{\mathbf{u}}_{0r}^{c(k)} = 1$	
ITERATION PROCEDURE	SEE EQ. (4.2.54) OR: $\mathbf{a}_G^{(k+1)} = (\mathbf{D}\mathbf{v}_2^{(k)T}\mathbf{G}_v\mathbf{D}\mathbf{v}_2^{(k)} + \mathbf{G}_e)^{-1} \cdot (\mathbf{D}\mathbf{v}_2^{(k)T}\mathbf{G}_v\mathbf{b}_2 + \mathbf{G}_e\mathbf{e}),$ $k = 0,1,\ldots, \quad \mathbf{a}_G^{(k+1)} \longrightarrow \hat{\mathbf{a}}_G$ WITH $\mathbf{D}\mathbf{v}_2^{(k)} = \begin{bmatrix} g_{g11}^{c(k)}, \ldots, g_{g1I}^{c(k)} \\ \cdots\cdots\cdots \\ g_{gN1}^{c(k)}, \ldots, g_{gNI}^{c(k)} \end{bmatrix},$ $\mathbf{b}_2^T = (\hat{\kappa}_{01}, \ldots, \hat{\kappa}_{0N}),\ g_{gr\iota}^{c(k)} = g_{gr\iota}(\mathbf{a}_G) = \hat{\mathbf{u}}_{0r}^{cT}\mathbf{M}\mathbf{G}_\iota\mathbf{M}\hat{\mathbf{u}}_{0r}^c$	(4.3.14)
SIMPLIFICATION	SEE TABLE 4.2	EMPHASIZING DOF WITH SMALL EIGEN-FREQUENCIES WITHOUT SIMPLIFICATION: THE EIGENVECTORS WILL ALSO BE ADJUSTED (ADAPTIVE PROCEDURE)

Table 4.4. Partial residuals: adjustment of the inertia matrix by the use of measured eigenvalues (eigenfrequencies)

DESCRIPTION	FORMULA	NOTE (EQ. NO.)
PARAMETRIZATION	$\mathbf{M}^c = \mathbf{M}(\mathbf{a}_M) = \sum_{\sigma=1}^{s} a_{M\sigma}\mathbf{M}_\sigma$	(4.3.15)
MEASUREMENTS	$\hat{\lambda}_{0r} \neq 0, \quad r = 1(1)\,N$	
RESIDUAL VECTOR	$\mathbf{v}_3^T(\mathbf{a}_M) = (\frac{1}{\lambda_{01}(\mathbf{a}_M)} - \frac{1}{\hat{\lambda}_{01}}, \ldots, \frac{1}{\lambda_{0N}(\mathbf{a}_M)} - \frac{1}{\hat{\lambda}_{0N}})$	
MATRIX EIGENVALUE PROBLEM	$\kappa_{0r} := \frac{1}{\lambda_{0r}}$,	(4.3.16)
	$(-\kappa_{0r}^{(k)}\mathbf{K} + \mathbf{M}^c)\hat{\mathbf{u}}_r^{c(k)} = \mathbf{0}$' k-ITERATION INDEX DUE TO $\mathbf{a}_M^{(k)}$ WITH $\mathbf{a}_M^{(0)} = \mathbf{e}, \mathbf{M}^{c(k)} = \mathbf{M}(\mathbf{a}_M^{(k)})$ etc.	(4.3.17)
	NORMALIZATION: $\hat{\mathbf{u}}_{0r}^{NcT}\mathbf{K}\hat{\mathbf{u}}_{0r}^{Nc} = 1, \quad \hat{\mathbf{u}}_{0r}^{N} = \frac{\hat{u}_{0r}}{\lvert\sqrt{\lambda_{0r}}\rvert}$, $m_{gr\sigma}^{Nc} = m_{gr\sigma}^{N}(\mathbf{a}_M) = \hat{\mathbf{u}}_{0r}^{NcT}\mathbf{M}_\sigma\hat{\mathbf{u}}_{0r}^{Nc}$	THE FORMALISM OF TABLE 4.2 IS VALID WITH THIS NORMALIZATION AND THE SUBSTITUTION OF THE GENERALIZED STIFFNESSES BY GENERALIZED MASSES $m_{gr\sigma}^{N}$
ITERATION PROCEDURE	SEE EQ: (4.2.54) OR $\mathbf{a}_M^{(k+1)} = (\mathbf{D}\mathbf{v}_3^{(k)T}\mathbf{G}_v\mathbf{D}\mathbf{v}_3^{(k)} + \mathbf{G}_e)^{-1} \cdot(\mathbf{D}\mathbf{v}_3^{(k)T}\mathbf{G}_v\mathbf{b}_3 + \mathbf{G}_e\mathbf{e})$, $k = 0,1,\ldots, \quad \mathbf{a}_M^{(k+1)} \longrightarrow \hat{\mathbf{a}}_M$ WITH $\mathbf{D}\mathbf{v}_3^{(k)} = \begin{bmatrix} m_{g11}^{Nc(k)}, \ldots, m_{g1S}^{Nc(k)} \\ \cdots\cdots \\ m_{gN1}^{Nc(k)}, \ldots, m_{gNS}^{Nc(k)} \end{bmatrix}$, $\mathbf{b}_3^T = (\hat{\kappa}_{01}, \ldots \hat{\kappa}_{0N})$.	IT IS AGAIN AN ADAPTIVE PROCEDURE WITH RESPECT TO THE EIGENVECTORS

Equation errors. The matrix eigenvalue problem with the identified eigenvalues will give the equation errors instead of the null vector for the adjustment of the stiffness matrix,

$$[-\hat{\lambda}_{0r}\mathbf{M} + \mathbf{K}(\mathbf{a}_K)]\hat{\mathbf{u}}_{0r}(\mathbf{a}_K) = \mathbf{v}_{4r}, \quad r = 1(1)N. \tag{4.3.18}$$

The total residual vector is

$$\mathbf{v}_4^T := (\mathbf{v}_{41}^T, \ldots, \mathbf{v}_{4N}^T) \tag{4.3.19}$$

which is dependent on $\mathbf{a}_K$. In order to perform the sensitivity matrix $D\mathbf{v}_4$, Eq. (4.3.18) will be partially differentiated with respect to $a_{K\iota}$:

$$\frac{\partial \mathbf{v}_{4r}}{\partial a_{K\iota}} = \mathbf{K}_\iota \hat{\mathbf{u}}_{0r}(\mathbf{a}_K) + [-\hat{\lambda}_{0r}\mathbf{M} + \mathbf{K}(\mathbf{a}_K)]\frac{\partial \hat{\mathbf{u}}_{0r}(\mathbf{a}_K)}{\partial a_{K\iota}}, \quad \iota = 1(1)I. \tag{4.3.20}$$

Equation (4.3.20) contains the partial derivatives of the eigenvectors to be adjusted with respect to the parameters to be estimated. If these derivatives cannot be assumed to be equal to zero, then two possibilities exist for their determination. One uses the differentiated eigenvalue problem, while the other is based on the spectral decomposition of the differentiated eigenvectors. The abbreviation

$$\boldsymbol{\xi}^c_{r\iota} \equiv \boldsymbol{\xi}_{r\iota}(\mathbf{a}_K) := \frac{\partial \hat{\mathbf{u}}^c_{0r}}{\partial a_{K\iota}} \tag{4.3.21}$$

may be introduced. The superscript will now generally be used in order to express the dependence on the parameters to be estimated. The matrix eigenvalue problem

$$(-\lambda^c_{0r}\mathbf{M} + \mathbf{K}^c)\hat{\mathbf{u}}^c_{0r} = \mathbf{0} \tag{4.3.22}$$

differentiated with respect to $a_{K\iota}$ yields the linear algebraic system of equations

$$(-\lambda^c_{0r}\mathbf{M}+\mathbf{K}^c)\boldsymbol{\xi}^c_{r\iota} = (k^c_{gr\iota}\mathbf{M}-\mathbf{K}_\iota)\hat{\mathbf{u}}^c_{0r}, \quad r = 1(1)N, \quad \iota = 1(1)I, \tag{4.3.23}$$

with the singular system matrix (rank $n-1$ for disjoint eigenvalues) dependent on $\mathbf{a}_K$. Therefore $\boldsymbol{\xi}^c_{r\iota}$ exists and can be calculated in every iteration step (by choosing one component arbitrarily). The other possibility mentioned is spectral decomposition

$$\boldsymbol{\xi}^c_{r\iota} = \sum_{k=1}^{n} \alpha^{(\iota)}_{rk}\hat{\mathbf{u}}^c_{0k}, \tag{4.3.24}$$

which requires all the eigenvectors to be adjusted. The result is [5]

$$\boldsymbol{\xi}^c_{r\iota} = \sum_{\substack{k=1\\k\neq r}}^{n} \frac{\hat{\mathbf{u}}^{cT}_{0k}\mathbf{K}_\iota\hat{\mathbf{u}}^c_{0r}}{\lambda^c_{0r} - \lambda^c_{0k}}\hat{\mathbf{u}}^c_{0k}. \tag{4.3.25}$$

This is a special solution, to which the eigenvector $\hat{\mathbf{u}}^c_{0r}$ can be added when multiplied with an arbitrary constant c. However, we are interested in a normalized eigenvector, and therefore $c = 0$ is chosen.

The procedure can be formulated with these equations. It is a nonlinear one in the parameters to be estimated. Because the eigenvectors also have to be adjusted, the computational expenditure is large. In the cases where the eigenvector derivatives are negligible (the partial eigenvector modifications are of second order compared with those of the eigenvalues), the prior eigenvectors can be used, and this produces a linear procedure with (see Eq. (4.2.56))

$$D\tilde{\mathbf{v}}_4 := \begin{bmatrix} \mathbf{K}_1\hat{\mathbf{u}}_{01} & \dots & \mathbf{K}_I\hat{\mathbf{u}}_{01} \\ \dots\dots & \dots & \dots\dots \\ \mathbf{K}_1\hat{\mathbf{u}}_{0N} & \dots & \mathbf{K}_I\hat{\mathbf{u}}_{oN} \end{bmatrix}, \quad \mathbf{b}_4^T := (-\hat{\lambda}_{01}\hat{\mathbf{u}}_{01}^T\mathbf{M}, \dots, -\hat{\lambda}_{0N}\hat{\mathbf{u}}_{0N}^T\mathbf{M}). \tag{4.3.26}$$

The formalism is summarized in Table 4.5.

For the adjustment of the flexibility matrix see [5, 6].

Table 4.5. Equation error: adjustment of the stiffness matrix by the use of measured eigenvalues (eigenfrequencies)

DESCRIPTION	FORMULA	NOTE (EQ. NO.)
PARAMETRIZATION	$\mathbf{K}^c = \mathbf{K}(\mathbf{a}_K) = \sum_{\iota=1}^{I} a_{K\iota}\mathbf{K}_\iota$	(4.3.7)
MEASUREMENTS	$\hat{\lambda}_{0r}, \quad r = 1(1)N$	
RESIDUAL VECTOR	$\mathbf{v}_4^T(\mathbf{a}_K) = (\mathbf{v}_{41}^T, \dots, \mathbf{v}_{4N}^T)$	
MATRIX EIGENVALUE PROBLEM	$(-\lambda_{0r}^{(k)}\mathbf{M} + \mathbf{K}^{c(k)})\hat{\mathbf{u}}_{0r}^{c(k)} = \mathbf{0}$, $(-\hat{\lambda}_{0r}\mathbf{M} + \mathbf{K}^{c(k)})\hat{\mathbf{u}}_{0r}^{c(k)} = \mathbf{v}_4^{(k)}$ k-ITERATION INDEX DUE TO $\mathbf{a}_K^{(k)}$ WITH $\mathbf{a}_K^{(0)} = \mathbf{e}$, $\mathbf{K}^{c(k)} = \mathbf{K}(\mathbf{a}_K^{(k)})$ etc. NORMALIZATION: $\hat{\mathbf{u}}_{0r}^{cT}\mathbf{M}\hat{\mathbf{u}}_{0r}^c = 1$	(4.3.18)
SIMPLIFICATION	$\xi_{r\iota}^c = \xi_{r\iota}(\mathbf{a}_K) = \frac{\partial\hat{\mathbf{u}}_{0r}^c}{\partial a_{K\iota}} \doteq 0$ APPROXIMATION: $\tilde{\mathbf{a}}_K \doteq \mathbf{a}_K$	
SOLUTION	$\hat{\tilde{\mathbf{a}}}_K = (D\tilde{\mathbf{v}}_4^T\,\mathbf{G}_v\,D\tilde{\mathbf{v}}_4 + \mathbf{G}_e)^{-1} \cdot (D\tilde{\mathbf{v}}_4\,\mathbf{G}_v\mathbf{b}_4 + \mathbf{G}_e\mathbf{e})$ WITH $D\tilde{\mathbf{v}}_4 = \begin{bmatrix} \mathbf{K}_1\hat{\mathbf{u}}_{01}, \dots, \mathbf{K}_I\hat{\mathbf{u}}_{01} \\ \dots\dots\dots\dots \\ \mathbf{K}_1\hat{\mathbf{u}}_{0N}, \dots, \mathbf{K}_I\hat{\mathbf{u}}_{0N} \end{bmatrix}$, $\mathbf{b}_4^T = (-\hat{\lambda}_{01}\,\hat{\mathbf{u}}_{01}^T\,\mathbf{M}, \dots, -\hat{\lambda}_{0N}\hat{\mathbf{u}}_{0N}^T\mathbf{M})$.	(4.3.26) WITHOUT THE SIMPLIFICATION THE PROCEDURE IS ITERATIVE AND ADAPTIVE WITH RESPECT TO THE NON-MEASURED EIGENVECTORS (SEE(4.3.18) TO (4.3.25)

4.3.3 Adjustment with Identified Eigenvalues and Eigenvectors

Although partial residuals can be applied, with the disadvantage of the necessary one-to-one relationship between the iterated and measured eigenquantities, this requirement is automatically fulfilled by using the equation error. Therefore only this procedure will be discussed. For special cases, for example where only the eigenvalues are measured, see [5]. We have to distinguish between the identified eigenquantities of the undamped and the damped model.

The eigenquantities of the undamped model. Simultaneous adjustment of the inertia and the stiffness matrices leads to the residual vector $\mathbf{v}_{MNr}$, Eq. (4.2.19), with the sensitivities (4.2.20) with respect to the parameters. With

$$D\mathbf{v}_5 := [D\mathbf{v}_M, D\mathbf{v}_K] \tag{4.3.27}$$

and

$$D\mathbf{v}_M := [\frac{\partial \mathbf{v}_{MNr}}{\partial a_{M\sigma}}], \quad r \text{ the row-index and } \sigma \text{ the column-index,} \tag{4.3.28}$$

$$D\mathbf{v}_K := [\frac{\partial \mathbf{v}_{MNr}}{\partial a_{K\iota}}], \quad \iota \text{ the column-index,} \tag{4.3.29}$$

and if the matrices are partitioned in such a way that

$$\mathbf{M}^c = \mathbf{M}' + \sum_{\sigma=1}^{S} a_{M\sigma}\mathbf{M}_\sigma, \quad \mathbf{K}^c = \mathbf{K}' + \sum_{\iota=1}^{I} a_{k\iota}\mathbf{K}_\iota, \tag{4.3.30}$$

which means that the matrices with a prime remain non-adjusted (by prior knowledge), it follows the procedure linear in the parameters to be estimated:

$$\left.\begin{aligned} &\mathbf{v}_5 = D\mathbf{v}_M\mathbf{a}_M + D\mathbf{v}_K\mathbf{a}_K - \mathbf{b}_5 = D\mathbf{v}_5\mathbf{a} - \mathbf{b}_5, \\ &\mathbf{b}_5^T := -(\hat{\mathbf{u}}_{01}^T(-\hat{\lambda}_{01}\mathbf{M}' + \mathbf{K}'), \hat{\mathbf{u}}_{01}^T\mathbf{M}'\hat{\mathbf{u}}_{01} - 1, \\ &\qquad \ldots, \hat{\mathbf{u}}_{0N}^T(-\hat{\lambda}_{0N}\mathbf{M}' + \mathbf{K}'), \hat{\mathbf{u}}_{0N}^T\mathbf{M}'\hat{\mathbf{u}}_{0N} - 1). \end{aligned}\right\} \tag{4.3.31}$$

The particular cases when only the stiffness or the inertia matrix is to be adjusted will also follow from these equations. For $S = 0$, $I > 0$ the residual vector $\mathbf{v}_5$ has to be differentiated only with respect to $a_{K\iota}$, so that $D\mathbf{v}_5$ is transferred to $D\mathbf{v}_K$ and

$$\mathbf{b}_5^T = \mathbf{b}_K^T =$$

$$= -(-\hat{\lambda}_{01}\hat{\mathbf{u}}_{01}^{T}\mathbf{M},\ \hat{\mathbf{u}}_{01}^{T}\mathbf{M}\hat{\mathbf{u}}_{01} - 1, \dots, -\hat{\lambda}_{0N}\hat{\mathbf{u}}_{0N}^{T}\mathbf{M},\ \hat{\mathbf{u}}_{0N}^{T}\mathbf{M}\hat{\mathbf{u}}_{0N} - 1)$$
$$\text{for } \mathbf{K}' = \mathbf{0}. \tag{4.3.32}$$

It follows $D\mathbf{v}_M\mathbf{e} = \mathbf{b}_K$. Further special cases follow without difficulties. The procedure is summarized in Table 4.6.

Table 4.6. Equation error: adjustment of the stiffness matrix by the use of measured eigenvalues (eigenfrequencies) and eigenvectors (normal modes)

DESCRIPTION	FORMULA	NOTE (EQ. NO.)
PARAMETRIZATION	$\mathbf{K}^c = \mathbf{K}(\mathbf{a}_K) = \mathbf{K}' + \sum_{\iota=1}^{I} \mathrm{a}_{K\iota}\mathbf{K}_\iota$,	
	$\mathbf{M}^c = \mathbf{M}(\mathbf{a}_M) = \mathbf{M}' + \sum_{\sigma=1}^{S} \mathrm{a}_{M\sigma}\, \mathbf{M}_\sigma$	(4.3.30)
MEASUREMENTS	$\hat{\lambda}_{0r}$, $\hat{\mathbf{u}}_{0r}$, r = 1(1)N	
RESIDUAL VECTOR	$\mathbf{v}_{MNr} = (-\hat{\lambda}_{0r}\, \mathbf{M}^c + \mathbf{K}^c)\, \widehat{\hat{\mathbf{u}}}_{0r}$, r = 1(1)N,	
	$\mathbf{v}_5 = \mathrm{D}\mathbf{v}_M\, \mathbf{a}_M + \mathrm{D}\mathbf{v}_K \mathbf{a}_K - \mathbf{b}_5$,	(4.3.31)
	$\mathrm{D}\mathbf{v}_M = [\frac{\partial \mathbf{v}_{MNr}}{\partial a_{M\sigma}}]$, $\mathrm{D}\mathbf{v}_K = [\frac{\partial \mathbf{v}_{MNr}}{\partial a_{K\iota}}]$,	
	$\mathbf{b}_5^T = -(\widehat{\hat{\mathbf{u}}}_{01}^T(-\hat{\lambda}_{01}\mathbf{M}' + \mathbf{K}'),\ \widehat{\hat{\mathbf{u}}}_{01}^T\ \mathbf{M}'-1, \dots,$ $\widehat{\hat{\mathbf{u}}}_{0N}^T(-\hat{\lambda}_{0N}\mathbf{M}' + \mathbf{K}'), \widehat{\hat{\mathbf{u}}}_{0N}\ \mathbf{M}'-1)$	(4.3.31)
SOLUTION	$\hat{\mathbf{a}} = (\mathrm{D}\mathbf{v}_5^T \mathbf{G}_\nu\, \mathrm{D}\mathbf{v}_5 + \mathbf{G}_e)^{-1} \cdot (\mathrm{D}\mathbf{v}_5^T\ \mathbf{G}_\nu \mathbf{b}_5 + \mathbf{G}_e\mathbf{e})$,	
	$\mathbf{a} := \left\{ \begin{array}{c} \mathbf{a}_K \\ \mathbf{a}_M \end{array} \right\}$	
SIMPLIFICATIONS	FOR EXAMPLE S=0, I>0 AND	
	$\mathbf{K}' = \mathbf{0}$: $\mathbf{b}_5 = \mathbf{b}_K$,	(4.3.32)
	IT FOLLOWS $\mathrm{D}\mathbf{v}_M\mathbf{e} = \mathbf{b}_K$	
		NO ONE-TO-ONE RELATION BETWEEN CALCULATED EIGENQUANTITIES IS REQUIRED
		N.B.: THE BIAS OF THE ESTIMATES HAS TO BE REDUCED BY FILTERING (SEE SECTION 4.2.3)

The eigenquantities of the damped model. The residual vector is defined by Eq. (4.2.11) with (4.2.10), from which the sensitivity matrix follows. The parameter sensitivities are easily obtained by partial differentiation with respect to the parameters linear in the equations. Therefore the LS solution is a linear system of equations which, however, results in a biased estimate, and therefore it needs statistically based weighting, as is the case for the input residuals (see Sect. 4.2.3). The formalism is summarized in Table 4.7.

Table 4.7. Equation error: adjustment of the stiffness and inertia matrices by the use of measured eigenfrequencies, damping ratios and eigenvectors

DESCRIPTION	FORMULA	NOTE (EQ. NO.)
PARAMETRIZATION	$\mathbf{M}^c = \mathbf{M}(\mathbf{a}_M) = \sum_{\sigma=1}^{S} \mathbf{a}_{M\sigma}\mathbf{M}_\sigma$,	
	$\mathbf{B}^c = \mathbf{B}(\mathbf{a}_B) = \sum_{\varrho=1}^{R} \mathbf{a}_{B\varrho}\mathbf{B}_\varrho$,	(4.3.1)
	$\mathbf{K}^c = \mathbf{K}(\mathbf{a}_K) = \sum_{\iota=1}^{I} \mathbf{a}_{K\iota}\, \mathbf{K}_\iota$	
MEASUREMENTS	$\hat{\lambda}_{Br}, \widehat{\mathbf{u}}_{Br}$, r = 1(1)N	
RESIDUAL VECTORS	$\mathbf{v}_{MNr} = \left\{ \begin{array}{c} \mathbf{v}_{Mr}(\mathbf{a}) \\ 2\hat{\lambda}_{Br}\widehat{\mathbf{u}}^T_{Br}\mathbf{M}(\mathbf{a}_M)\widehat{\mathbf{u}}_{Br} + \widehat{\mathbf{u}}^T_{Br}\mathbf{B}(\mathbf{a}_B)\hat{u}_{Br} - 1 \end{array} \right\}$,	(4.2.11)
	$\mathbf{v}_{Mr} = [\lambda^2_{Br}\mathbf{M}(\mathbf{a}_M) + \lambda_{Br}\mathbf{B}(\mathbf{a}_B) + \mathbf{K}(\mathbf{a}_K)]\widehat{\mathbf{u}}_{Br}$,	(4.2.10)
	$\mathrm{D}\mathbf{v}_M := [\frac{\partial \mathbf{v}_{MNr}}{\partial a_{M\sigma}}], D\mathbf{v}_B := [\frac{\partial \mathbf{v}_{MNr}}{\partial a_{B\varrho}}], D\mathbf{v}_K := [\frac{\partial \mathbf{v}_{MNr}}{\partial a_{K\iota}}]$	
SOLUTION	$\hat{a} = \mathrm{Re}\{\mathrm{D}\mathbf{v}_6^T\ \mathrm{G}_\nu\ \mathrm{D}\mathbf{v}_6^T\}^{-1}\ \mathrm{Re}\{\mathrm{D}\mathbf{v}_6^T\ \mathrm{G}_\nu\ \mathbf{b}_6\}$,	
	$\mathrm{a} := \begin{bmatrix} \mathbf{a}_K \\ \mathbf{a}_B \\ \mathbf{a}_M \end{bmatrix}$	
	$\mathbf{v}_6 = \mathrm{D}\mathbf{v}_6\ \mathbf{a}\ \text{-}\mathbf{b}_6$	SEE [123]

4.3.4 Adjustment Based on Input/Output Measurements

The experimental determination of eigenquantities is subjected to estimation. It is an initial condensation of information from special input/output measurements [5]. Instead of following this path, the measured input and output quantities can be taken directly for adjustment. As already mentioned, averaged measurements should be taken for adjustment if possible.

The use of input/output measurements can be superior to that of incomplete eigenquantities if all the modes are contained in the related input/ouput measurements in a non-negligible manner, assuming that the random measurement errors are sufficiently small.

Output residuals. The output residuals are defined in Eq. (4.2.4), written out in full using the Laplace transforms,

$$\mathbf{v}_{Or}(\mathbf{a}) = [s_r^2\mathbf{M}(\mathbf{a}_M) + s_r\mathbf{B}(\mathbf{a}_B) + \mathbf{K}(\mathbf{a}_K)]^{-1}\mathbf{P}^m(s_r) - \mathbf{U}^m(s_r). \quad (4.3.33)$$

Eq. (4.2.24) contains the parameter sensitivities. The sensitivity matrix follows with the abbreviation

$$\mathbf{H}_r^c := \mathbf{H}_r(\mathbf{a}) \quad (4.3.34)$$

to

$$D\mathbf{v}_O = -\begin{bmatrix} s_1^2\mathbf{H}_1^c\mathbf{M}_1\mathbf{U}_1^c & \dots & s_1^2\mathbf{H}_1^c\mathbf{M}_S\mathbf{U}_1^c & s_1\mathbf{H}_1^c\mathbf{B}_1\mathbf{U}_1^c & \dots & s_1\mathbf{H}_1^c\mathbf{B}_R\mathbf{U}_1^c & \mathbf{H}_1^c\mathbf{K}_1\mathbf{U}_1^c & \dots & \mathbf{H}_1^c\mathbf{K}_I\mathbf{U}_1^c \\ \dots & \dots & \dots\dots & \dots & \dots & \dots & \dots & \dots & \dots \\ s_N^2\mathbf{H}_N^c\mathbf{M}_1\mathbf{U}_N^c & \dots & s_N^2\mathbf{H}_N^c\mathbf{M}_S\mathbf{U}_N^c & s_N\mathbf{H}_N^c\mathbf{B}_1\mathbf{U}_N^c & \dots & s_N\mathbf{H}_N^c\mathbf{B}_R\mathbf{U}_N^c & \mathbf{H}_N^c\mathbf{K}_1\mathbf{U}_N^c & \dots & \mathbf{H}_N^c\mathbf{K}_I\mathbf{U}_N^c \end{bmatrix}. \quad (4.3.35)$$

Taking into account that the quantities are complex and the estimates required are real, then the necessary condition (4.2.54) yields for the WLS

$$\left.\begin{aligned} \Re\left[\sum_{r=1}^N s_r^2\mathbf{U}_r^{cT}\mathbf{M}_\sigma\mathbf{H}_r^c\mathbf{G}_{vr}(\overline{\mathbf{U}_r^c - \mathbf{U}_r^m})\right]\Big|_{\mathbf{a}=\hat{\mathbf{a}}} &= \mathbf{0}, \quad \sigma = 1(1)S, \\ \Re\left[\sum_{r=1}^N s_r\mathbf{U}_r^{cT}\mathbf{B}_\rho\mathbf{H}_r^c\mathbf{G}'_{vr}(\overline{\mathbf{U}_r^c - \mathbf{U}_r^m})\right]\Big|_{\mathbf{a}=\hat{\mathbf{a}}} &= \mathbf{0}, \quad \rho = 1(1)R, \\ \Re\left[\sum_{r=1}^N \mathbf{U}_r^{cT}\mathbf{K}_\iota\mathbf{H}_r^c\mathbf{G}''_{vr}(\overline{\mathbf{U}_r^c - \mathbf{U}_r^m})\right]\Big|_{\mathbf{a}=\hat{\mathbf{a}}} &= \mathbf{0}, \quad \iota = 1(1)I, \end{aligned}\right\} \quad (4.3.36)$$

$\overline{(\dots)}$ means the conjugate complex value of $(\dots)$, and the weighting matrices $\mathbf{G}_{vr}$, $\mathbf{G}'_{vr}$, $\mathbf{G}''_{vr}$ are quadratic of order n and must be positive definite.

This iterative procedure has the advantage that it is asymptotically unbiased. The Hessian matrix (4.2.15) gives a lower bound of the parameter covariances [5]. However, it is difficult to find starting values for which the iteration converges, because the convergence domain can be very narrow. Simulations with $\mathbf{M} = \mathbf{M}^T$ show that

1. fast convergence and an extension of the convergence domain for the Newton-Raphson method can be achieved by the use of not too small additive damping (in the Laplace transforms instead of using the Fourier transform) for the dynamic responses due to impulse testing, and
2. the iteration may converge if an initial iteration for $\mathbf{a}_K$ is performed with fixed $\mathbf{a}_B = \mathbf{e}$. Then $\mathbf{a}_K$ and $\mathbf{a}_B$ will be iteratively determined,
3. the narrow valley of the loss function hyper-surface at the minimum can be widened by the use of state observers (see Sect. 4.3.5).

Recent research is directed to the application of interval arithmetics in the context of the linear (in the parameters to be estimated) input residual method (see next paragraph) in order to obtain an enclosure for the solution [136]. This enclosure, if not sufficiently accurate, can be improved by the nonlinear output residual method. In addition to the enclosure of the solution, the linear system of equations to be solved has the advantage of yielding the global minimum. Another direction of research is the combination of mapping applied to the output residual method with the selective sensitivity (see Sect. 2.1.3). The adaptive forces and the corresponding dynamic responses needed for this will be computed from existing input/output measurements [137].

Input residuals. As already stated, this procedure uses the adjustment of the input vector, described here in the image domain. The residual is defined in Eq. (4.2.7), and the parameter sensitivities are given in Eq. (4.2.22). The sensitivity matrix reads

$$D\mathbf{v}_I = \left[\begin{array}{cccccccccc} s_1^2\mathbf{M}_1\mathbf{U}_1^m & \dots & s_1^2\mathbf{M}_S\mathbf{U}_1^m & s_1\mathbf{B}_1\mathbf{U}_1^m & \dots & s_1\mathbf{B}_R\mathbf{U}_1^m & \mathbf{K}_1\mathbf{U}_1^m & \dots & \mathbf{K}_I\mathbf{U}_1^m \\ \dots & \dots & \dots & \dots & \dots & \dots & \dots & \dots & \dots \\ s_N^2\mathbf{M}_1\mathbf{U}_N^m & \dots & s_N^2\mathbf{M}_S\mathbf{U}_N^m & s_N\mathbf{B}_1\mathbf{U}_N^m & \dots & s_N\mathbf{B}_R\mathbf{U}_N^m & \mathbf{K}_1\mathbf{U}_N^m & \dots & \mathbf{K}_I\mathbf{U}_N^m \end{array}\right]. \tag{4.3.37}$$

It is a linear method in the parameters to be estimated, because $D\mathbf{v}_I$ is independent of $\mathbf{a}$. With parametrization like (4.3.30) for all the parameter matrices, the right side of the system of equations is

$$\mathbf{b}_I^T := -(\mathbf{P}_1^{mT} - \mathbf{U}_1^{mT}(s_1^2\mathbf{M}' + s_1\mathbf{B}' + \mathbf{K}'), \dots, \mathbf{P}_N^{mT} - \mathbf{U}_N^{mT}(s_N^2\mathbf{M}' + s_N\mathbf{B}' + \mathbf{K}')). \tag{4.3.38}$$

The particular cases follow easily.

Statistically based weighting is conditionally necessary, as stated in Sect. 4.2.3. The noise-to-signal ratio has to be very small (averaging).

Choice of frequencies The computations are carried out for ω_r, $r = 1(1)N$. The procedure discussed here has the property $\nu = N$. N should be chosen as large as possible. However, sensors have a relative measuring error dependent on their measurement interval, and therefore it is not advisable to measure and use very small values of the dynamic responses compared with the maximum values. Consequently, transformed measurements should be chosen adjacent to the resonance frequencies. It is necessary to take at least a minimum number of frequencies ω_r close to each resonance peak in the frequency response function. In general, only a few (≤ 15) frequencies per resonance peak are necessary; their actual number depends on the chosen frequency spacing and the damping ratios

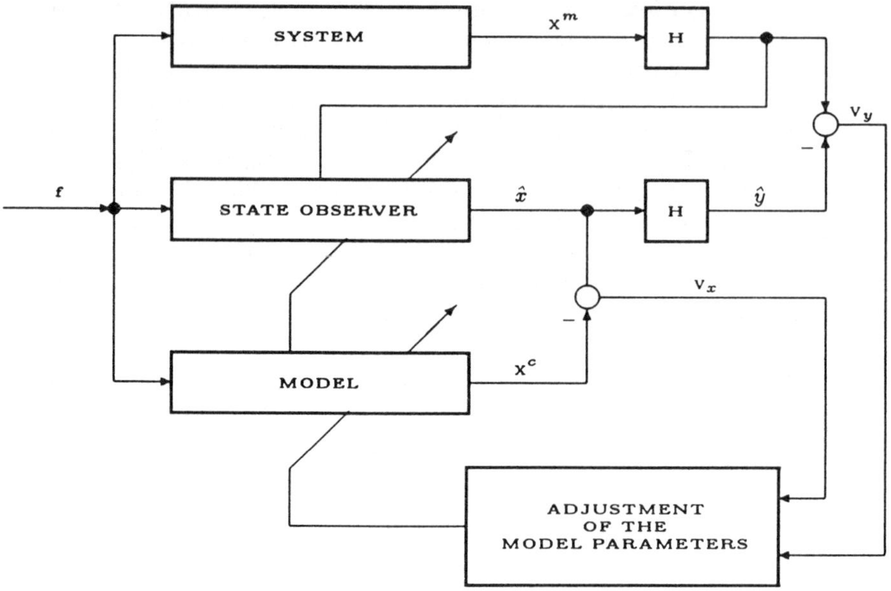

Fig. 4.4. The procedure using state observers

of the eigenmodes of the system [119]. Frequencies chosen too close will not extend the information of the measurements, because these measured values are (deterministically or) stochastically dependent on each other.

4.3.5 State Observers

The state observers, well-known in control theory, are generally used for the state reconstruction of non-measured states $\mathbf{x}(t)$. The advantage of state observers is that they can be formulated adaptively, which means that they are applicable in the case of large measuring errors. Their combination with parameter estimation smoothes the loss function hyper-surface and widens the often very small valley with respect to the (global) minimum [138].

The details will be very briefly summarized here following [138], without including the more recent developments of the additional error modelling of parameters and of measurements, as is mentioned at the beginning of Chap. 3. Figure 4.4 explains the procedure. The state observer is used in order to reconstruct all the system states with the help of the measured states. These "estimates" will then be used for parameter estimation: substituting non-measured states by reconstructed states, which means extending the number of equations used ($\rightsquigarrow$ regularization). It should be noted that the reconstruction uses the prior mathematical model, and therefore it may feign additional knowledge. In [138] it is shown that the output residual method combined with the LS estimation is a special case of the adaptive state observer.

The time-discrete formulation for the linear model (following Eq. (3.01) without noise and disturbances, see [106, 139]), $\mathbf{x}(t_k) =: \mathbf{x}_k$,

$$\mathbf{x}_{k+1} = \mathbf{A}\mathbf{x}_k + \mathbf{B}\mathbf{f}_k, \tag{4.3.39}$$

leads to the Luenberger observer

$$\tilde{\mathbf{x}}_{k+1} = \mathbf{A}_1\tilde{\mathbf{x}}_k + \mathbf{A}_2\mathbf{y}_k + \mathbf{A}_3\mathbf{f}_k, \tag{4.3.40}$$

where $(\tilde{\ldots})$ designates the reconstructed state vector. $\mathbf{y}_k$ are the measurements. The requirement

$$\Delta\mathbf{x}(t) := \mathbf{x}(t) - \tilde{\mathbf{x}}(t) \to \mathbf{0} \tag{4.3.41}$$

leads to

$$\Delta\mathbf{x}_{k+1} = \mathbf{A}_1\Delta\mathbf{x}_k + (\mathbf{A} - \mathbf{A}_1 - \mathbf{A}_2\mathbf{H})\mathbf{x}_k + (\mathbf{B} - \mathbf{A}_3)\mathbf{f}_k, \tag{4.3.42}$$

where $\mathbf{H}$ stems from the measuring equation

$$\mathbf{y}(t) = \mathbf{H}\mathbf{x}(t). \tag{4.3.43}$$

The requirement with respect to the error (4.3.41) turns into the observer design

$$\left.\begin{array}{lcl} \mathbf{A}_1 & = & \mathbf{A} - \mathbf{A}_2\mathbf{H}, \\ \mathbf{A}_3 & = & \mathbf{B}. \end{array}\right\} \tag{4.3.44}$$

It follows

$$\Delta\mathbf{x}_{k+1} = \mathbf{A}_1\Delta\mathbf{x}_k, \tag{4.3.45}$$

which means that the reconstruction error obeys the same dynamics as the state observer.

If $\mathbf{H}$ is given, and in practice it should be chosen dependent on the needs, only $\mathbf{A}_1$ or $\mathbf{A}_2$ can be chosen. In order to make sure that the procedure converges and that the solution is stable, the observer dynamics should be chosen by the choice of its eigenvalues. For the computation of the observer matrices see [140, 141].

If it is provided that the measurement noise has statistical properties, then the Kalman filter follows (see the note in Sect. 4.2.3). From the point of view of the possibility of the observer design, it has the disadvantage that the design of the filter is automatically determined by the statistical properties of the errors. The reader can find the theory of Kalman filtering in [133], and a description of the system identification including the parameter estimation by Kalman filter[11] and its application to concrete structures in [142].

[11] Kalman filtering can be seen as a recursive Bayesian approach [4], p.426 ff.

The special case of the output error residual is obtained with $\mathbf{A}_1 = \mathbf{A}$ and $\mathbf{A}_2 = \mathbf{0}$.

It should be noted that this procedure needs additional theoretical investigation. The same applies to the computational expenditure. When the existing disturbances are understood as external but unknown inputs, as written in Eqs. (3.0.1) and (3.0.2), then the filter (observer) can be designed so that these inputs are decoupled from system modifications [143, 144]. Consequently, the residual is robust against unknown inputs and only sensitive to system modifications; this method is called a robust method. It is obvious that this procedure can also be used for failure isolation (each failure requires a filter).

4.3.6 ARMA Models

Approximations of the governing differential equations by finite differences or the corresponding integral equations by numerical integration methods lead to algebraic equations. Knowledge of the method of approximation permits a physical interpretation of the parameters [139, 145]. Independent of these approximations, a system of algebraic equations can be assumed as a process model a priori. If the common notation in ARMA modelling is used, the symbols for the input and output quantities previously used in this book now deviate from the following ones; the model then reads

$$\begin{aligned} &y_k + a_{n-1}y_{k-1} + a_{n-2}y_{k-2} + \ldots a_0 y_{k-n} \\ &\quad = b_{n-1}u_{k-1} + b_{n-2}u_{k-2} + \ldots + b_0 u_{k-n}, \end{aligned} \tag{4.3.46}$$

with $y_k := y(t_k)$, $u_k := u(t_k)$, $t_k = kh$, h the time increment, $k = 1, \ldots, n$. In order to calculate y_k, n previous values of the output and n previous values of the input (initial conditions) must be given.

Equation (4.3.46) is called an ARMA model with the autoregressive (AR) part $\{y_i\}$ and the moving average (MA) part $\{u_i\}$. The coefficients $a_{n-1}, \ldots, a_0$ and $b_{n-1}, \ldots, b_0$ are constant for linear and time-invariant systems.

The solution of Eq. (4.3.46) can be obtained by applying the z-transform. With the shift operator q,

$$q^i y_k := y_{k+i}, \tag{4.3.47}$$

Eq. (4.3.46) turns into the equation

$$(q^n + a_{n-1}q^{n-1} + \ldots + a_0)y_k = (b_{n-1}q^{n-1} + b_{n-2}q^{n-2} + \ldots + b_0)u_k. \tag{4.3.48}$$

By introducing the z-transforms

$$Y(z) := \mathfrak{Z}\{y_k\}, \;\; U(z) := \mathfrak{Z}\{u_k\}, \;\; \mathfrak{Z}\{f_{k+n}\} = z^n F(z), \tag{4.3.49}$$

with initial conditions equal to zero, one obtains

$$(z^n + a_{n-1}z^{n-1} + \ldots + a_0)Y(z) = (b_{n-1}z^{n-1} + b_{n-2}z^{n-2} + \ldots + b_0)U(z). \tag{4.3.50}$$

The transfer function immediately follows:

$$H(z) = \frac{b_{n-1}z^{n-1} + b_{n-2}z^{n-2} + \ldots + b_0}{z^n + a_{n-1}z^{n-1} + \ldots + a_0}. \tag{4.3.51}$$

The solution of the homogeneous Eq. (4.3.46) is

$$y_{hk} = y_h(k) = c_1 p_1^k + c_2 p_2^k + \ldots + c_n p_n^k \tag{4.3.52}$$

with constants c_i and poles of the transfer functions p_i equal to the roots of the characteristic polynomial

$$z^n + a_{n-1}z^{n-1} + \ldots + a_0 = \Pi_{k=1}^n (z - p_k) = 0.$$

If the particular integral is designated by $y_p(k)$, then the total solution of the linear model is

$$y_k = y(k) = y_h(k) + y_p(k). \tag{4.3.53}$$

The reader can find a detailed discussion in [106], for example. The coefficients of the ARMA model can be estimated by the use of process measurements. The definition of residuals and the application of a chosen estimator lead to the required estimates (see, for example [69, 146]). This can be done in the time domain or the z-domain. The order determination requires special attention; here the reader is additionally referred to [147]. Several ARMA models permitted for vibrating systems with their orders are summarized in [148]. Order determination of ARMA models is crucial. Order estimation of AR and ARMA models can be done based on the system-balancing theory. The approach is balanced if it consists of the determination of state coordinates, when the system is controllable and observable in the same degree [149]. If some state variables are weakly controlled, and, at the same time, weakly observed, they can be reduced.

As already mentioned, ARMA models represent time series; they are process models. As can be seen from Eq. (4.3.51), the AR part determines the poles of the model which are related to the eigenfrequencies of the system which causes the process [106]. The solution (4.3.52) of the difference equation can be compared approximately with the solution of the corresponding differential equation, which consists of the fundamental terms $\exp(\delta_i + j\omega_i)kh$ at the time kh with the time step h and its multiple. Comparison of the fundamental solutions p_i^k and $\exp(\delta_i + j\omega_i)kh$ yields

$$p_i = e^{(\delta_i + j\omega_i)h}.$$

Distinction of the various possible cases of the poles by means of the ln-operation provides a first guess concerning the required eigenquantities. As can be imagined, it can be difficult to decide which eigenfrequencies are related to the object under consideration and which ones are caused by noise. But, consequently, the AR coefficients of the ARMA model contain the information about the system modifications via the modified process (time series); they indicate possible damage.

Another physical interpretation of these estimates needs further calculation steps: backward transformation. This transformation can be performed by calculating the eigenquantities (singular values and vectors) of the ARMA model, and by inserting them in a suitable way into the spectral decompositions of the parameter matrices of the assumed system model. This can be a difficult task, when one considers the possible incompleteness of the eigenquantities and the possible physically non-interpretable modes due to the noisy data used for modal identification. Another method of physically interpreting the ARMA coefficients is given by the forward discretization of the equations of motion in the form of differential equations by pre-defined finite difference expressions, and integral equations by pre-given numerical integration methods [139, 145]. In addition to deterministic handling, a statistical criterion can be taken which results in a covariant equivalent formulation [150].

What is writtten above for a single input/single output (SISO) system can be written, of course, for a multi-input/multi-output (MIMO) system. For an n DOF model only the second order AR parts are necessary in order to determine the system quantities. The various types of modelling are illustrated in Fig. 4.5, following [151]. The abbreviation "X" stands for eXogeneous excitation. Lattice filters are digital filters which are based on ARMA models and realized by the ARMA z-transform transfer function $H(z)$. The MA coefficients, however, determine the identification methods. The reader can find a summary of the possible models and identification methods in [152].

The $2n$-th order MIMO state-space model corresponds to Eqs. (3.01, 3.02) without the failure terms,

$$\dot{\mathbf{x}}(t) = \mathbf{A}\mathbf{x}(t) + \mathbf{B}\mathbf{u}(t),$$

$$\mathbf{y}(t) = \mathbf{C}\mathbf{x}(t) + \mathbf{D}\mathbf{u}(t),$$

where $\mathbf{x}(t)$ is $2n$-dimensional, $\mathbf{u}(t)$ is l-dimensional, and $\mathbf{y}(t)$ is m-dimensional, and $\{\mathbf{A}, \mathbf{B}, \mathbf{C}, \mathbf{D}\}$ are dimensioned compatibly. When the discretization methods mentioned above are followed, or when an intervalwise approximation of the forcing function is implied (e.g. by taking the averaged sample value), the discrete time state-space model corresponding to the above continuous time formulation reads (it can be compared with Eq. (4.3.39))

$$\mathbf{x}(k) = \mathbf{A}_d\mathbf{x}(k-1) + \mathbf{B}_d\mathbf{u}(k-1), \tag{4.3.54}$$

$$\mathbf{y}(k) = \mathbf{C}_d\mathbf{x}(k) + \mathbf{D}_d\mathbf{u}(k-1), \tag{4.3.55}$$

where the matrices result dependent on the type of approximation, and the abbreviation

$$\mathbf{u}(k) = \mathbf{u}(t), \quad t \epsilon (kh, kh+h] \tag{4.3.56}$$

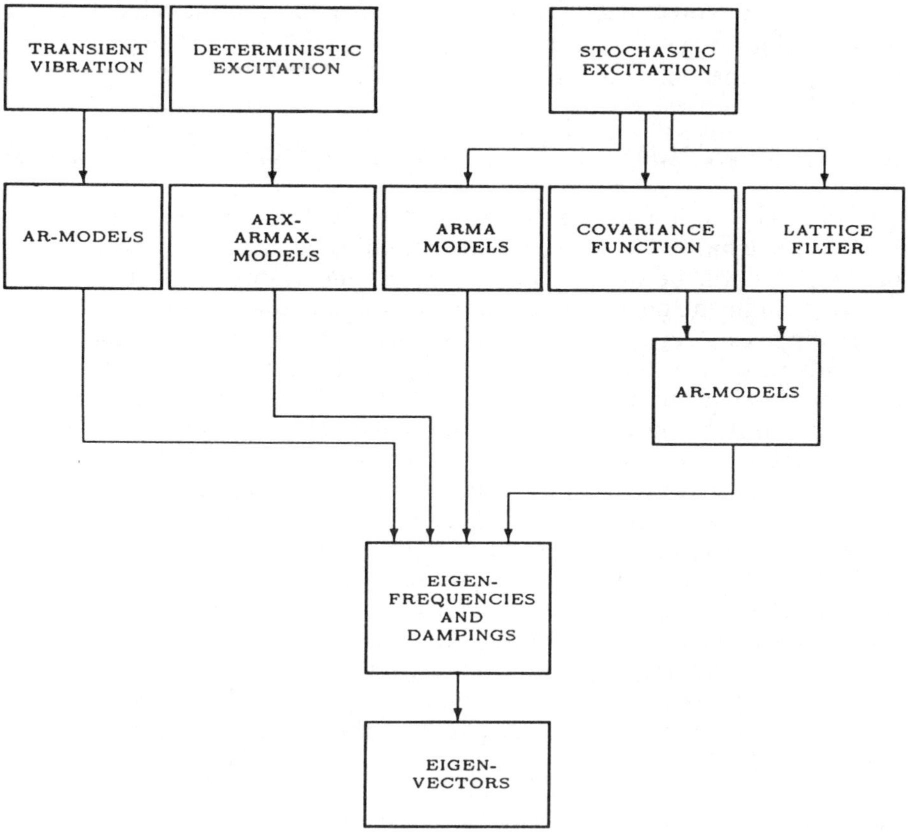

Fig. 4.5. Time domain modelling of linear systems

is used. In the above equations no distinction is made between the exact quantities $\mathbf{x}, \mathbf{y}$ and their approximations.

4.3.7 Additional Methods

No attempt is made to present a review of adjustment methods. Here the reader is referred to reviews and surveys in addition to the books on system identification already cited: [150] to [153]. But the reader's attention must be drawn to two methods that are interesting in adjustment: one method is interesting due to its closed form solution, and the other method is a real time identification method.

It is assumed that $m < n$ measured eigenfrequencies and natural modes are measured. The stiffness matrix adjustment problem is formulated as

$$\left.\begin{array}{rl} & \min \| \mathbf{M}^{-1/2}(\mathbf{K}^c - \mathbf{K})\mathbf{M}^{-1/2} \|_F, \\ \text{subject to} & \mathbf{K}^c\hat{\mathbf{U}}_0 = \mathbf{M}\hat{\mathbf{U}}_0\hat{\mathbf{\Lambda}}_0 \text{ and } \mathbf{K}^c = \mathbf{K}^{cT}. \end{array}\right\} \tag{4.3.57}$$

The identified modal matrix $\hat{\mathbf{U}}_0$ is an (n, m)-matrix, and the related matrix of eigenvalues $\hat{\mathbf{\Lambda}}_0$ is a quadratic matrix of order m. The problem formulated in [154] has the closed-form solution given in [155]:

$$\mathbf{K}_c = \mathbf{K} + \mathbf{M}\hat{\mathbf{U}}_0\hat{\mathbf{\Lambda}}_0\hat{\mathbf{U}}_0^T\mathbf{M} - \mathbf{K}\hat{\mathbf{U}}_0\hat{\mathbf{U}}_0^T\mathbf{M} - \mathbf{M}\hat{\mathbf{U}}_0\hat{\mathbf{U}}_0^T\mathbf{K} + \mathbf{M}\hat{\mathbf{U}}_0\hat{\mathbf{U}}_0^T\mathbf{K}\hat{\mathbf{U}}_0\hat{\mathbf{U}}_0^T\mathbf{M}. \quad (4.3.58)$$

The adjusted stiffness matrix is positive definite if the non-adjusted matrix is positive definite. However, if the non-adjusted matrix $\mathbf{K}$ has a sparse pattern, this will be lost in adjustment. As mentioned in Sect.4.4.3, this sparsity can be included as prior information in the adjustment process by including it as a constraint. In [156] the case of the associated undamped matrix eigenvalue problem and the case of statics are dealt with taking into account missing measurements, this means with missing eigenquantities and reduced static measurements, respectively. The missing quantities, and to some extent the measured ones, are introduced as hidden functions of the structural parameters. Consequently, the Euclidean norm (equivalent to the LS) of the matrix to be minimized leads to a nonlinear system of equations with more than one (local) minimum.

System identification methods can be classified as off-line and on-line methods. Among the on-line methods those are of particular interest which are real-time methods. Real-time identification methods are important when systems undergo degrading behaviour under large forces. In [157] state observers are applied for vibration monitoring and damage detection. Spectral decomposition and truncation of the higher modes are used in order to reduce the computational expenditure. Another real-time identification method is published in [158]. It is a time domain method based on a second order linear ODE with time-varying model parameters. The dynamic response of the model is obtained by application of the Newmark time-integration method. The LS method is applied for estimation of the physical model parameters (e.g. stiffness, damping). Since the current system state is used for adjustment at each identification time instant, no identification error is accumulated. It has to be noted that the time identification interval must be short enough to capture the parameter variations in order to yield reliable estimates, and yet it must be long enough to allow necessary numerical computations. The demonstration is done with an MDOF model up to 3 DOF. For an n-DOF model, about $60n^2$ multiplications and additions have to be made to estimate n model parameters at each time step.

4.4 Algorithms

The adjustment procedures lead to linear and nonlinear algebraic equations in the parameters to be estimated. These systems of algebraic equations have to be solved while bearing in mind that the data used are erroneous.

4.4.1 Linear Equation Solvers

Equations linear in the parameters can advantageously be solved numerically by SVD[12] [159]. If the system matrix is designated by A, and assuming it is real, then the product $A^T A$ does not have to be calculated, and the number of digits needed to compute the solution of the normal equations with the same accuracy is half because it holds true [160] that

$$\begin{aligned} 1 &\leq condition\ number(A) \\ &= \sqrt{condition\ number(A^T A)} \leq condition\ number(A^T A). \end{aligned}$$

As is already known, the minimum number of digits of the accuracy necessary for solving a system of linear equations is equal to $log[condition\ number(A)]$.

One should choose a numerical algorithm which avoids the amplification of the errors inhibited in the system of equations. For example, orthogonal triangularization combined with the QR-algorithm by the use of finite arithmetic (which serves as the regularization method by modifying the mathematical operator, yielding an adjacent but stable solution, see Sect. 4.4.3) is a method of this kind [72, 69]. The QR method is expensive in computations (twice the number of computations as a direct solver of LS equations), but it is not sensitive to rounding errors.

Once again, the interval arithmetic with the Gauss-Seidel algorithm applied to the input residual method [136] is mentioned. This yields an enclosure of the solution.

4.4.2 Nonlinear Equation Solvers

If the resulting system of equations is nonlinear in the parameters, then iterative methods have to be applied which are characterized by the choice of initial values, generation of the improved parameters and the choice of a termination criterion. The Newton-Raphson methods and their modifications, and also the gradient methods, are well-known. For these methods the choice of initial values dependent on the shape of the unknown loss function can be very difficult for the LS procedures when a large number of parameters have to be estimated. The convergence region of the iteration procedure can be very small. If insufficient prior knowledge of the values to be estimated exists, starting values for iteration have to be constructed. This can be done by methods that are linear in the parameters (for instance, based on input residuals and subsystem modelling with approximated statistical weighting, as described in Sect. 4.2.3, or by taking the solution of the linearized system of equations as starting values). Another approach uses the EWLS in order to construct an iteration procedure that is always convergent [127, 5] (and see footnote 9 of this section). It

[12] Singular Value Decomposition

is a parameter perturbation method, starting with predominant weighting of the penalty term with the known solution **e** from system analysis (see Eq. (4.2.53)). Then the ratio of the weighting is modified in such a way that the solution of the previous iteration step serves as initial values within the convergence region of the iteration under computation. The convergence will be certain, but the solution can be a local minimum.

Additionally, the Levenberg-Marquardt method should be mentioned as a combination of the Gauss -Newton and gradient methods. It contains a parameter ε in the form $\varepsilon\mathbf{I}$, which serves as a penalty term for regularization.

Termination criteria are also part of regularization [71]. All the above methods behave well if the loss function has a sufficiently "smooth" surface. If not, iteration procedures may fail, e.g. when parameter values are significantly changed after a particular iteration step, with an increase of the loss function. Appropriate provisions thus have to be installed into the iteration scheme to overcome such difficulties (e.g. in the above-mentioned case by repeatedly halving the step size).

An interesting publication in this context is [161], which contains a discussion of estimators including numerical algorithms.

4.4.3 Role of Regularization

Applications and theoretical investigations of large-scale models show that, in general, the resultant inverse problems are ill-posed in the definition by Hadamard [71]. He took the existence, uniqueness and stability of solutions to be the characteristics of the well-posedness of problems. Spatially discretized problems are ill-conditioned. A well-known example is estimation of the impulse response function by deconvolution. The application of regularization means the substitution of the solution operator by an adjacent operator to provide a unique and stable solution. A generalized solution based on a minimum norm requirement is unique with respect to the chosen minimum energy considerations (corresponding to the norm). Additional manipulation is then needed in order to obtain a stable solution. The modified operator can be obtained by taking additional information into account. This can be knowledge of mathematical properties (symmetry, definiteness of matrices) and of system properties, such as connectivity, it also can concern prior results of reconstructed states, of standard deviations of measurements and prior estimates, or of parameter sensitivities with respect to measurement errors [162] etc. The latter is the reduction of the measurement error effect by special weighting. This is performed by including the data sensitivity of the normal solution (LS-solution) with respect to the measuring errors [166]. The inverse problem is solved, for example, by applying the LS with the measured data, and then additionally known randomly varying errors are superimposed on the measurements and the problem is solved again. This sensitivity is determined in a similar way to the Monte Carlo method, using additional known randomly varying errors which are superimposed on the already erroneous data, i.e. the weighting matrix is chosen as

$$\mathbf{G}_e := s \sum_{i=1}^{N} [\bar{\mathbf{x}}(\mathbf{y}) - \bar{\mathbf{x}}(\mathbf{y} + \boldsymbol{\eta})][\bar{\mathbf{x}}(\mathbf{y}) - \bar{\mathbf{x}}(\mathbf{y} + \boldsymbol{\eta})]^T , \tag{4.4.1}$$

where $\mathbf{y}$ is the data vector, $\boldsymbol{\eta}$ the known random vector, and $\bar{\mathbf{x}}$ denotes the normal solution. s is a scalar which, for instance, can be estimated by cross-validation. The improvement is described in [166]. Additional regularization methods are discussed in the same thesis. Instead of measured values one can extend the knowledge by the calculated (by the use of the validated model) dynamic responses of a changed mathematical model of system analysis; the changes can concern the boundary conditions [163] and the theoretically introduced "anti-resonances" [164]. An overview of regularization methods with some of the authors' own results is presented in [72]. A review of the literature is also given, and therefore only two methods are mentioned in some detail. It will be noted here that one can combine global with local residuals in order to enlarge the information content [165].

Bayesian approach. Normally distributed random variables for model parameters as well as for measurement errors lead to the EWLS (see Eqs. (4.2.53) to (4.2.56)). Here the possibility is given of regularizing the operator by the penalty term.

Additionally, submodelling with the adjustment factors far fewer in number than the number of matrix elements to be estimated means the introduction of a coarser topology, which is also a type of regularization: the share of information content of the finite measurement data set that is related to each single parameter is enlarged when the total number of the parameters is diminished, which means that the number of equations is now greater and the number of parameters is smaller than before.

As already mentioned, statistical weighting by $\mathbf{G}_e$ can be a problem. The Tikhonov-Phillips procedure can be applied as a deterministic alternative to this.

Tikhonov-Phillips procedure. If one changes the parameters to be estimated from $\mathbf{a}$ to $\Delta\mathbf{a} := \mathbf{a} - \mathbf{e}$, then the prior knowledge for $\Delta\mathbf{a}$ is the zero vector $\mathbf{0}$. By substituting the weighting matrix $\mathbf{G}_e$ of the penalty term in Eq. (4.2.54) by $\gamma^2\mathbf{Y}^T\mathbf{Y}$ with a pre-given matrix $\mathbf{Y}$, the problem is reduced to determining the regularization parameter γ. This can be done a priori under specified circumstances [71, 70], or a posteriori by cross-validation [128]. The additional expenditure is not negligible, but examples show that the expense is almost justified [165, 166].

4.5 Practical Hints

The essential statements have already been made with regard to subsystems modelling and prior knowledge, so that only general comments on mathematical handling and experience will be made here. The adjustment procedures also have to be seen in the context that the order of the initial mathematical model (number of DOF) is unequal to the number of measuring points. Additionally, the inputs often cannot be measured while the outputs are known.

4.5.1 General

In parametric identification the mathematical model is the starting point, and it is a very important subject. The model structure must be adequate to the problem, which means, for example, that if the load configuration of the system has changed a check must be made to determine whether the model structure is still valid, because the energy intensity fluxs can have changed etc. (change of the physical model [5] and see Fig. 1.3 and Sect. 2.1.4). Any error in the model structure affects the parameter estimates in the sense that they become biased. Spatial discretization (e.g. by finite elements) generally influences the structure as well as the parameter values. Therefore the discretization error has to be taken into account, and here, especially, the error in the model structure (including connectivity), while with an adequate model structure the statistical parameter error will be reduced by the estimators. The modelling of boundary conditions is already included in this discussion when one thinks of the flexibility formulation.

The choice of parameters to be estimated is most important for continuous models as well as for discrete (with respect to the spatial coordinates) ones. Error localization[13] [167] can be done deterministically and statistically and needs a lot of information, including the fact that the measurements are incomplete and erroneous. Here prior knowledge and regularization again play a major role.

Parameter sensitivity helps one to find out the parameters influenced mainly by the measurements. In addition, measurement error sensitivity [162], [166, 168] helps one to choose those parameters which are affected most by measurement errors and should therefore be emphasized less or eliminated in the estimation. As can already be seen, the user can formulate the problem at the beginning (e.g. by parameter choice) in a well-posed manner.

Parameter reduction has already been mentioned in connection with the sensitivity investigations. Another approach is parameter topology by submodel formulation, also already discussed.

[13] Here, of course, error localization is identical with the localization of model modifications, whereby the models describe different state conditions dependent on life times.

With regard to the convergence of iterative procedures, the idea is to reduce the number of estimates during iteration. Multi-level estimation can be performed sequentially with subsets of the parameters [169]. Another procedure is adaptive testing based on selective sensitivity with respect to particular parameters or a subset of parameters [170, 171]. These latter methods mentioned here are parts of the regularization methods, because they reduce the number of equations to be solved and diminish the respective condition numbers.

The adjustment methods discussed work well for a number of $J < 100$ of parameters to be estimated and with $n < 10^3$ if

1. all the information needed is contained in the measured data set in a non-negligible manner, for example the DOF affected by the system modifications
2. the structure of the (prior bandlimited) mathematical model is equivalent to the system behaviour
3. the parametrization is chosen "suitably" (for which several tools are provided)
4. as much information as possible is used (e.g. prior knowledge)
5. in the case of the application of the (E)WLS statistically based weighting is used (in order to reduce the random error correlation and therefore the resulting bias)
6. a stable algorithm is used (in order to avoid amplified error propagation)
7. regularization methods are applied in the case of ill-conditioning.

4.5.2 Incomplete Measurements

Incomplete measurements are a severe problem. They affect the identification possibilities (methods) as well as the quality of the parameter estimates.

Missing measuring points. Missing measurements are, for example, expressed by Eq. (4.3.43). They can be substituted by simulation with the prior validated mathematical model, as already mentioned a number of times. Additionally, they can be estimated together with the parameters (Kalman filter, see Sect. 4.2.4). Missing dynamic responses can also be reconstructed (see Sect. 4.3.5 on state observers). At least missing components can be accepted by the use of the measuring equation within the EWLS, instead of using all the components of the related output and/or input vectors in the corresponding residuals: the sums of the loss functions are built up using fewer terms.

Non-measurable input quantities. The use of available modal data avoids this problem. For example, signal processing can easily result in identified eigenfrequencies.

The excitation of some operating systems can be assumed as random with constant spectral power density within a frequency band, and therefore white noise excitation can be provided with a variance that is also to be estimated in addition to the parameters.

At least, if the input can be parametrized, these input parameters can then be estimated simultaneously with the other parameters of the model, as is done in [172].

4.5.3 The Use of Order-Reduced Models

Another problem which has to be mentioned here is the reduction of the order of the discretized model itself. The incomplete modal transformation can be emphasized, which is advantageous in addition to other properties, because it does not change the parameter set to be estimated [68, 135]. The parameters are defined for the non-reduced model, and estimated within the reduced model. Therefore they can be applied to the total model, but one has to prove whether they also update the total model (dependent on the information, the frequency contents of the measurements, the chosen parametrization). Various types of reduction transformations and the incomplete measurements and their errors including the parametrization are discussed in [173] and [68]. This means that the measuring matrix **H** and the control matrix have to be taken into account in addition to the error models. The main results to be taken into account here are

- that the submodel parametrization has to fulfil special requirements with the application of the frequency-independent incomplete modal transform (fewer eigenvectors than the number of DOF of the prior model, however, complete within the band-limitation) with $m < n$ measuring points, and
- that the reduced mathematical model means a loss of information.

The $n' < n$ primary (essential) eigenvectors of the associated undamped model are determined by the band-limitation, and they will yield a relatively complete decomposition of the dynamic response due to the related band-limited excitation. In the image domain the transformation[14] reads

$$\mathbf{U}(s) = \mathbf{T}\mathbf{Q}(s), \tag{4.5.1}$$

with the vector of generalized coordinates $\mathbf{Q}(s)$. The constant transformation matrix $\mathbf{T}$ can be, for example, the incomplete modal matrix $\hat{\mathbf{U}}_{0p}$. Further transforms are contained in [135]. With the complete prior modal matrix

$$\hat{\mathbf{U}}_0 =: [\hat{\mathbf{U}}_{0p}, \hat{\mathbf{U}}_{0s}] \tag{4.5.2}$$

(the index p stands for primary coordinates and s for secondary coordinates) the summand matrices have to fulfil additional conditions. Otherwise the estimates of the correction factors may be biased, though the

[14] It can also be performed in the time domain.

reducing modal transformations used are exact for simulation purposes. The (rather restrictive) conditions

$$\left.\begin{array}{lll}\hat{\mathbf{U}}_{0s}^T\mathbf{M}_\sigma\hat{\mathbf{U}}_{0p} & = & \mathbf{0} \text{ for } \sigma = 1(1)S, \\ \hat{\mathbf{U}}_{0s}^T\mathbf{B}_\rho\hat{\mathbf{U}}_{0p} & = & \mathbf{0} \text{ for } \rho = 1(1)R, \\ \hat{\mathbf{U}}_{0s}^T\mathbf{K}_\iota\hat{\mathbf{U}}_{0p} & = & \mathbf{0} \text{ for } \iota = 1(1)I\end{array}\right\} \quad (4.5.3)$$

are sufficient to prevent biased estimates.

The necessary conditions (4.3.3) now change [68] to

$$\left.\begin{array}{llll}\text{rank }\{(\hat{\mathbf{U}}_{0p}\otimes\hat{\mathbf{U}}_{0p})^T[cs\mathbf{M}_1,\dots,cs\mathbf{M}_S] & = S & \leq & (n')^2, \\ \text{rank }\{(\hat{\mathbf{U}}_{0p}\otimes\hat{\mathbf{U}}_{0p})^T[cs\mathbf{B}_1,\dots,cs\mathbf{B}_R] & = R & \leq & (n')^2, \\ \text{rank }\{(\hat{\mathbf{U}}_{0p}\otimes\hat{\mathbf{U}}_{0p})^T[cs\mathbf{K}_1,\dots,cs\mathbf{K}_I] & = I & \leq & (n')^2\end{array}\right\} \quad (4.5.4)$$

where $\otimes$ denotes the Kronecker product.

4.6 Use of the Knowledge Base for Diagnosis

The far-reaching information now available is the validated mathematical model for every past life time required. If the measured data used also contain sufficient information on the system condition, and if the estimation performed approximates the required expectation well, then the errors of the validated mathematical model are not only known, but they are also sufficiently small (usability of the model).

The estimated variances and covariances of the estimates are thus first given for the recommended EWLS. They serve for checking the usability, and secondly, as is discussed in the next chapter, for decision-making: the test of the significance of model modifications (on time θ_i compared with the model for time θ_j, $j < i$). Diagnosis is then performed by the use of reliable and usable static and dynamic models.

4.6.1 Parameter Errors: Covariance Matrix and Confidence Regions

The EWLS is recommended. The reasons for this recommendation can be found in Sects. 4.2.4 and 4.4.3. We will mainly follow [174] for determination of the covariance matrix. Additional information can be obtained elsewhere, for example in [126].

Using the Bayesian estimator as described in Sect. 4.2.4, one gains a posteriori the most probable parameter values $\hat{\mathbf{a}}$ for estimates that minimize the loss functional Eq. (4.2.52). Taking the Taylor series expansion $\mathcal{J}_B(\mathbf{a})$ for parameter values $\mathbf{a}$ not too far from $\hat{\mathbf{a}}$, one thus obtains

$$\mathcal{J}_B(\mathbf{a}) \doteq \mathcal{J}_B(\hat{\mathbf{a}}) + \frac{1}{2}(\mathbf{a}-\hat{\mathbf{a}})^T\mathbf{H}_{\hat{a}}(\mathbf{a}-\hat{\mathbf{a}})$$

as a second order approximation. $\mathbf{H}_{\hat{a}}$ is the Hessian matrix (4.2.55) taken for the estimates $\hat{\mathbf{a}}$. Hence, assuming an a posteriori normal pdf for the parameters (as random variables) one can take approximately

$$p_{\mathbf{a}}(\mathbf{a}|\hat{\mathbf{a}}(\mathbf{v})) \propto \exp[-\frac{1}{2}(\mathbf{a}-\hat{\mathbf{a}})^T \mathbf{H}_{\hat{a}}(\mathbf{a}-\hat{\mathbf{a}})]. \tag{4.6.1}$$

(4.6.1) is a conditional normal pdf with mean $E\{\mathbf{a}\} = \hat{\mathbf{a}}$ and covariance matrix

$$cov\{\mathbf{a}\} = \mathbf{H}_{\hat{a}}^{-1} = \left[\left. \frac{\partial \mathcal{J}_2(\mathbf{a})}{\partial \mathbf{a} \partial \mathbf{a}^T} \right|_{\mathbf{a}=\hat{\mathbf{a}}} + \mathbf{G}_e \right]^{-1}. \tag{4.6.2}$$

If (4.6.1) is applicable, the quadratic form

$$Q(\mathbf{a}) = (\mathbf{a}-\hat{\mathbf{a}})^T \mathbf{H}_{\hat{a}}(\mathbf{a}-\hat{\mathbf{a}})$$

is distributed as χ^2 with J degrees of freedom (J the total number of parameter estimates). For the parameter values $\mathbf{a}$ to be a posteriori with probability γ within the region (γ-confidence region) $Q(\mathbf{a}) \leq c$ one obtains

$$P[Q(\mathbf{a}) \equiv (\mathbf{a}-\hat{\mathbf{a}})^T \mathbf{H}_{\hat{a}}(\mathbf{a}-\hat{\mathbf{a}}) \leq c] = \gamma, \tag{4.6.3}$$

with $\gamma = P(\chi^2|J)|_{\chi^2=c}$ from the χ^2-distribution with J degrees of freedom.

For a pre-given γ (posterior probability), the confidence regions for each single parameter (as a random variable) can be estimated by

$$\hat{a}_r - \Delta a_r \leq a_r \leq \hat{a}_r + \Delta a_r, \tag{4.6.4}$$

where

$$\Delta a_r := \sqrt{\frac{c}{(\mathbf{H}_{\hat{a}})_{r,r}}}, \quad r = 1(1)J. \tag{4.6.5}$$

From Eq. (4.6.2) one can obtain the associated correlation matrix for the parameters. The statistical correlation between the parameters a_r and a_s is given by the correlation coefficient

$$\rho_{rs} := \frac{(\mathbf{H}_{\hat{a}}^{-1})_{rs}}{(\mathbf{H}_{\hat{a}}^{-1})_{rr}^{\frac{1}{2}} (\mathbf{H}_{\hat{a}}^{-1})_{ss}^{\frac{1}{2}}}. \tag{4.6.6}$$

Here the diagonal elements of the inverse Hessian matrix are obviously taken for the variances of the parameters.

When applying the formulas above, one should take into account that the inverse Hessian matrix is in general only a lower bound for the covariance matrix (Rao-Cramér inequality) [126].

In a similar manner one can estimate the variances of the parameter estimates [126] by taking into account several different measurement data sets (samples) for the parameter estimation. For the covariance matrix of the parameter estimates, with the usual assumptions concerning the measuring errors [126, 174] one obtains

$$\mathbf{V}_{\hat{a}} = cov\{\hat{\mathbf{a}}\} = \mathbf{H}_{\hat{a}}^{-1} \left[\left. \frac{\partial \mathcal{J}_2(\mathbf{a})}{\partial \mathbf{a} \partial \mathbf{a}^T} \right|_{\mathbf{a}=\hat{\mathbf{a}}} \right] \mathbf{H}_{\hat{a}}^{-1}. \tag{4.6.7}$$

If the influence of the weighting matrix $\mathbf{G}_e$ on the (at least asymptotically unbiased) parameter estimates $\hat{\mathbf{a}}$ is negligible, and assuming that the sampling distribution is (approximately) normal, one can also estimate confidence intervals for the distances between the parameter estimates and the *true* values like (4.6.3) [126, 174]. (If the sampling distribution cannot be assumed to be normal, the confidence intervals have to be estimated by means of Chebychev's inequality.)

4.6.2 Dynamic Models

The resulting mathematical model dependent on a discrete life time $\theta_i =$ const. serves for the simulation and prediction of dynamic quantities including symptoms of condition. These results serve for the studies of causes of possible faults and therefore are the basis for diagnosis. The deviations resulting from the comparison of the corresponding quantities of the previous mathematical models with the recently adjusted mathematical model, together with the damage catalogue and the quantities which lead to it, serve for assessment purposes. The significance of the resulting deviations in the model and the derived quantities, and the severity of the located faults resulting from assessment then lead to a decision as mentioned in the Introduction and described in Chap. 4.

Fig. 3.1 describes the knowledge flow schematically. It is noted that the loading may have changed during the life time. The causes can be due to modified forces. This fact is also shown in Fig. 3.1. If necessary, weak point analysis, sensitivity studies etc. have to be revised.

4.6.3 Static Models

As already mentioned, if, for example, dependent on the problem under consideration, stress distributions with their maximum values are decisive, then an additional static model has to be included in the investigation. An interface to the static model (FE program) should therefore be provided in the knowledge-based system (program).

Often it is unnecessary to correct the total static model. This is the case if local damage has to be considered. The corresponding submodel must be corrected in detail while the remaining model may be corrected globally if necessary. The advantage is obvious: only a few measurements are necessary and the expenditure is relatively small. This procedure, however, must be developed and checked in each particular case.

4.6.4 Trend

After some applications of model-supported diagnostics, several life time-dependent adjusted mathematical models $\mathcal{M}(\theta_i)$, $i = 1(1)N_1$, are available which permit the performance of a trend analysis by establishing $\mathcal{M}(\theta_{N_1+1})$. This means that a fault, or damage evolution model, can be obtained in addition. Regression analysis combined with extrapolation can be used. If a sufficiently large number of measurements are available, some type of pdfs can be introduced. The prior knowledge has to be taken into account, of course. This subject is discussed in Sect. 2.3. In brief, except when rapid changes occur in the characteristics, the Weibull, Fréchet, and Pareto type models are suitable methods for trend prediction (model of evolution), and for calculating the shape coefficient γ as illustrated in Fig. 2.24 and given in Eq. (2.3.27). The trend of the parameters can be calculated with Eq. (2.3.30), and the symptom trend by (2.3.60) for periodic observations. The results allow the user inspection on request, instead of uneconomical continuous or periodical inspection.

As already mentioned, the model-based procedures can be understood and formulated as self-learning. One approach is to formulate them recursively. One part of self-learning is the adjustment, the second part of self-learning is the trend prediction based on performing $\mathcal{M}(\theta_{N_1+1})$.

4.7 Summary

Model supported diagnosis is based on model adjustment via parameter estimation . This means that the model describing the previous state condition is corrected in such a way that the recent state condition will now be represented. Systematic measurement errors have to be avoided or detected and removed. Irregular measurement errors are modelled stochastically. Then estimators have to be applied in order to obtain the corrected, adjusted parameter values including their covariances.

This part of system identification is described with its application to diagnosis. It is emphasized that prior knowledge is essential in order to deal with a parameter estimation problem with a minimum number of parameters to be estimated. Submodelling and the application of the extended weighted least squares is recommended. Additionally, practical hints are given (algorithms, regularization, measurements, various models) for obtaining success in adjustment.

The associated problem of connectivity, which is under research, is not discussed explicitly. On one hand, it concerns the mathematical properties of the related matrices including their minors [175, 176] for reconstruction, and on the other hand – excluding connectivity identification – it concerns the preservation of the connectivities in the model during parameter estimation. This is not evident, since, for example in FE-modelling, the parameter matrices are a priori summands of element matrices, and estimation in the LS sense will smear various physical effects dependent on the quality criterion (residuals) used.

In the last section preparations are made for the use of the resulting knowledge base. This concerns the errors of the estimates, the ability and use of the dynamic and static models, and trend prediction.

CHAPTER 5

Decision, Assessment, and System Modification Procedures

Again Fig. 3.1 is the starting point. Therefore it is repeated here as Fig. 5.1, and the decision problems are outlined in bold type. As can be seen from the decision diamonds, four types of decisions have to be made:

1. the first type of decision is concerned with the weak point observations by deciding whether certain pre-given thresholds are exceeded by observed symptoms,
2. the second type of decision is based on the significance test of changes with respect to particular values of symptoms as functions of life time,
3. the third type of decision problem consists of the significance test of changes of parameter estimates. Model structure modifications are obviously significant, and
4. the fourth type of decision concerning action required has to be made under risk, dependent on the previous assessment.

Rapoport [177] distinguished the following types of decision:

- decisions under uncertainty
- decisions under risk, and
- multi-objective decisions.

Decisions under uncertainty, if modelled probabilistically, are characterized by the unknown relative likelihoods with which the various states of nature are obtained at the time. Additionally, a finite number of alternatives are given which are the basis of the outcome of the decision as a course of action among these alternatives. However, the adjusted mathematical model with its estimates also gives the information on the covariances of the estimates. Consequently, significance tests taking into account confidence intervals can be performed. Additionally, the dynamic responses (for example accelerations, stresses) can be calculated with the corresponding errors.

Decisions under risk require the knowledge of pdfs: a clear formulation of the sample space with the a priori probability. The decision tree for sequential decisions under probable information is well known. Decisions and actions also are taken under unequality; this means that different beneficiaries have to be taken into account, and that, in addition to the intended consequences, previously unknown and not intended consequences often appear. Risk here is defined as the product of the damage measure and the probability of the occurrence of the damage.

Multi-objective decision: if multi-objective criteria of values are given, the preference relation is sought. It follows an estimation (optimization) procedure.

These topics are briefly introduced in the following. Instead of a deep and broad discussion of decision-making, the tools will be provided for our purposes, and some illustrations of their use will be given. Finally, the application of fuzzy logic in decision-making will be mentioned.

One action resulting from decision-making may be to modify the system. Design engineers generally know where and how to change the construction with respect to the requirements. But these requirements have to be formulated, and within model-based diagnosis there exist a powerful knowledge base which, of course, should be used for the purpose of system modification. Consequently, modification procedures are discussed at the end.

5.1 Fundamentals of Decision

The testing of hypotheses is the basis of decision-making. The simplest way is to test the exceedance of a threshold. The threshold can be chosen (problem-dependent) deterministically or statistically (e.g. multiples of a standard deviation). If statistical hypotheses have to be tested, we ask for the significance level. The application of fuzzy logic also results in a measure which serves the purpose of decision-making. The reader is reminded that decision-making concerns, or is based on, significant deviations of chosen features (scalar or vector) from their previous values. Within parameter estimation significant modifications of the estimates dependent on the life time are connected with confidence intervals and permissible deviations.

5.1.1 Pseudo-deterministic Decision

In addition to the remarks made in the Introduction we must remember that the features we consider consist of symptoms $S(\theta)$, i.e. quantities sensitive to expected failures. We also have to take into account alert-thresholds: thresholds which may be tolerated, as well as those which cannot be tolerated [178]. Economic and safety thresholds cannot be forgotten. The sensitivity of the symptoms is decisive for various early warning-times. This will be demonstrated in Sect. 5.2.

Theoretical studies, for example, can result in a fault classification (Fig. 5.2). This shows the concept of a test using two symptoms. First a check has to be made as to where the point is located (area). If it lies outside the healthy area (allowable range) it has to be classified (fault 1 or fault 2). Generally, the centre of gravity points of the related sets are computed and compared with the corresponding points of simulated, historical failure data or previous data. Distances then serve as a basis for decision.

Another measure has already been introduced: it is the local and global damage measure defined in Sect. 2.3.2.

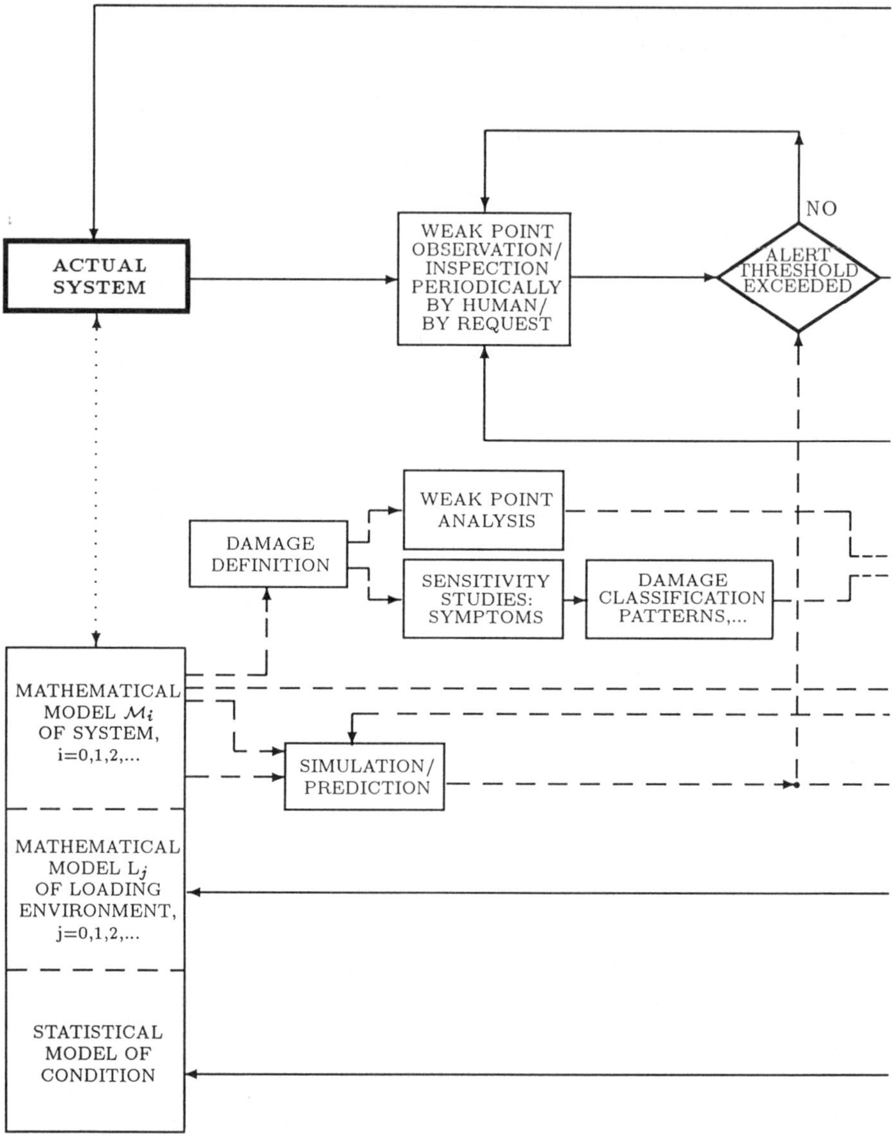

Fig. 5.1. Flow-chart of model-supported diagnosis emphasizing the decision problem included

The health index [179] is a very simple measure. It is defined as the sum of the squares of residuals, each consisting of a current parameter value und its nominal value, and divided by a corresponding warning deviation of the related parameter. It is a quadratic loss function (global) measure which emphasizes large deviations:

$$h_i := [(p_i - p_{inom})/(\Delta_{pi})]^2, \quad h := \frac{1}{N}\sum_{i=1}^{N} g_i h_i, \tag{5.1.1}$$

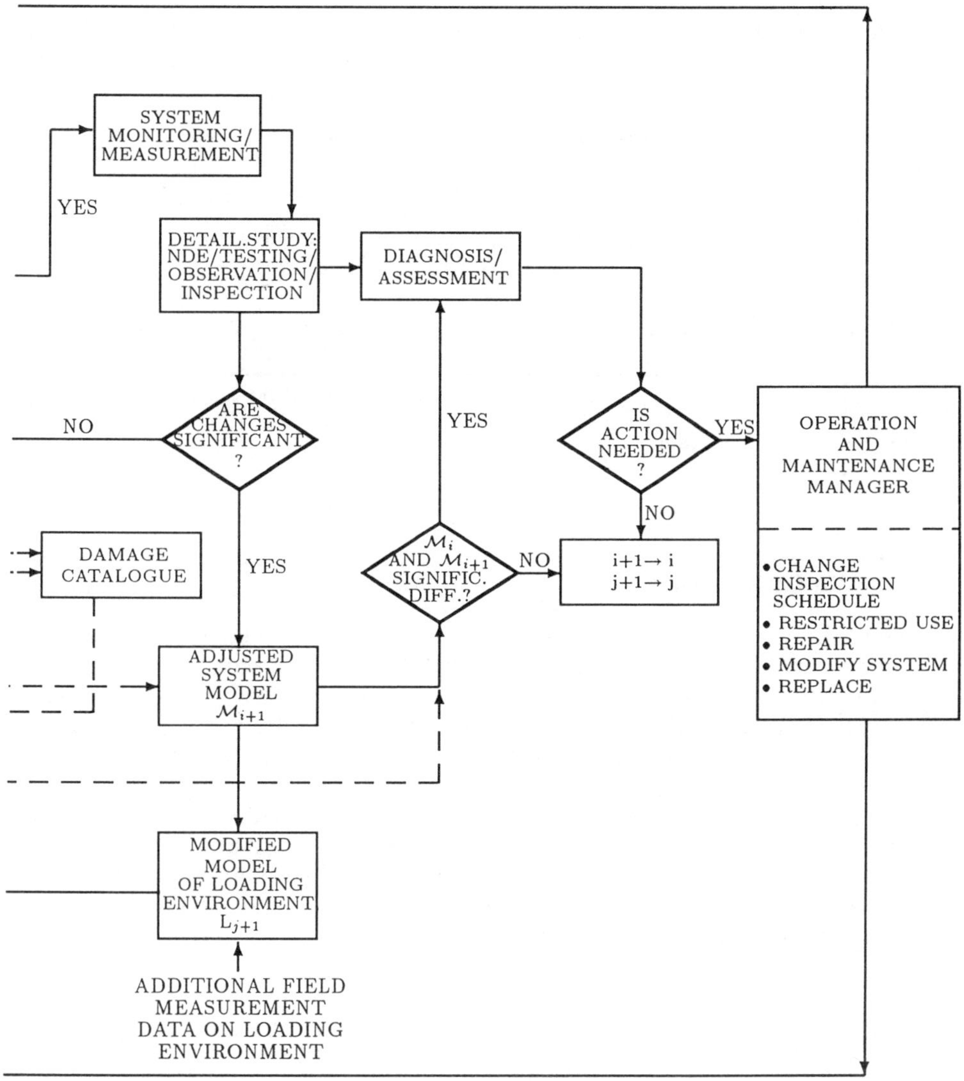

Fig. 5.1. (continued)

g_i is a suitably (with respect to critical symptoms) chosen weighting. As can be seen, h_i and/or h can be plotted against life time, and its limit value can also be assessed.

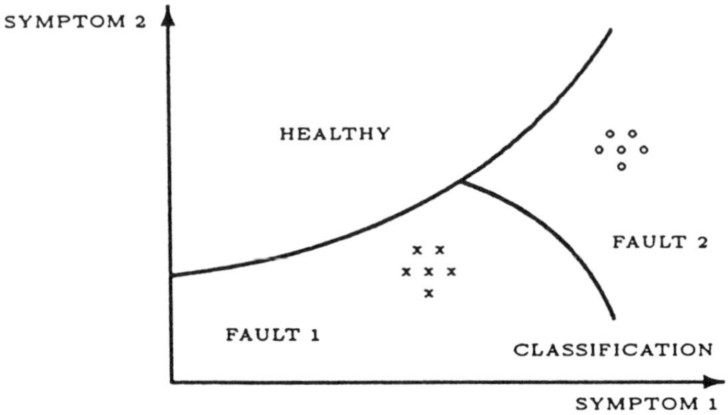

Fig. 5.2. Fault classification

5.1.2 Bayesian Decision Theory

The following discussion is close to that given in [180]. One measure of the performance of a decision is the probability of making an incorrect decision; this is also called the probability of error. We will follow a heuristic approach. Measured features are designated by $\mathbf{y}^m$. They can be symptoms (e.g. dynamic responses) assembled in a vector. The feature classes are denoted by ω_i, $i = 1, 2, ..., n_\omega$. They are defining assumed hypotheses. We will restrict ourselves to two hypotheses: good and faulty conditions. They will be designated by $\omega_1 = \omega_G$ and $\omega_2 = \omega_F$. The measured vector $\mathbf{y}^m$ has to be classified. The posterior probability of each class is $P(\omega_i|\mathbf{y}^m)$. For the two classes chosen, the decision rule using Bayes rule (see footnote 7 in Chap. 4) is defined as

$$\frac{p_{y|\omega_G}(\mathbf{y}^m|\omega_G)P(\omega_G)}{p_y(\mathbf{y}^m)} \begin{cases} > & \frac{p_{y|\omega_F}(\mathbf{y}^m|\omega_F)P(\omega_F)}{p_y(\mathbf{y}^m)} \quad \text{choose } \omega_G, \\ \text{else} & \quad \text{choose } \omega_F. \end{cases} \tag{5.1.2}$$

The weighting of the posteriori conditional pdfs is done with the prior probabilities $P(\omega_i)$. Consequently, the decision rule can be formulated as the likelihood ratio

$$l(\mathbf{y}^m) := \frac{p_{y|\omega_G}(\mathbf{y}^m|\omega_G)}{p_{y|\omega_F}(\mathbf{y}^m|\omega_F)} \begin{cases} > & \frac{P(\omega_F)}{P(\omega_G)} \quad \text{choose } \omega_G, \\ < & \frac{P(\omega_F)}{P(\omega_G)} \quad \text{choose } \omega_F. \end{cases} \tag{5.1.3}$$

The equality sign which defines the decision boundary can be chosen dependent on the problem for the one class or the other (hypothesis).

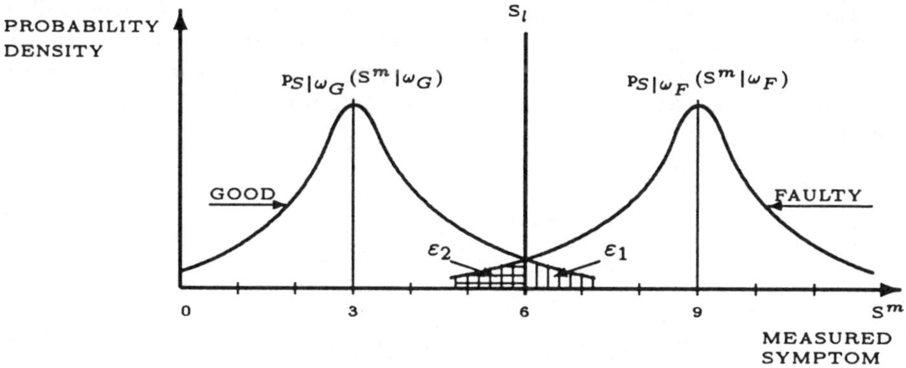

Fig. 5.3. Class conditional densities and decision regions

The following example is given to clarify the meaning of this ratio test. A scalar symptom measurement S^m at a pre-given time θ of the symptom S has the conditional pdfs

$$p_{S|\omega_G}(S^m|\omega_G) = \frac{1}{\sqrt{2\pi}} e^{(S^m-3)^2/2},$$

$$p_{S|\omega_F}(S^m|\omega_F) = \frac{1}{\sqrt{2\pi}} e^{(S^m-9)^2/2}.$$

The densities are shown in Fig. 5.3. In addition, we assume that the two classes are equally likely:
$P(\omega_G) = P(\omega_F) = 1/2$. Then Eq. (5.1.3) gives

$$l(S^m) = \frac{e^{(S^m-3)^2/2}}{e^{(S^m-9)^2/2}} \begin{cases} > & 1, \text{ choose } \omega_G, \\ < & 1, \text{ choose } \omega_F. \end{cases}$$

The decision can be simplified by taking the ln-operation:

$$(S^m - 3)^2 - (S^m - 9)^2 = 12S^m - 72 \begin{cases} > & 0, \text{ choose } \omega_F, \\ < & 0, \text{ choose } \omega_G. \end{cases}$$

The symptom limit value is $S_l = 6$, and the decision rule gives (see Fig. 5.3)

$$S^m \begin{cases} > & 6, \text{ choose } \omega_F, \\ < & 6, \text{ choose } \omega_G. \end{cases}$$

In the example a symmetrically located intersection point $S_l = 6$ exists which is the case when the prior probabilities are equal. Generally, the intersection as the decision boundary is defined by the equality of the weighted class conditional pdfs of Eq. (5.1.2):

$$p_{y|\omega_G}(y_1|\omega_G)P(\omega_G) = p_{y|\omega_F}(y_1|\omega_F)P(\omega_F). \tag{5.1.4}$$

The performance of a decision rule can be measured by its probability of error ε: although the measurements $\mathbf{y}^m$ require the choice of, for example, ω_G, the other region is chosen due to calculation (with uncertain measurements). Various formulations can be chosen for the error probability:

$$P(\varepsilon) = \int_{-\infty}^{\infty} P(\varepsilon|\mathbf{y}) p_y(\mathbf{y}) d\mathbf{y}, \tag{5.1.5}$$

or

$$P(\varepsilon) = P(\varepsilon|\omega_G)P(\omega_G) + P(\varepsilon|\omega_F)P(\omega_F). \tag{5.1.6}$$

The probabilities of error for each class are

$$\varepsilon_i := P(\varepsilon|\omega_i), \ \{1,2\} = \{G, F\}. \tag{5.1.7}$$

For the two classes considered they are defined as

$$\varepsilon_G := P(\varepsilon|\omega_G) = P(\text{choose } \omega_F|\omega_G) = \int_{R_F} p_{y|\omega_G}(\mathbf{y}|\omega_G) d\mathbf{y}, \tag{5.1.8}$$

$$\varepsilon_F := P(\varepsilon|\omega_F) = P(\text{choose } \omega_G|\omega_F) = \int_{R_G} p_{y|\omega_F}(\mathbf{y}|\omega_F) d\mathbf{y}. \tag{5.1.9}$$

In the example it holds true: for the symptom $y = S\epsilon(-\infty, \infty)$ and the limit value S_l: $\omega_G\epsilon(-\infty, S_l)$, $\omega_F\epsilon(S_l, \infty)$. It follows

$$\begin{aligned} P(\varepsilon) &= \varepsilon_G P(\omega_G) + \varepsilon_F P(\omega_F) \\ &= P(\omega_G) \int_{R_F} p_{y|\omega_G}(\mathbf{y}|\omega_G) d\mathbf{y} + P(\omega_F) \int_{R_G} p_{y|\omega_F}(\mathbf{y}|\omega_F) d\mathbf{y}. \end{aligned} \tag{5.1.10}$$

The class error probabilities of the previous example are shaded in Fig. 5.3.

It can be shown that these decision rules are optimum in the sense of minimizing the error probability or quantities related to them. For this purpose it is useful to introduce costs (cost functionals, loss functionals). The cost is denoted by J_{ij} if it is decided on ω_i when $\mathbf{y}^m$ is actually from $\omega_j, i \neq j$. The cost deriving from an incorrect decision is assumed to be greater than that of making a correct decision,

$$J_{ij} > J_{jj}. \tag{5.1.11}$$

The Bayes risk $\mathcal{R}$ is defined as the expectation of the total cost, consisting of the sum of the weighted costs, where the weightings are the probabilities of deciding ω_i **and** $\mathbf{y}$ from ω_j. With the abbreviations introduced previously, mathematical manipulations lead to

$$\mathcal{R} = P(\omega_G)J_{GG} + P(\omega_F)J_{FF} + P(\omega_G)(J_{FG} - J_{GG})\varepsilon_G + P(\omega_F)(J_{GF} - J_{FF})\varepsilon_F. \tag{5.1.12}$$

Binary costs,

$$J_{ij} = \begin{cases} 0 \text{ if } i = j, \\ 1 \text{ if } i \neq j, \end{cases} \tag{5.1.13}$$

lead to (5.1.11) [180] by setting $\mathcal{R} = P(\varepsilon)$.

5.1.3 Neyman-Pearson Rule

One of the class error probabilities is fixed within this rule. However, starting with the Bayes risk mentioned in the previous section, minimization of the error probability again yields the likelihood ratio test for hypothesis testing.

Let us assume that

$$\varepsilon_F = \int_{R_G} p_{y|\omega_F}(\mathbf{y}|\omega_F)d\mathbf{y} = c \tag{5.1.14}$$

holds true. The value c is the probability of observing a point $\mathbf{y}$ within the class ω_G, assuming that the hypothesis concerned is true. With c the second error probability ε_G will be minimized:

$$\varepsilon_G = \int_{R_F} p_{y|\omega_G}(\mathbf{y}|\omega_G)d\mathbf{y}. \tag{5.1.15}$$

The minimization of ε_G, for example, means minimizing the number of breakdowns of a system (decision GOOD when the system is getting near to becoming FAULTY). The minimization of (5.1.15) subject to the constraint (5.1.14) is equivalent to minimizing

$$J_\varepsilon = \varepsilon_G + \lambda(\varepsilon_F - c), \tag{5.1.16}$$

where λ is the Lagrange multiplier. Substitution of the related integrals (5.1.15) and (5.1.14), taking into account

$$\int_{R_G} p_{y|\omega_i}(\mathbf{y}|\omega_i)d\mathbf{y} + \int_{R_F} p_{y|\omega_i}(\mathbf{y}|\omega_i)d\mathbf{y} = 1,$$

leads to the minimization of an integral with the integrand

$$g(\mathbf{y}) = \lambda p_{y|\omega_F}(\mathbf{y}|\omega_F) - p_{y|\omega_G}(\mathbf{y}|\omega_G).$$

Consequently, for the minimization of the integral the region R_1 must be defined where $g(\mathbf{y})$ is negative. The decision rule

$$l(\mathbf{y}^m) = \frac{p_{y|\omega_G}(\mathbf{y}|\omega_G)}{p_{y|\omega_F}(\mathbf{y}|\omega_F)} \begin{cases} > & \lambda \text{ for } \mathbf{y} \epsilon R_G \\ < & \lambda \text{ for } \mathbf{y} \epsilon R_F \end{cases} \tag{5.1.17}$$

then follows. The determination of λ, which is called the threshold of the decision, can be done as follows: the ratio (5.1.17) is a function of the random variable $\mathbf{y}$, and consequently $l(\mathbf{y})$ is a random variable. The decision regions $R_{1,2}$ in the $\mathbf{y}$-space are mapped into the intervals (λ, ∞) and $(-\infty, \lambda)$ in the l-space, respectively. The condition (5.1.14) can therefore be written in the form

$$c = \int_{\lambda}^{\infty} p_{l|\omega_F}(l|\omega_F)dl. \tag{5.1.18}$$

This is an implicit equation in l.

Comparison of (5.1.17) with the likelihood ratio (5.1.3) shows the essential difference between the two rules. The Neyman-Pearson rule uses the probability c based on the error probabilities rather than the class probabilities.

Only the probability $P(\omega_G)$ and an approximation of the pdf $p_{S|\omega_G}(S|\omega_G) = p_G(S)$ are available for symptom condition monitoring. It is thus impossible to calculate the risk function (5.1.12) due to a lack of information. Taking into account the decision error (5.1.10), we can minimize the probability of the system's breakdown,

$$P(\omega_F) \int_{R_G} p_{S|\omega_F}(S|\omega_F)dS = min, \tag{5.1.19}$$

which means that the limit value S_l should be chosen so as to fulfil Eq. (5.1.19). This requirement means we will "pay for it" when stopping and repairing, even if the system is in a GOOD condition. So we allow ourselves to commit the error of the wrong classification of a GOOD condition as being FAULTY on a probability level A. It is

$$A = P(\omega_G) \int_{R_F} p_{S|\omega_G}(S|\omega_G)dS. \tag{5.1.20}$$

Transforming this equation in the previous simplified notation (see Sect. 2.3.5 and Fig. 5.4) for the symptom S and the symptom limit value S_l, it follows

$$A = G \int_{S_l}^{\infty} p_G(S)dS, \tag{5.1.21}$$

where A is the allowable probability of needless repair, G is the availability of the systems in operation/service, and $p_G(S)$ is the pdf with respect to the GOOD condition. It must be remembered that the above integral is nothing other than the previously defined symptom reliability (see Eq. (2.3.44)):

$$R(S) = \int_{S} p_G(s)ds,$$

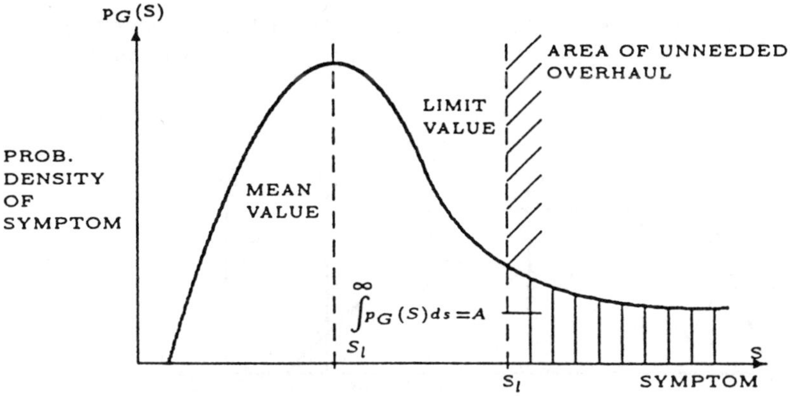

Fig. 5.4. Illustration of Neyman-Pearson rule for vibration condition monitoring

with s the dummy variable for S. Hence, the final relationship for the symptom limit value calculation is (see Eq. (2.3.55))

$$A = G \int_{S_l}^{\infty} p_G(S) dS = GR(S_l). \tag{5.1.22}$$

In diagnostics [24, 95] the probability of a wrong classification of a GOOD condition is usually fixed at some low level, for example $A = 0.05$. It means that one is ready to pay for needless repairs to 5% of the machinery stock when a wrong assessment is made, in order not to permit the breakdown of a system in operation. The allowable probability of wrong classification A is illustrated in Fig. 5.4 as the tail of the shown probability density of a GOOD condition $p_G(S)$.

5.1.4 Fuzzy Logic and Decision

Fuzzy sets can be useful in making these decisions, because very often the probability information is not available. Then the membership of, for example, symptoms has to be considered. As examples, the recent papers [18] and [19] are briefly reviewed here.

Frank [18] discussed the roles of the following four parts of a knowledge-based observer:

1. Qualitative Model – a mathematical model for the determination of the expected behaviour.
2. Discrepancy Detector – a procedure for the determination of the discrepancy between measured and calculated symptoms using quadruple membership functions.
3. Candidate Generator – a catalogue to suggest the fault candidate on the basis of discrepancies.

B = [b_1, b_2, b_3, b_4,] ⟹ CALCULATED STATES
G = [g, g, 0, 0,] ⟹ MEASURED STATES

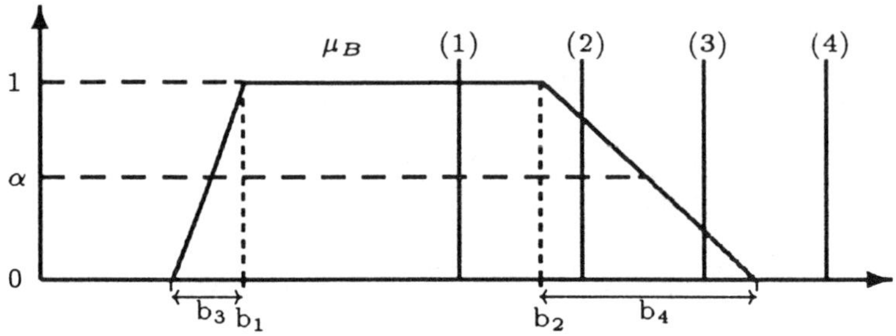

G INTERSECTS WITH **B**

(1) IF g ε [b_1, b_2,] ⟹ THEN **G** & **B** ARE "MATCHED"

(2) IF g ε [b_2, b_2+ b_4(1-α)], $\alpha \geq 0.5$, THEN "MATCHED"

(3) IF g ε [b_2+b_4(1-α), b_2+ b_4], $\alpha < 0.5$, THEN
g-(b_2 + b_4(1-α)) = TYPE I DISCREPANCY

(4) IF **G** DOES NOT INTERSECT WITH **B**, THEN
g-(b_2 + b_4(1-α)) = TYPE II DISCREPANCY

SIMILARLY FOR **g** < $\mathbf{b_1}$

Fig. 5.5. Decision-making strategy (following Frank 1994)

4. Diagnostic Strategist – an algorithm to coordinate the entire integrative search process in order to ensure that the mathematical model matches the time evolution of the symptoms of the actual process.

The observed/calculated residual symptoms are fuzzified through the use of the appropriate membership functions. A rule-based inference engine was used to obtain results, which were then defuzzified. In the decision-making strategy, a quadruple membership function as shown in Fig. 5.5 is used in [18]. In addition, an adaptive threshold using fuzzy relations is recommended in order to minimize the occurrence of false alarms.

Isermann [19] suggested calculating the mean $\bar{S}_i$ and standard deviation $\bar{\sigma}_i$ of the symptoms $S_i(t)$[1] as reference values prior to the occurrence of any fault. After a fault appears, any changes in S_i in the form of $\Delta S_i = E\{S_i(t) - \bar{S}_i\}$, $\Delta\sigma_i = E\{\sigma_i(t) - \bar{\sigma}_i\}$ can be evaluated using fuzzy sets, as shown in Fig. 5.6. In this particular example, fuzzy sets representing large decrease, small decrease, normal, small increase, and large increase

[1] The symptoms are measured at a fixed life time θ_j; consequently the notation is $S_i(\theta_j, t)$. The first argument is suppressed in the following.

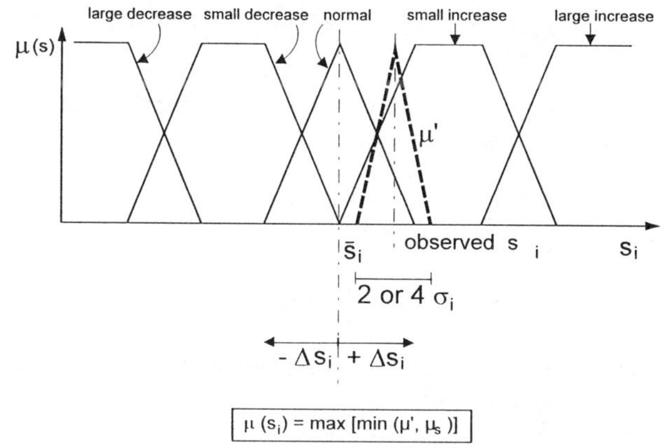

Fig. 5.6. Evaluation of observed symptom (following Isermann 1994)

are used to determine whether the observed S_i has changed. He also used a fault tree with iterative forward and backward chaining in order to determine faults.

The theory of fuzzy sets has been well developed during the past three decades (e.g. see [181, 16]). Fuzzy rules and their applications are summarized in the following.

Consider the following sample rules,

IF $S(t)$ is A_j and $\Delta S(t)$ is B_l, THEN $L(t)$ is $C_{j,l}$.
IF $S(t)$ is A_{j+1} and $\Delta S(t)$ is B_l, THEN $L(t)$ is $C_{j+1,l}$.
IF $S(t)$ is A_{j+1} and $\Delta S(t)$ is B_{l+1}, THEN $L(t)$ is $C_{j+1,l+1}$.
IF $S(t)$ is A_j and $\Delta S(t)$ is B_{l+1}, THEN $L(t)$ is $C_{j,l+1}$.

Membership functions for $A_j, A_{j+1}, B_l, B_{l+1}, C_{j,l}, C_{j+1,l}, C_{j,l+1}, C_{j+1,l+1}$ are illustrated in Fig. 5.7. We can find the membership function C_{ik} by using the following formula:

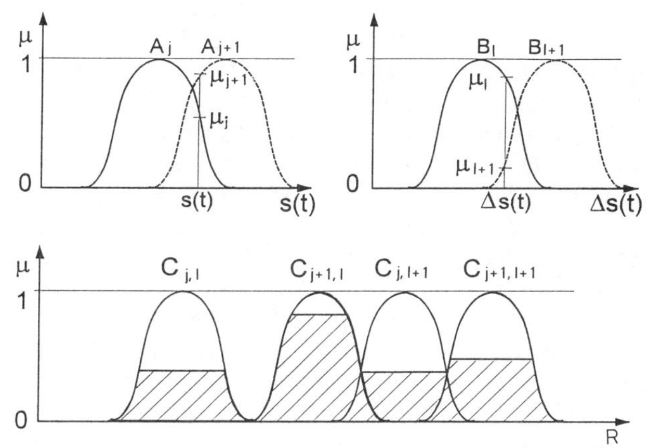

Fig. 5.7. Membership function for fuzzy rules

$$\mu_C(S) = \max_{jj=j,j+1,ll=l,l+1} \{\min[\min(\mu_{A_{jj}}(S(t), \mu_{B_{ll}}(\Delta S(t)), \mu_{C_{jj,ll}}(S)]\}. \quad (5.1.23)$$

In addition to the min/max operations, the other fundamental t-norms as listed in the following equation may be used:

$$t_S(a,b) = \begin{cases} \min(a,b), & \text{if } S=0 \\ a \cdot b, & \text{if } S=1 \\ \max(0, a+b-1), & \text{if } S=\infty \\ \log_S[1+(S^a-1)(S^b-1)/(S-1)], & \text{otherwise.} \end{cases} \quad (5.1.24)$$

The defuzzification process can use either the moment of the maximum method or the centroid of the area method.

A simple example may serve for illustration. The structural reliability function [182], $R(\theta)$, is defined as the probability of the event that the useful life θ is at least

$$R(\theta) \equiv P(\theta_b > \theta). \quad (5.1.25)$$

It can also be expressed in terms of symptoms as follows (see Eqs. (2.3.38) and (3.3.42)):

$$R_{S_b}(S^m) \equiv P(S_b > S^m) = \int_{S^m}^{\infty} p_{S_b}(S)dS, \quad (5.1.26)$$

where S_b is a random variable denoting the symptom limit state of the system, and S^m is an observed symptom. The pdf used in Eq. (5.1.26) will be equal to the pdf in Eq. (5.1.22) if the symptom readings taken for the elaboration of $p_{S_b}(S)$ are only from systems in operation.

When there are several types of symptoms, assuming that distributions of their respective limit states are available, we may calculate reliabilities R_i based on the symptom of type i, $i = 1, \ldots, n$. Different symptoms describe different faults. Consequently, the reliabilities generally have different values. A decision can then be made on the basis of these reliability measures. Frequently, however, sufficient statistical data are not available for the construction of probability density functions $p_{S_b}(S)$. In such cases, the fuzzy rules may be applied as illustrated in the following example.

Consider membership functions for A_1 ($S(t)$ Small), A_2 ($S(t)$ Large), B_1 (No change in $S(t)$), B_2 (Increase in $S(t)$, C_{11} (Good condition), C_{12} (Caution), C_{21} (Poor), and C_{22} (Failure), where $S(t)$ denotes a symptom measure at time t. Suppose that $S(t)$ is found to be $0.54S_{\max}$, such that $\mu_{A_1}(0.54S_{\max}) = 0.8$, and $\mu_{A_2}(0.54S_{\max}) = 0.2$. In addition, $\Delta S(t)$ is found to be $0.035S_{\max}$, such that $\mu_{B_1}(0.035S_{\max}) = 0.3$, $\mu_{B_2}(0.035S_{\max}) = 0.7$. Using min/max operations ($S = 0$ in t_S as given in Eq. 5.1.24, we find that $C_{11} = 0.3$, $C_{12} = 0.7$, $C_{21} = 0.2$, $C_{22} = 0.2$ as listed in Table 5.1, in which results are given for several other values of S in the t_S calculations. For this particular example (Fig. 5.8), the finally assessed condition of this system is 'caution' using various t-norms because the calculated reliability values R_I (using moment of maximum method) and R_{II} (using centroid of area method) range from 0.546 to 0.624.

Table 5.1. Calculations for the reliability and condition assessment example

	$S=0$	$S=1$	$S=e$	$S=10$	$S=\infty$
C_{11}	0.300	0.240	0.223	0.201	0.100
C_{12}	0.700	0.560	0.544	0.527	0.500
C_{21}	0.200	0.006	0.044	0.027	0.000
C_{22}	0.200	0.140	0.123	0.101	0.000
R_I	0.546	0.582	0.590	0.602	0.658
R_{II}	0.570	0.590	0.592	0.591	0.624

Notes: R_I is calculated by using the moment of maximum (MOM) method.
R_{II} is calculated by using the centroid of area (COA) method.

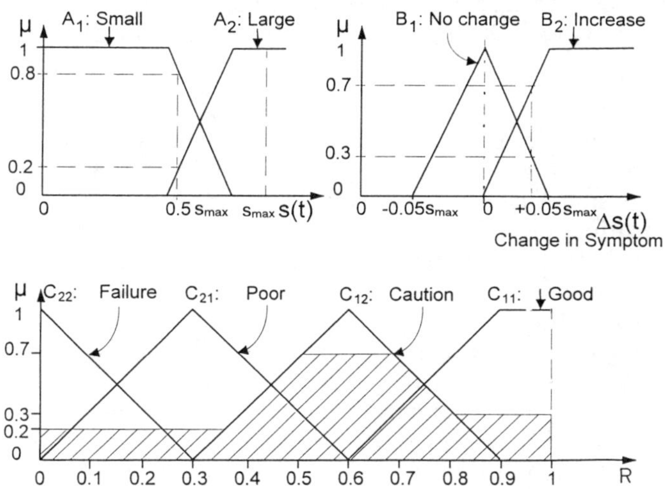

Fig. 5.8. Example of reliability and condition assessment

As was suggested in [18], fuzzy relations can be useful in obtaining adaptive alarm thresholds in order to minimize 'false' alarms. The fuzzy fault tree as applied by Isermann [19] is also a useful tool in holistic dynamics.

5.2 Symptom-based Decision

Decision-making based on symptoms can be done simply, for example by the use of thresholds and also statistically-based.

5.2.1 Deterministic Decision

Symptoms such as scalar or vectorial quantities, or discriminants, can serve for decision-making. Examples of this are

- resonance (or eigen-) frequencies f_{ref} in the form

$$\frac{f_{ref} - \Delta f}{f_{ref}}$$

  with the modification Δf with respect to a parameter modification, for example a stiffness modification $\Delta k/k$,
- spectra ratio (new spectrum/previous or reference spectrum),
- amplitude and phase or real and imaginary components of a dynamic quantity.

The choice of symptoms has already been discussed in Sect. 2.2.2. The goal is important. For example, the early warning time is essential if the symptom serve for monitoring. This effect is indicated in Fig. 5.9. The frequency of the measurement is also important. In the frequency domain the frequency modification in the time unit $\Delta f/\Delta\tau$ and the frequency resolution B are essential for the optimum mesuring frequency $(\Delta f/\Delta\tau)/B$.

An example of decision-making is shown in Fig. 5.10. The relative eigenfrequency deviation versus the relative parameter modification is plotted as a fault indicator. Figure 5.10 is self-explanatory. It should be noted that, for example, the fault evolution of structures and machinery rotors is quite different. The eigenfrequency changes of structures are often of the magnitude of $0.1 Hz/year$. With a frequency resolution of $0.125Hz$ a monitoring of $1\times$ to $3\times$ per year is optimum [183]. In contrast to these mod-

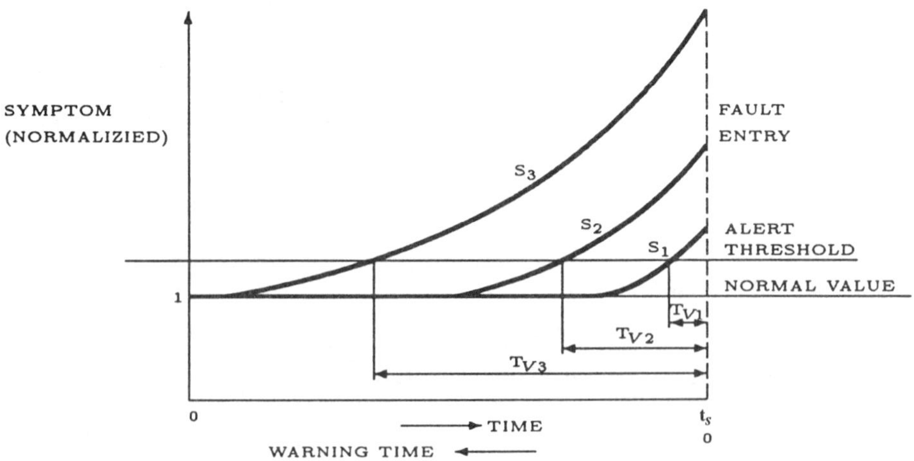

Fig. 5.9. Symptoms or fault indicators with their various sensitivities with respect to early warning times

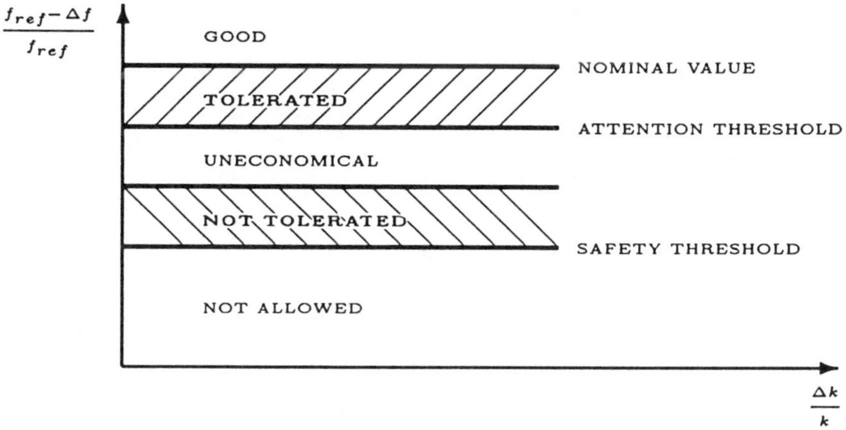

Fig. 5.10. Fault indicator (relative frequency) versus relative parameter modification

ifications, rotor damage can evolve within hours; this means that several measurements per minute are necessary etc. One has to optimize symptom sensitivity and the frequency of measurements in order to minimize the amount of data.

5.2.2 Statistical Decision Based on Histograms

One can use empirical statistics within the symptom-based decision. Figure 5.2.2 shows an example of the averaged symptom life curve $S(\theta)$ [24]. As can be seen, it is possible to obtain the pdf of GOOD condition $p_G(S)$ and the pdfs of the breakdown $p(S_b)$ and $p(\theta_b)$, even though on the basis of very few events, and the pdf of the initial symptom $p(S_0)$.

In a first approximation, assuming Gaussian distribution of the FAULTY condition, $p(S_b)$, with the corresponding standard deviation σ_{S_b} the symptom limit value S_l can then be calculated as the difference between the averaged breakdown value $\bar{S}_b$ and the standard deviation σ_{S_b} (see Fig. 5.2.2):

$$S_l = \bar{S}_b - \sigma_{S_b}.$$

As a result, theoretically we will have 16% of possible breakdowns. Of course, the symptom alarm value, for example $S_a = \bar{S}_b - 3\sigma_{S_b}$ can also be calculated, which will give 0.13% of possible breakdowns [24].

Another method for calculating the symptom limit value S_l is the use of the Neyman-Pearson rule (see Chap. 2). It is based on the minimum number of breakdowns (see Eq. (5.1.19), classification of the FAULTY condition as GOOD), having specified the allowable probability A (see Eq. (5.1.20)) of unnecessary repair (i.e. classification of a GOOD condition as FAULTY). This procedure is in use in machine condition monitoring. Examples are given in Chap. 2.

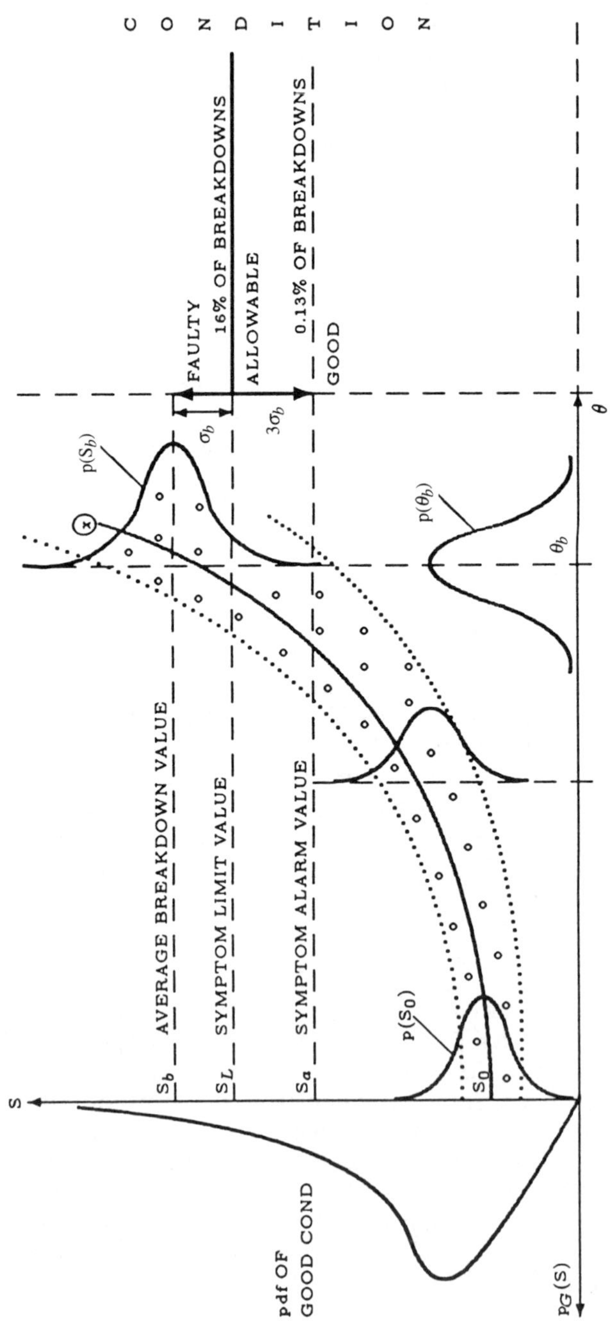

Fig. 5.11. Statistics and decisions connected with the averaged symptom life curve

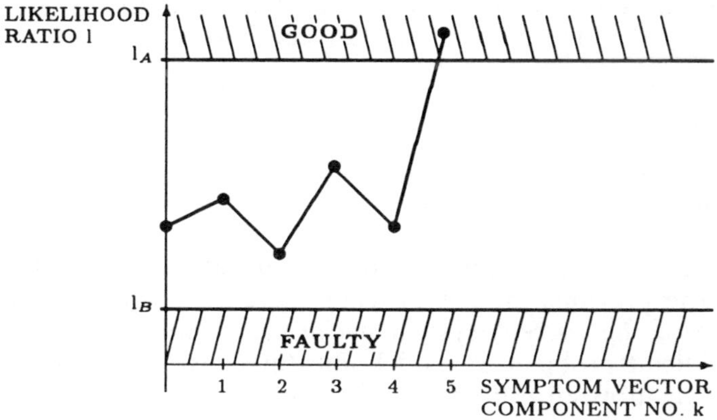

Fig. 5.12. Sequential probability ratio test using successive components of the symptom vector

An additional method will be mentioned here due to its simplicity. This is the sequential probability ratio test (SPRT). If the symptom vector components are ordered according to descending information contents, or ascending costs of measurements, the use of SPRT can be advantageous due to lower costs for the measurements. The idea behind this is the assessment of the probability ratio of a GOOD condition to a FAULTY condition for each component [24, 180]:

$$l_K(S_K^m) = \Pi_{k=1}^{K} \frac{P_{S_k|G}(S_k|\omega_G)}{P_{S_k|F}(S_k|\omega F)} \begin{cases} > A \text{ choose } \omega_G, \\ < B \text{ choose } \omega_F. \end{cases}$$

As is seen, monitoring and measurements can be stopped when the likelihood ratio exceeds the limits l_A or l_B, as illustrated in Fig. 5.12.

5.2.3 Statistical Decision Using Significance Tests

Statistical decision implies that symptoms are assumed to be random variables. The sample mean, the sample variance and moments are well known. The quality of the estimates is described by particular criteria, for example by the bias, variances etc. (see Sects. 4.2.4 and 4.6.1). Significance tests in these simple cases are based on assumed pdfs, mostly normal distributions, and considering the finite sample size, on Student's t-distribution. Relations of random quantities often follow the χ^2-distribution. For details see elsewhere, for example [12]. The probability c of error (type 1) is referred to as the significance level. Which significance level should be adopted in the special case, of course depends on the particular problem. Common values for c are 0.001, 0.01, and 0.05. A value of c between 5% and 1% is considered as *almost significant*, a value between 1% and 0.1% as *significant*, and a value below 0.1% as *highly significant*.

Significance testing is advantageous, too, when more than two classes are considered, but without a proper statistical description. For example, a good statstical description of a GOOD condition may be available, and many faults have to be distinguished in FAULTY conditions.

5.3 Model-based Decision

The models used here are models adjusted to the recent state θ_i defining the system condition. The previous models provide reference models, which allow a comparison of the current system performance or a diagnostic mode. The most difficult problem in model adjustment is to establish a usable model. This requires an estimation of variances of the parameter estimates which are small enough for the purpose.

As already stated, a dynamic model with a few DOF can represent the dominant dynamic condition of the system: a low-order model. The difficulty is to find the characteristics of the system which are sensitive with respect to the system modifications, and which allow mapping in order to construct the model. For diagnostic purposes a more complex model may be required, for example a static model with many DOF.

Tests of significance are discussed within the fundamental statistics. There are model fit tests using hypotheses testing, for example [12, 69]. In many situations, especially in the case in question, the decision would be made better by quoting confidence intervals of the parameter estimates. Often one obtains detailed information by treating it as a pure estimation problem. The estimation of parameter errors and how to perform confidence intervals is discussed in Sect. 4.6.1. Significant parameter modifications therefore can be determined and used as a basis for decision-making in the particular case.

5.4 Assessment

Assessment is now possible with the various adjusted mathematical models describing the state conditions of the system under investigation at the chosen life times. Consequently, one can use all the advantages of system analysis. Additionally, these mathematical models are verified, validated and (hopefully) usable. This is the best known knowledge base concerning the system. According to Figs. 3.1 and 5.1, these models enable the user

- to make an assessment of the state condition by simulation (maximum stresses etc.)
- to find causes in the case of significant modifications
- to make a trend prediction of the system's state condition, and
- to make decisions on further actions if necessary.

These points are dealt with in detail in the following.

5.4.1 Assessment of State Condition by Simulation

All the features of system analysis are available. Various loading conditions and various faults with their effects on dynamic and static performances can be simulated. Limit stresses, pre-given acoustic levels, limit displacements etc. lead to an assessment when compared with the simulated corresponding quantities. As already mentioned, one has to take into account the characteristics of the models used. Dynamic models are global, energy-equivalent models with generally a few DOF which are not able to investigate local properties if required. In the latter case static models have to be taken. However, if the system modification is restricted to local causes, then static subsystem modelling restricted to the submodel with the local modification is sufficient.

It is noted here that eigenquantities are generally seldom suitable for serving as symptoms. Eigenfrequencies are proportional to the energies of the respective DOF, therefore they are global quantities with some local properties (dependent on the number of DOF). The mode shapes with their local properties are not very sensitive with respect to local system modifications, and additionally, measured mode shapes are corrupted by noise. However, as already mentioned in Sect. 2.1.2, the second and third spatial derivatives of the modal shapes are very sensitive to local stiffness modifications. Consequently, they are appropriate in this sense, but one has to avoid numerical differentiation (by taking the equations of motion, see Sect. 2.1.2 and [9]). But the advantage of the eigenquantities is their use in model adjustment to obtain a validated mathematical model, which means that the structure of the model can be corrected directly by the known eigenstructure given by the identified modal quantities.

If, for example, a crack is located, then a special crack element can be introduced in the corresponding place of the FE-model [184] in order to estimate the crack depth for assessment.

5.4.2 Cause Finding

Evolving modifications of the system are detected by deviations of symptoms, states, and/or parameters. Significant deviations compared with the previous values can lead to the causes of these modifications. The causes can be due to weak points of the system, and to erroneous assumptions of the loadings. The tools for cause finding are given by prior knowledge of system analysis (including, of course, the symptoms chosen, sensitivity analysis, weak point analysis etc.) which results in the damage catalogue (see Fig. 5.1), by simulation, because various assumptions have to be checked.

In many cases the causes of failures are primary causes[2] described by patterns in the symptom domain. In such a case, pattern recognition

[2] Sometimes called root causes, i.e. principle causes which are common to a class of systems.

techniques or neural networks can be used in order to recognize on-line the emerging primary cause of faults.

In this context mention must be made of input identification as a systematic approach when the loading has changed during the system's life time. If, as already mentioned, it is possible to find a parametric model for the forcing, then the problem is reduced to parameter estimation, and the known methods of estimation can be applied.

Additionally, system parameter modifications resulting from a comparison of the recently adjusted parameters with the previously adjusted ones indicate related properties of the system responsible for the modifications. This recognition provides motives for thinking over system changes.

5.4.3 Trend Prediction

The knowledge base consisting of the various adjusted mathematical models at the different life times θ_i, $i = 1(1)N$, makes it possible to investigate the fault/damage evolution, and to establish a mathematical model for θ_{N+1} by the application of deterministic (extrapolation) or stochastical methods (see Sects. 2.3 and 4.6.4). The future state condition can also be predicted and assessed with this predicted model (see above).

Many prediction techniques can be used, even some taken from econometry like Brown's exponential smoothing, which gives satisfactory results in machine diagnostics [24]. However, the best approach is to find a proper evolutionary model of the system under consideration as an ETS. Weibull and Fréchet distributions are suitable for this purpose [185]. They result in minimum prediction error and are easy to program for computing (see Sect. 2.3, Eqs.(2.3.30) and (2.3.60)).

5.4.4 Actions

Based on the assessment, various decisions can be made. These can consist of a changed type of monitoring and shortened inspection intervals, via restricted use, right to a modification of the system and phase-out. These possible actions are indicated in the extended system identification methodology (Fig. 1.2) and in the flow-chart 5.1.

Of course, the actions are connected with costs, risks etc., and therefore require suitable organisation and management methods which, however, are outside the scope of this book.

5.5 System Modification Procedures

System modification may be required as a result of assessment and decision-making. The existing knowledge base can help to find suitable modifications. Linearized or finite sensitivity analysis thus shows which parameters of which parts (subsystems, elements) of the system produce a large effect in the desired direction of modification.

If direct modification of physical parameters due to the result of diagnosis is required, then system analysis will give the resulting dynamic response. If the diagnosis requires modified dynamic behaviour, then the parameter modifications are asked what will produce these responses. Only this problem will be discussed in the following.

As already stated, sensitivity analysis can help to find the most sensitive parameters with respect to the required response. One can obtain more information with the use of the spectral decompositions of the parameter matrices if eigenquantities are decisive. The best procedure to be applied is to follow the problem formulation itself: it is a design problem, which means an optimization problem with constraints. What is required is a minimum modification (in location and quantification) subject to pre-defined constraints. The latter procedure will be discussed only in principle.

5.5.1 Sensitivity Analysis

Differential sensitivity analysis yields amplification factors/functions with respect to model parameters which influence the state condition of the system to a great extent. Eq. (2.1.40) for example, describes the amplification factor in subsystem modelling for the eigenfrequency shift with respect to a change of the related factor a_{Ki} of the submodel stiffness. Here, however, in this example, what is required is the change of a_{Ki} with respect to a required change in the eigenfrequency λ_{0r}:

$$\Delta a_{Ki} \doteq \frac{1}{\hat{\mathbf{u}}_{0r}^T \mathbf{K}_i \hat{\mathbf{u}}_{0r}} \Delta\lambda_{0r}. \tag{5.5.1}$$

However, the coupling, and this means the effect of the modification (5.5.1) on the other eigenquantities ($\neq r$), has to be taken into account.

The reader is referred to Sect. 2.1.3, where various examples are discussed. As already stated: parametrization, for example within subsystem modelling (see Eqs. (2.1.37) or (2.1.45)) and differential analysis or finite modification investigations, including opportune modifications of the model structure, will give information on the system parts (subsystems, elements) which have essential effects on the quantities required.

5.5.2 Spectral Decomposition

If modal quantities are decisive for system modifications, then the spectral decompositions of the parameter matrices will give the answer concerning which parameter matrix elements should be changed. It may be required that

- some eigenfrequencies should be shifted,
- in order to generate a pre-defined nodal point of a mode etc.

We will restrict ourselves to the associated undamped model. Models of damped systems can be handled correspondingly in the state space formulation.

The spectral decompositions are given in Sect. 2.1, Eqs.(2.1.16), (2.1.17):

$$\mathbf{K} = \sum_{i=1}^{n} \lambda_{0i} \mathbf{M} \hat{\mathbf{u}}_{0i} \hat{\mathbf{u}}_{0i}^T \mathbf{M}, \quad \mathbf{M} = \mathbf{M}^T, \tag{5.5.2}$$

$$\mathbf{G} = \sum_{i=1}^{n} \frac{1}{\lambda_{0i}} \hat{\mathbf{u}}_{0i} \hat{\mathbf{u}}_{0i}^T, \tag{5.5.3}$$

with the normalization $m_{gi} \equiv 1$. In addition,

$$\mathbf{M}^{-1} = \sum_{i=1}^{n} \hat{\mathbf{u}}_{0i} \hat{\mathbf{u}}_{0i}^T \tag{5.5.4}$$

holds true for the inverse inertia matrix due to the orthonormalization condition of the eigenvectors.

As can be seen, pre-given modified eigenquantities determine the modifications in the parameter matrices which have to be realized in order to obtain a modified system. The explicit formulation of how the stiffness matrix has to be modified for pre-given changes of the modal quantities (2.1.23) is given by Eqs. (2.1.24) to (2.1.27).

However, this method has to be applied cautiously, because the eigenquantity modifications cannot be chosen arbitrarily (due to the requirements of operation/service). They must be consistent with the system properties from the point of view of mechanics, otherwise the pre-given quantities will result in non-physical interpretable results and/or non-realizable system modifications. Thus, for example, for beam-like systems the order of the normal modes (number of nodal points in comparison with the increasing eigenfrequencies) is not permitted to be changed, the generalized matrices must remain positive definite etc. It is hard to fulfil all the required conditions for pre-given values.

5.5.3 Minimum Modification by the Application of Optimization

Instead of direct modification one can ask for *minimum* modification subject to pre-given parts (submodels) of the system and due to dynamic requirements. The first step is the preselection of those submodels which are free for modifications. This is done by prior knowledge based on the diagnostics already done, by sensitivity analysis and, if necessary, by recalculations. The second step is then the definition of design parameters. The third step consists of the definition of the required dynamic behaviour. As stated above, the dynamic behaviour cannot be defined arbitrarily.

As can be seen, the problem is formally identical to the adjustment problem, where the measured data of the identification problem are substituted by predefined data. The measured data of existing systems automatically fulfil the requirements of dynamics and realization (existence). However, pre-given data have to be checked carefully with respect to their realizations in the particular case.

Various optimum modifications (stiffnesses, inertias, dampings) with various constraints are discussed in [186]. The solutions presented in [187] for the various formulations can be completed by the explicit solutions published by Baruch [155, 154]. The solution (4.3.57) can be supplemented by

$$\mathbf{K}_c = \mathbf{K} - (\mathbf{K}_1 + \mathbf{K}_1^T), \tag{5.5.5}$$

$$\mathbf{K}_1 = [\mathbf{I} - \frac{1}{2}(\mathbf{U}_s^T\mathbf{U}_s)^{-1}\mathbf{U}_s^T](\mathbf{K}\mathbf{U}_s - \mathbf{P}_s)(\mathbf{U}_s^T\mathbf{U}_s)^{-1}\mathbf{U}_s^T, \tag{5.5.6}$$

using static measurements: $\mathbf{U}_s$ is the measured and corrected (e.g. symmetrized) displacement matrix, and $\mathbf{P}_s$ is the force matrix. Of course, all the matrices should be composable with each other. The latter solution minimizes the Euclidean norm of $\mathbf{K}_1$. These inverse problems seem to be well-posed, because the solutions avoid inverse matrices, or they contain already generalized inverses.

If weightings are implied in the loss functions of the minimum formulations, then the weightings for the physical parameter changes have to be chosen carefully. If masses are modified, then relatively large mass modifications, assembled in the matrix $\Delta\mathbf{M}$, are allowed where large masses exist: with the Cholesky decomposition of the inertia matrix $\mathbf{M} = \tilde{\mathbf{M}}^T\tilde{\mathbf{M}}$ the minimization problem can be formulated as

$$\|(\tilde{\mathbf{M}}^{-1})^T\Delta\mathbf{M}(\mathbf{a}_M)\tilde{\mathbf{M}}^{-1}\| \rightarrow min(\mathbf{a}_M).$$

Large changes may not be significant, where large stiffnesses exist. However, if there are large masses, the weighting can be chosen as

$$\|(\tilde{\mathbf{M}}^{-1})^T\Delta\mathbf{K}(\mathbf{a}_K)\tilde{\mathbf{M}}^{-1}\| \rightarrow min(\mathbf{a}_K).$$

With regard to the weighting of the damping modification without any additional knowledge, it can be chosen with the damping matrix $\mathbf{B} = \tilde{\mathbf{B}}^T\tilde{\mathbf{B}}$:

$$\|(\tilde{\mathbf{B}}^{-1})^T\Delta\mathbf{B}(\mathbf{a}_B)\tilde{\mathbf{B}}^{-1}\| \rightarrow min(\mathbf{a}_B).$$

Since the goals by the Euclidian norm mentioned above are not independent, hierarchical optimization should be done [188]. Of course, constraints must accomplish the above aims for non-trivial solutions.

5.6 Summary

The chapter concerns decision-making. As can be seen from the table of contents, some fundamental introductions are given. Then the symptom-based decision is discussed. As a first step, decision-making uses thresholds (see Fig. 5.1). If the modifications of the symptoms are significant, then the measuring capabilities have to be used in order to check the system's integrity. This check is again the basis for decision-making to determine whether the changes of the system require an adjustment of the mathematical model to the recent state. If the resulting model fulfils the requirements of verification, validation, and if it is usable (see Sects. 1.3.1 and 4.2.2), then it permits decision-making with respect to physical model parameters. If these modifications are significant (in the meaning of statistics) and serious with respect to performance, operation/service and comfort conditions, the latter known from assessment investigations, decision-making is necessary with regard to further actions.

These decisions can be made deterministically, pseudo-deterministically, statistically or fuzzily. Which rule or method will be chosen depends on the available type of information and the particular problem under consideration.

As a result of decision-making, and due to the assessment performed, modification of the system itself can be necessary. Consequently, a systematic procedure may be required instead of a trial-and-error one. Therefore, finally, there is a discussion of various modification methods which can take constraints into account.

CHAPTER 6

Examples and Remarks on Applications

The examples discussed here will illustrate some of the approaches that have been presented in the book only in principle. We do not intend to demonstrate the model-based procedure completely or in detail.

The reader can find many academic examples with respect to damage detection and localization in the technical literature. The examples are mainly based on simulated data; some references are [189] to [198]. [189] and [190] identify the extent of a crack by various (spring) models, and [189, 184], and [142] emphasize the Kalman filter technique, while [192, 193] work with ARMA models. The application of Kalman filtering[1] has the advantage that missed state measurements are estimated in addition to the parameters. [194, 195] take into account the structure of the matrices (called connectivity) and some invariants (as additional information for regularization!).

The principles of model adjustment are discussed in Sects. 6.1 and 6.3, and the latter section contains an application from civil engineering, while applications within mechanical engineering are mentioned in Sect. 6.2. In addition to the references already cited, paper [196] will be cited in connection with model-supported diagnosis, although nonlinearities play a role here.

6.1 An Academic Example: The FE-Model of a Satellite

6.1.1 The System and its Model

The model is a beam model which is a simplified map of a satellite base structure [199]. This beam model (Fig. 6.1) will be taken as the system under consideration. It is then modelled by finite elements. This is done by beam elements as portrayed in Fig. 6.1. Each beam element is defined by two nodes orientated as described in Table 6.1. The simulated damage locations are characterized by capital letters and symbols: B for bending stiffness, T for tensile stiffness, with a square (□) for changes in lumped masses and with a circle (○) for a change in moments of inertia. The nodes of the finite element model are indicated in Fig. 6.1 with unbracketed

[1] Again it is noted that filtering/estimation techniques are averaging procedures which work against the intention of dealing with local effects. This means that modelling in general (for

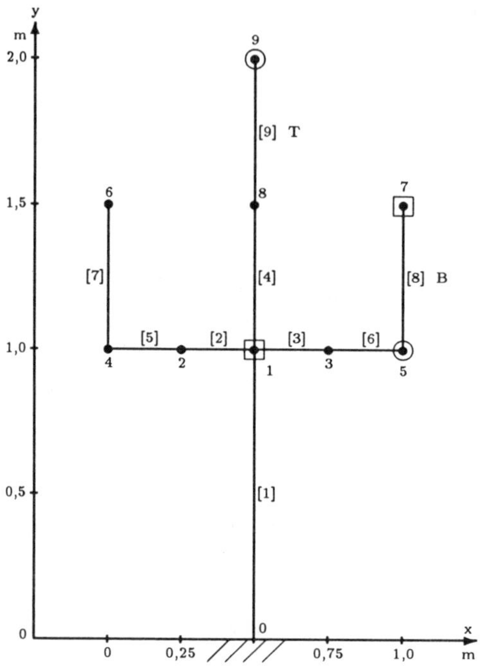

Fig. 6.1. The beam model of the satellite

Table 6.1. Orientation of beam elements in the finite element model

element	1	2	3	4	5	6	7	8	9
orientation	y	$-x$	x	y	$-x$	x	y	y	y
node numbers j, k	0,1	1,2	1,3	1,8	2,4	3,5	4,6	5,7	8,9

numbers (0–9) and the beam elements are labelled with square-bracketed numbers ([1]–[9]). The 0th node is fixed, while nodes 1 to 9 each have two translational degrees and one rotational degree of freedom, all in the plane of the system.

Each element of the nine beam elements is characterized by four physical parameters, and in order to complete the physical description of the structure with each of the nine active nodes (node 0 is fixed and therefore irrelevant) two physical parameters are associated, as shown in Table 6.2. For the ith ($i = 1(1)9$) beam element ρ_i, A_i, I_i and ℓ_i (see Fig. 6.1) represent the density, the cross-section area, the area moment of inertia and the length, while for the ith node m_i and $I_{\alpha i}$ represent the lumped mass and the rotational moment of inertia, respectively.

global identification) and subsystem modelling (for local identification) combined with prior knowledge of the system behaviour and that of damage effects are very important.

Table 6.2. Physical parameters of the initial model

i	$\rho_i[10^5\text{kg/m}^3]$	$A_i[10^{-4}\text{m}^2]$	$I_i[10^{-6}\text{m}^4]$	$m_i[\text{kg}]$	$I_{\alpha i}[\text{kgm}^2]$
1	0.5384	7.4283	369.52	1	42
2	1.051	5.7103	5.2872	180	42
3	1.401	6.8524	7.1377	165	42
4	1.077	9.2854	527.88	1	42
5	1.051	5.7103	5.2872	1	42
6	1.401	4.5682	6.609	14	1.8
7	5.045	0.1487	1.9457	39	4.2
8	10.09	0.1487	2.8198	360	37.5
9	0.0	0.08566	0.46396	12	1.0

In accordance with the generalized displacement vector (see Fig. 6.1)

$$\mathbf{u}^T := (w_{x1}, w_{y1}, \alpha_1, w_{x2}, w_{y2}, \alpha_2, \ldots, w_{x27}, w_{y27}, \alpha_{27}), \tag{6.1.1}$$

where w_{xi}, w_{yi}, α_i denote in-plane bending deflections and torsion angles, the inertia matrix of the 27-degrees-of-freedom FE-model is diagonal and consists of three additive parts:

$$\mathbf{M} = \sum_{i=1}^{9} \text{diag}(\mathbf{m}_a^i + \mathbf{m}_m^i + \mathbf{m}_I^i) \tag{6.1.2}$$

$$=: \text{diag}(\mathbf{m}_0). \tag{6.1.3}$$

$\mathbf{m}_a^i$, $\mathbf{m}_m^i$, $\mathbf{m}_I^i$ and $\mathbf{m}_0$ are (27,1)-vectors whose structure is specified as follows:

- *The rotational moment of inertia of lumped mass of the ith node:* The (27,1)-vector $\mathbf{m}_I^i$ contains the dimensional parameter $I_{\alpha i}$ in position $3i$, $i = 1(1)9$, and zeros elsewhere. That is, $\mathbf{m}_I^i = I_{\alpha i}\mathbf{e}_{3i}$, where $\mathbf{e}_n$ is the nth standard basis vector of dimension 27, which consists of zeros except for a unit in the nth component.
- *The lumped mass of the ith node:* The (27,1)-vector $\mathbf{m}_m^i$ contains the dimensional parameter m_i in positions $3i - 2$ and $3i - 1$, and zeros elsewhere. That is, $\mathbf{m}_m^i = m_i(\mathbf{e}_{3i-2} + \mathbf{e}_{3i-1})$, with $i = 1(1)9$.
- *The element mass contribution:* The index i ranges over $1 \ldots 9$, denoting the nine beam elements of the model. The ith beam element is now defined by the ordered pair of nodes with indices (j, k), $j < k$. For the nine elements the defining nodes are listed in Table 6.1. If $i = 1$ and therefore $j = 0, k = 1$, the (27,1)-vector $\mathbf{m}_a^1$ contains the dimensional parameter $\ell_1\rho_1 A_1/2$ in positions 1 and 2 and zeros elsewhere. That is, $\mathbf{m}_a^1 = \frac{\ell_1\rho_1 A_1}{2}(\mathbf{e}_1 + \mathbf{e}_2)$. For all other beam elements the (27,1)-vector $\mathbf{m}_a^i (i > 1)$ contains the dimensional parameters $\ell_i\rho_i A_i/2$ in positions

$3j-2, 3j-1, 3k-2$ and $3k-1$ and zeros elsewhere. That is, $\mathbf{m}_a^i = \frac{\ell_i \rho_i A_i}{2}(\mathbf{e}_{3j-2} + \mathbf{e}_{3j-1} + \mathbf{e}_{3k-2} + \mathbf{e}_{3k-1})$.

For each of the nine beam elements the stiffness matrix consists of two parts, representing the tensile and the bending terms

$$\mathbf{K} = \sum_{i=1}^{9} [\mathbf{K}_t^i + \mathbf{K}_b^i], \tag{6.1.4}$$

where $\mathbf{K}_t^i$ and $\mathbf{K}_b^i$ are (27, 27)-matrices associated with the ith beam element. With the functions

$$\begin{aligned} s_{1i} &= 12 \cdot E_{\text{dyn}} \cdot I_i/\ell_i^3 \\ s_{2i} &= 6 \cdot E_{\text{dyn}} \cdot I_i/\ell_i^2 \\ s_{3i} &= 4 \cdot E_{\text{dyn}} \cdot I_i/\ell_i \\ s_{4i} &= 2 \cdot E_{\text{dyn}} \cdot I_i/\ell_i \\ s_{5i} &= E_{\text{dyn}} \cdot A_i/\ell_i, \end{aligned} \tag{6.1.5}$$

of the parameters of the ith beam element, where $E_{\text{dyn}} = 7 \cdot 10^{10}$ $[\text{N/m}^2]$ denotes the Young's modulus, the stiffness matrices $\mathbf{K}_t^i$ and $\mathbf{K}_b^i$ of the ith beam element can now be defined. They depend on the orientation of the beam element. The positive x and y directions are indicated in Fig. 6.1. For node-index $k \neq 0$, define the triplet: $n_k = (3k-2, 3k-1, 3k)$. Let the ith beam element be orientated in the $+x$ direction from node j to node k. Then all the elements in the (27, 27)-matrices $\mathbf{K}_t^i$ and $\mathbf{K}_b^i$ are zero, except the elements with the row index n_j and column index n_k. The structures of the non-zero 6×6 submatrices of $\mathbf{K}_t^i$ and $\mathbf{K}_b^i$ with row and column indices (n_j, n_k) are:

$$\mathbf{K}_{t,(n_j,n_k)\otimes(n_j,n_k)}^i = \begin{pmatrix} s_{5i} & 0 & 0 & -s_{5i} & 0 & 0 \\ 0 & 0 & 0 & 0 & 0 & 0 \\ 0 & 0 & 0 & 0 & 0 & 0 \\ -s_{5i} & 0 & 0 & s_{5i} & 0 & 0 \\ 0 & 0 & 0 & 0 & 0 & 0 \\ 0 & 0 & 0 & 0 & 0 & 0 \end{pmatrix}, \tag{6.1.6}$$

$$\mathbf{K}_{b,(n_j,n_k)\otimes(n_j,n_k)}^i = \begin{pmatrix} 0 & 0 & 0 & 0 & 0 & 0 \\ 0 & s_{1i} & -s_{2i} & 0 & -s_{1i} & -s_{2i} \\ 0 & -s_{2i} & s_{3i} & 0 & s_{2i} & s_{4i} \\ 0 & 0 & 0 & 0 & 0 & 0 \\ 0 & -s_{1i} & s_{2i} & 0 & s_{1i} & s_{2i} \\ 0 & -s_{2i} & s_{4i} & 0 & s_{2i} & s_{3i} \end{pmatrix}. \tag{6.1.7}$$

If the ith beam element is orientated in the $-x$ direction, then $\mathbf{K}_t^i$ is modified by replacing s_2 by $-s_2$. If the ith beam element is orientated in the $+y$ direction, then the 6×6 non-zero submatrices of $\mathbf{K}_t^i$ and $\mathbf{K}_b^i$ become:

$$\mathbf{K}^i_{t,(n_j,n_k)\otimes(n_j,n_k)} = \begin{pmatrix} 0 & 0 & 0 & 0 & 0 & 0 \\ 0 & s_{5i} & 0 & 0 & -s_{5i} & 0 \\ 0 & 0 & 0 & 0 & 0 & 0 \\ 0 & 0 & 0 & 0 & 0 & 0 \\ 0 & -s_{5i} & 0 & 0 & s_{5i} & 0 \\ 0 & 0 & 0 & 0 & 0 & 0 \end{pmatrix}, \tag{6.1.8}$$

$$\mathbf{K}^i_{b,(n_j,n_k)\otimes(n_j,n_k)} = \begin{pmatrix} s_{1i} & 0 & s_{2i} & -s_{1i} & 0 & s_{2i} \\ 0 & 0 & 0 & 0 & 0 & 0 \\ s_{2i} & 0 & s_{3i} & -s_{2i} & 0 & s_{4i} \\ -s_{1i} & 0 & -s_{2i} & s_{1i} & 0 & -s_{2i} \\ 0 & 0 & 0 & 0 & 0 & 0 \\ s_{2i} & 0 & s_{4i} & -s_{2i} & 0 & s_{3i} \end{pmatrix}. \tag{6.1.9}$$

The first beam element is orientated in the y direction, and since its first node (index 0) is inactive, the non-zero entries of its stiffness matrix are contained in the second (3, 3)-diagonal block of Eq. (6.1.8) and Eq. (6.1.9) for $i = 1$:

$$\mathbf{K}^1_{t,n_1\otimes n_1} = \begin{pmatrix} 0 & 0 & 0 \\ 0 & s_{51} & 0 \\ 0 & 0 & 0 \end{pmatrix}, \tag{6.1.10}$$

$$\mathbf{K}^1_{b,n_1\otimes n_1} = \begin{pmatrix} s_{11} & 0 & -s_{21} \\ 0 & 0 & 0 \\ -s_{21} & 0 & s_{31} \end{pmatrix}. \tag{6.1.11}$$

In order to avoid singularities in the dynamic flexibility matrix a proportional damping matrix is defined as

$$\mathbf{C} := c_0\mathbf{M}, \quad c_0 = \ln(100), \tag{6.1.12}$$

where c_0 is chosen arbitrarily. Combining Eqs. (6.1.3) to (6.1.9) yields the dynamic stiffness matrix of the initial model

$$\mathbf{S}(j\omega) := -\omega^2\mathbf{M} + j\omega\mathbf{C} + \mathbf{K} \tag{6.1.13}$$

which completes the formulation of the finite element model of the example considered.

6.1.2 The Damaged System and the Simulation of the Dynamic Response

The damage is simulated in the FE-model, and it is assumed that this system state occurred at a particular life time without denoting it.

The model of the damaged system. The damage of the system is simulated by deterministic changes in some physical parameters of the initial model. For the particular example presented here, the damage assumed is due to an increase of inertia in nodes 5 and 9 of 40% and 10%, respectively, and due to an increase of moments of inertia in nodes 1 and 7 of 10% and 20%, respectively. Thus the matrix of inertia $\mathbf{M}^d = \text{diag}(m_0^d)$ of the model of the damaged system differs from that of the initial model (see Eq. (6.1.3)) only by the six components

$$\begin{aligned} m_{0\,3}^d &= 1.1 \cdot m_{0\,3} \\ m_{0\,13}^d &= 1.4 \cdot m_{0\,13} \\ m_{0\,14}^d &= 1.4 \cdot m_{0\,14} \\ m_{0\,21}^d &= 1.2 \cdot m_{0\,21} \\ m_{0\,25}^d &= 1.1 \cdot m_{0\,25} \\ m_{0\,26}^d &= 1.1 \cdot m_{0\,26}. \end{aligned} \tag{6.1.14}$$

Moreover, a decrease in the bending stiffness of beam element 8 of 20%, and a decrease in the tensile stiffness of beam element 9 of 10% is assumed. Therefore the stiffness matrix $\mathbf{K}^d$ of the model of the damaged system differs from that of the initial model only in two of the 18 element stiffness matrices (see Eq. (6.1.4))

$$\begin{aligned} \mathbf{K}_b^{d\,8} &= 0.8 \cdot \mathbf{K}_b^8 \\ \mathbf{K}_t^{d\,9} &= 0.9 \cdot \mathbf{K}_t^9. \end{aligned} \tag{6.1.15}$$

Using Eqs. (6.1.3) and (6.1.4) together with Eqs. (6.1.14) and (6.1.15), and using the same damping matrix (6.1.12) as within the initial model, the dynamic stiffness matrix of the model of the damaged system becomes

$$\mathbf{S}^d(j\omega) = -\omega^2\mathbf{M}^d + j\omega\mathbf{C} + \mathbf{K}^d. \tag{6.1.16}$$

This completes the definition of the model of the damaged system.

Simulation of the dynamic responses. Using the dynamic stiffness matrix of the model of the damaged system defined in Eq. (6.1.16), the response of the system in the frequency domain is given by

$$\mathbf{U}^d(j\omega) = \left[\mathbf{S}^d(j\omega)\right]^{-1} \mathbf{P}(j\omega), \tag{6.1.17}$$

where the 27-dimensional complex excitation vector $\mathbf{P}(j\omega)$ denotes the Fourier transform of the vector of the input signals. The optimum choice of the excitations is due to a method introduced by Cottin [119]. In this example only translational excitations are chosen, i.e.

$$\mathbf{P}(j\omega) = \mathbf{B}_0\widetilde{\mathbf{P}}(j\omega), \tag{6.1.18}$$

with the (27, 18)-control matrix

$$\mathbf{B}_0 = [\mathbf{e}_1, \mathbf{e}_2, \mathbf{e}_4, \mathbf{e}_5, \cdots, \mathbf{e}_{23}, \mathbf{e}_{25}, \mathbf{e}_{26}]. \tag{6.1.19}$$

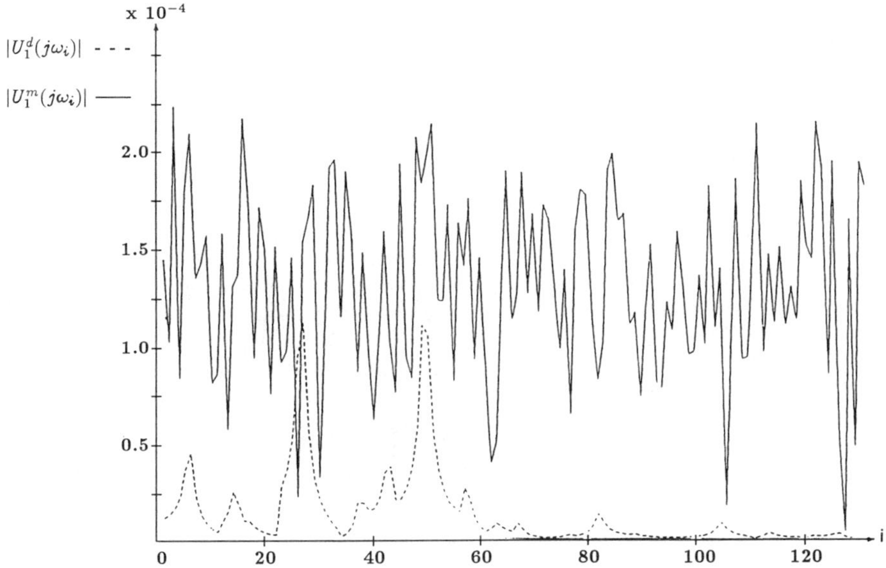

Fig. 6.2. Effect of simulated measurement errors on the first component of the response of the model of the damaged system

This restriction leads to a suboptimum excitation. In addition to each of the first 12 eigenfrequencies of the associated undamped initial model, 10 excitation frequencies in the neighbourhood of each eigenfrequency are chosen for a total of 132 excitation frequencies. The output measurement is then simulated by corrupting the responses with an additive noise of non-correlated complex random numbers with zero mean values and with a magnitude of 3 % of the maximum absolute value of the components of the response vectors, i.e.

$$\mathbf{U}^m(j\omega_i) = \mathbf{U}^d(j\omega_i) + \underbrace{S \cdot \boldsymbol{\xi}_i}_{\boldsymbol{\eta}_i}, \tag{6.1.20}$$

$$S := \frac{3}{100} \cdot \max_{\substack{l=1(1)27\\ k=1(1)132}} |U_l^d(j\omega_k)|, \tag{6.1.21}$$

$$E\{\boldsymbol{\xi}_i\} = 0, \tag{6.1.22}$$

$$E\{\Re\boldsymbol{\xi}_i \Re\boldsymbol{\xi}_i^T\} = \mathbf{I}_{27} = E\{\Im\boldsymbol{\xi}_i \Im\boldsymbol{\xi}_i^T\}. \tag{6.1.23}$$

The extreme effects of this measurement error simulation are depicted in Figs. 6.2 and 6.3. The dashed line shows the dynamic response components U_1^d and U_{26}^d, respectively, without measurement errors, while the solid line represents the same response components including the measurement errors defined above, i.e. U_1^m and U_{26}^m.

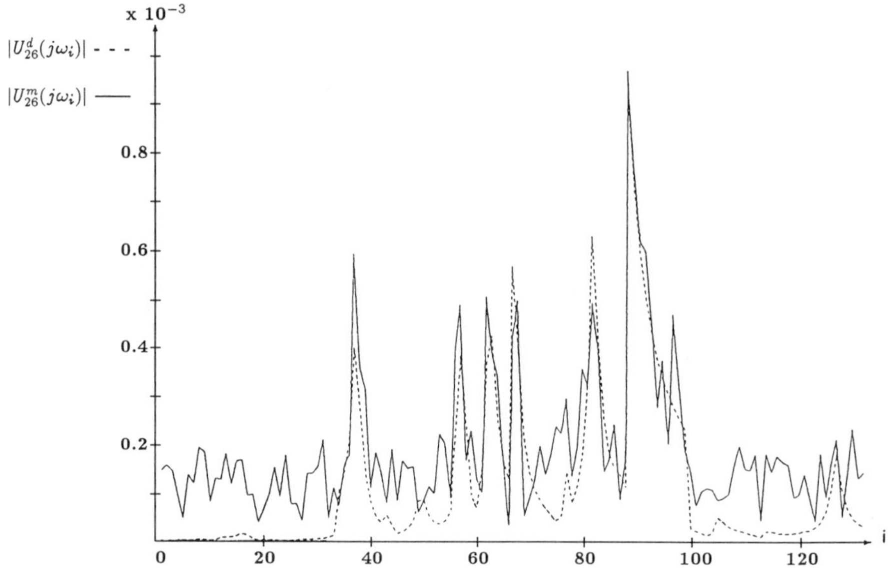

Fig. 6.3. Effect of simulated measurement errors on the last translational component of the response of the model of the damaged system

6.1.3 The Localization of Faults

The localization of the faults will be performed directly via parameter estimation and subsystem modelling. This means that we are looking for parameter estimates which correctly describe the local modification properties of the system, while the modifications of the system are required only qualitatively.

The input residuals are used together with the submodel decomposition method (see Sect. 4.3.4) for the localization, and this provides affine linear model parametrization

$$\mathbf{S}(j\omega, a) = \sum_{\alpha=1}^{r} \mathbf{S}_\alpha(j\omega)a_\alpha + \mathbf{S}^c(j\omega), \tag{6.1.24}$$

where $\mathbf{S}^c(j\omega)$ denotes a constant part of the dynamic stiffness matrix with respect to the life time of the system, i.e. it is independent of the dimensionless model adjustment parameters $\mathbf{a} = (a_1, \ldots, a_r)^T$. Within the example considered here this constant part is given by

$$\mathbf{S}^c(j\omega) = j\omega\mathbf{C} = j\omega c_0\mathbf{M}. \tag{6.1.25}$$

The submodels, represented by the matrices $\mathbf{S}_\alpha(j\omega)$, $\alpha = 1(1)r$, have been chosen as indicated in Table 6.3. They can always be defined in such a way that the following holds for the prior model (see Eq. (6.1.13))

$$\mathbf{S}(j\omega) = \mathbf{S}(j\omega, \mathbf{e}) \tag{6.1.26}$$

with the r-dimensional vector $\mathbf{e} = (1, \cdots, 1)^T$.

Table 6.3. The dynamic stiffness matrices of the submodels

α	$\mathbf{S}(j\omega)_\alpha$
1	$-\omega^2 \operatorname{diag}(m_{0\,1}\mathbf{e}_1 + m_{0\,2}\mathbf{e}_2)$
$\vdots$	$\vdots$
9	$-\omega^2 \operatorname{diag}(m_{0\,25}\mathbf{e}_{25} + m_{0\,26}\mathbf{e}_{26})$
10	$-\omega^2 \operatorname{diag}(m_{0\,3}\mathbf{e}_3)$
$\vdots$	$\vdots$
18	$-\omega^2 \operatorname{diag}(m_{0\,27}\mathbf{e}_{27})$
19	$\mathbf{K}_b^1$
$\vdots$	$\vdots$
27	$\mathbf{K}_b^9$
28	$\mathbf{K}_t^1$
$\vdots$	$\vdots$
36	$\mathbf{K}_t^9$

Extending the transformed input residual

$$\underbrace{\mathbf{S}^{-1}(j\omega, \mathbf{e})\mathbf{S}(j\omega, \mathbf{a})}_{=:\, \mathbf{Z}(j\omega, \mathbf{a})} \mathbf{U}(j\omega, \mathbf{a}) = \mathbf{S}^{-1}(j\omega, \mathbf{e})\mathbf{P}(j\omega) = \mathbf{U}(j\omega, \mathbf{e}) \tag{6.1.27}$$

for all the excitation frequencies $\omega \to \omega_i$, $i = 1(1)132$, and rewriting the result in a real form by doubling the order yields

$$\mathbf{Z}(\mathbf{a})\mathbf{U}(\mathbf{a}) = \mathbf{U}(\mathbf{e}), \tag{6.1.28}$$

where the generalized real output vectors $\mathbf{U}(\mathbf{e}), \mathbf{U}(\mathbf{a})$ of the initial model and of the parametrized model respectively are both of dimension $27 \cdot 132 \cdot 2 = 7128$. By introducing the model error vector $\mathbf{x} = \mathbf{a} - \mathbf{e}$, Eq. (6.1.28) can be rewritten yielding

$$\mathbf{Z}(\mathbf{x})\mathbf{U}(\mathbf{a}) = \mathbf{U}(\mathbf{e}) - \mathbf{U}(\mathbf{a}). \tag{6.1.29}$$

After defining the (7128, 36)-real matrix $\mathbf{A} = \mathbf{A}(\mathbf{U})$ through $\mathbf{A}(\mathbf{U})\mathbf{x} = \mathbf{Z}(\mathbf{x})\mathbf{U}$, which holds for all $\mathbf{x}, \mathbf{U}$, and after abreviating the right-hand side of Eq. (6.1.29) by $\mathbf{f}(\mathbf{U}(\mathbf{a}))$, the substitution of $\mathbf{U}(\mathbf{a})$ by the measured generalized output $\mathbf{U}^m$ which results from Eq. (6.1.20) by the same extension,

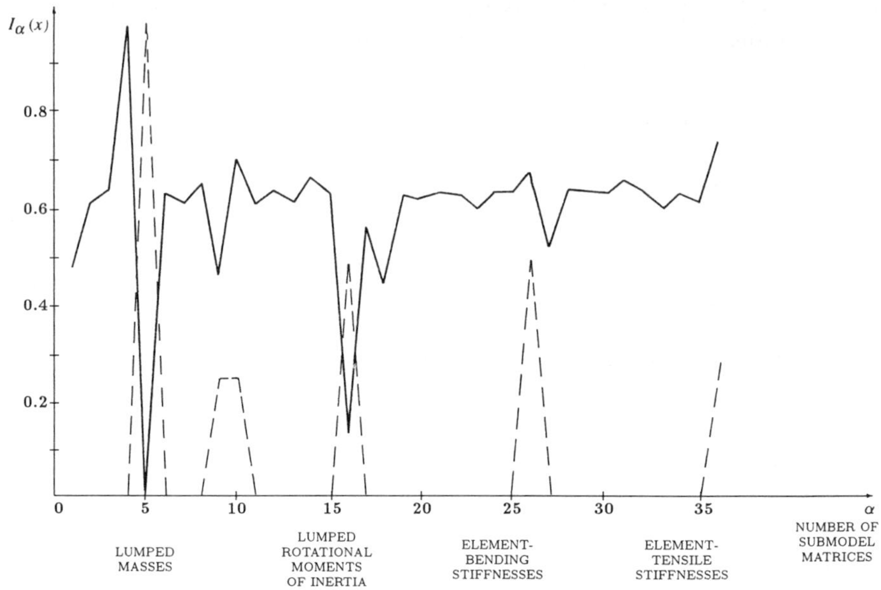

Fig. 6.4. The values of the indicator using the normal solution (see Table 6.3)

as explained above, leads to the basic equation of the inverse problem of localization

$$\mathbf{A}(\mathbf{U}^m)\mathbf{x} = \mathbf{f}(\mathbf{U}^m). \tag{6.1.30}$$

LS solution without regularization. The LS solution of Eq. (6.1.30) is called the normal solution

$$\bar{\mathbf{x}}(\mathbf{U}^m) = \mathbf{A}^+(\mathbf{U}^m)\mathbf{f}(\mathbf{U}^m), \tag{6.1.31}$$

where the superscript + denotes the Moore-Penrose inverse. Figure 6.4 shows the indicator with respect to the solution $\bar{\mathbf{x}}$, i.e. a measure of modification locations,

$$\mathcal{I}_\alpha(\mathbf{x}) := \frac{|x_\alpha|}{\max\limits_{\beta=1(1)r} |x_\beta|}, \tag{6.1.32}$$

where x_α, x_β are components of $\mathbf{x}$, calculated at $\mathbf{x} = \bar{\mathbf{x}}$.

The dashed curve is the exact solution. A comparison of the LS solution with the exact one shows that the normal solution is not usable due to the ill-conditioning of the inverse problem.

LS solution with Tikhonov-Phillips regularization. Using the Tikhonov-Phillips regularization, the regularized solution

$$\mathbf{x}_\gamma = \left[\mathbf{A}^T(\mathbf{U}^m)\mathbf{A}(\mathbf{U}^m) + \gamma\mathbf{I}_r\right]^{-1} \mathbf{A}^T(\mathbf{U}^m)\mathbf{f}(\mathbf{U}^m) \tag{6.1.33}$$

solves the modified problem

$$\min_x \left\{ \|\mathbf{Ax} - \mathbf{f}\|^2 + \gamma\|\mathbf{x}\|^2 \right\}, \tag{6.1.34}$$

where the regularization parameter γ is calculated via the Lagrange multiplicator method using pre-given error bounds[2]

$$\begin{aligned} \|\boldsymbol{\eta}\| &\leq \frac{\epsilon}{\sqrt{2}}, \\ \|\mathbf{A}(\boldsymbol{\eta})\| &\leq \frac{\delta}{\sqrt{2}}, \quad \epsilon,\ \delta > 0. \end{aligned} \tag{6.1.35}$$

The result is depicted in Fig. 6.5 with $\gamma = 2.3\ 10^{-4}$ together with the exact solution (dashed). As can be seen, the regularized solution is now better than the normal solution, but it still does not fit the demand of localization.

LS solution with regularization through data sensitivity. Figure 6.6 shows the values of the indicator using the weighted regularized solution

$$\mathbf{x}_\gamma(\mathbf{G}) = \left(\mathbf{A}^T(\mathbf{U}^m)\mathbf{A}(\mathbf{U}^m) + \gamma\mathbf{G}\right)^{-1} \mathbf{A}^T(\mathbf{U}^m)\mathbf{f}(\mathbf{U}^m). \tag{6.1.36}$$

This solution satisfies the problem

$$\min_{\mathbf{x}} \left\{ \|\mathbf{Ax} - \mathbf{f}\|^2 + \gamma\mathbf{x}^T\mathbf{Gx} \right\}, \tag{6.1.37}$$

where the weighting matrix $\mathbf{G}$ is an estimate of the measure of the data sensitivities of the components of the normal solution [166], and the regularization parameter γ is calculated as before (6.1.35). This data sensitivity is to be understood in the following sense: if the normal solution $\overline{\mathbf{x}} = \overline{\mathbf{x}}(\mathbf{U}^m)$ is visualized as a nonlinear function of the generalized output $\mathbf{U}^m$, the measurement errors are small changes in the argument of the normal solution. Thus a measure for the data sensitivity of the normal solution can be defined empirically by calculating the following expectations[3]

$$\mathbf{G} = s \cdot E\left\{ \left(\overline{\mathbf{x}}(\mathbf{U}^m + \mathbf{Z}) - \overline{\mathbf{x}}(\mathbf{U})\right) \left(\overline{\mathbf{x}}(\mathbf{U}^m + \mathbf{Z}) - \overline{\mathbf{x}}(\mathbf{U}^m)\right)^T \right\}, \tag{6.1.38}$$

where $\mathbf{Z}$ is a vector consisting of noncorrelated random numbers with zero mean values and s is a scalar, which is chosen here in such way that the largest diagonal element of $\mathbf{G}$ is equal to one.

---

[2] $\boldsymbol{\eta}$ is the extension of the error vectors $\boldsymbol{\eta}_i$ defined in Eq. (6.1.20).

[3] Its performance is similar to the Monte Carlo method.

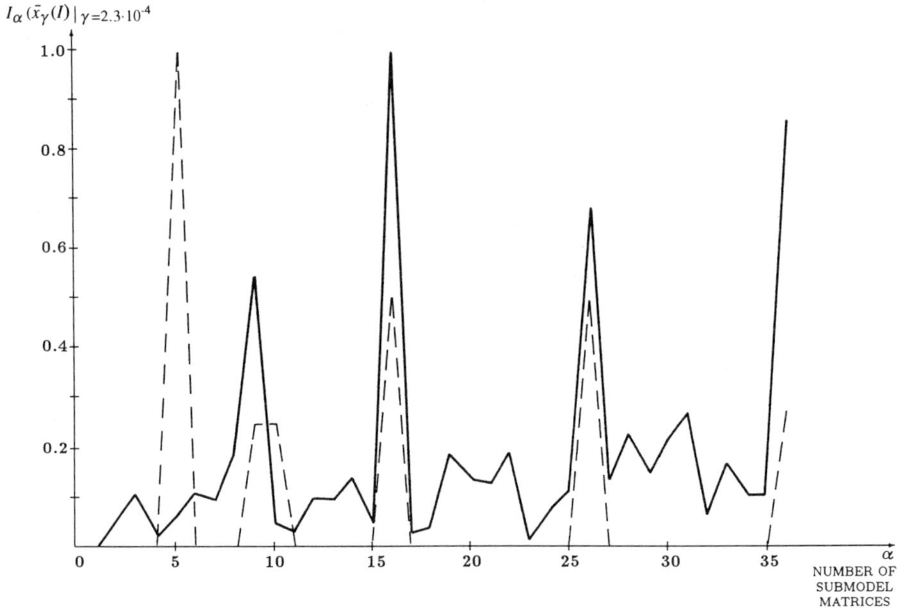

Fig. 6.5. The values of the indicator using the Tikhonov-Phillips regularized solution (see Table 6.3)

All 6 parameter errors are localized correctly by choosing a decision level of 0.15 (see the dashed curve which represents the exact solution): it is the correct qualitative solution.

6.1.4 The Adjusted Mathematical Model

A reduction of the dimension of the solution space to these 6 components leads to the reduced problem [166]

$$\mathbf{A}^{\text{red}}\mathbf{x}^{\text{red}} = \mathbf{f}. \tag{6.1.39}$$

The normal solution of the reduced problem can be taken approximately as the solution of the problem. The regularization is done here by introducing the reduced parameter space, which is known from the result of localization done in the previous section. A WLS solution can be performed in order to improve this result. The corresponding covariances can be estimated, or approximated by the Hessian matrix. The various estimates for the model modifications are compared in Table 6.4. The first row shows the model modifications introduced for simulation of the damage. The second row shows the corresponding components of the normal solution (6.1.31). The third row contains the result of the normal solution of the reduced problem (6.1.39). The relative errors are given in rows four and five.

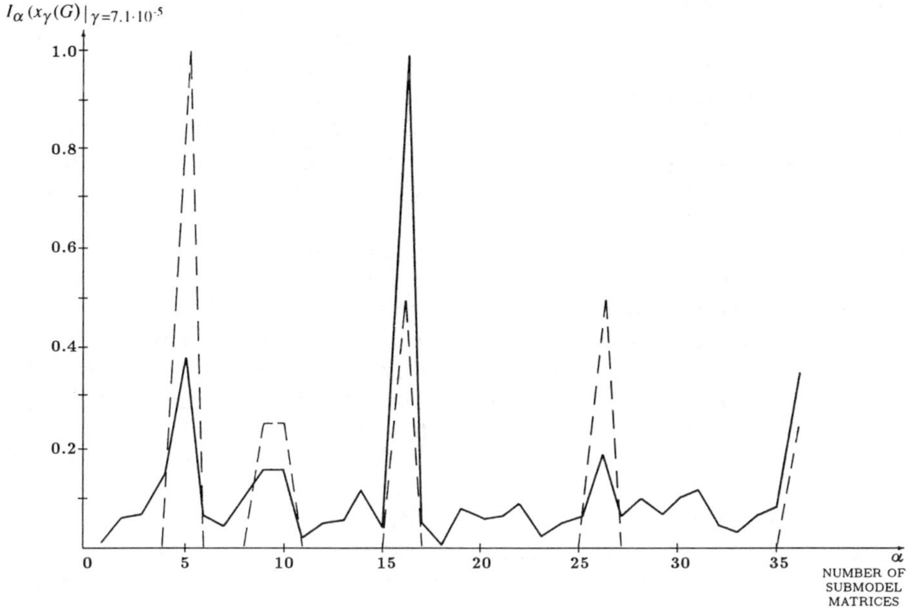

Fig. 6.6. The values of the indicator using the weighted Tikhonov-Phillips regularized solution, with the data sensitivities of the normal solution as the weighting matrix (see Table 6.3)

Table 6.4. The model errors used in order to simulate the damage, the results of model error estimation with (superscript red) and without reduction of the dimension of the solution space

α	5	9	10	16	26	36
x_α	0.4	0.1	0.1	0.2	−0.2	−0.1
$\overline{x}_\alpha$	0.01	−0.29	−0.45	−0.08	−0.43	−0.48
$\overline{x}_\alpha^{\mathrm{red}}$	0.17	0.10	0.05	0.43	−0.05	−0.10
$\dfrac{x_\alpha - \overline{x}_\alpha}{x_\alpha}$	0.99	3.94	5.53	1.41	1.16	3.77
$\dfrac{x_\alpha - \overline{x}_\alpha^{\mathrm{red}}}{x_\alpha}$	0.58	0.003	0.46	−1.14	0.75	0.04

6.2 Application in Mechanical Engineering

Some differences exist between diagnostics and condition monitoring in mechanical and civil engineering. One difference lies in the fact that, in addition to the structural parameters, operational parameters like rotation speed and pressure play an important role for systems of mechanical engineering. Another difference is to be seen in the following. Condition monitoring for some critical machines, like turbosets, main drive engines etc., started some 50 years ago, with the simple measurements of vibrations and other symptoms (see, for example, [200, 201]. Therefore a substantial amount of information has been gathered concerning the faults of machine element, subassemblies, etc. The same statement holds true concerning the methodology of the observation of related faults in machines. One can refer, for example, to the special table prepared by IRD Mechanalysis (see [24]), where one can find which type of machinery fault can be observed using a suitable symptom and with a pregiven probability of fault detection. This is even distinguished for the given locations of signal and symptom observation. For example, the detection of shaft misalignment can be done by observing the shaft's relative vibration with the probability 0.8, but when observed at the bearing support the probability of detection will be as low as 0.1. The same applies to machine casing. With respect to the direction of vibration observation for this fault, the axial direction has the probability 0.5, the horizontal one 0.3, and the vertical one 0.2.

Based on these data one can state that for machine diagnostics the fault location problem is currently not as urgent as for civil engineering systems (with the exception of special cases of turbosets, for example). This may be also due to the much smaller dimensional scale of machines in comparison to buildings (approximately ten times smaller). However, this does not mean that model-supported diagnostics is not needed for machines and equipment. Apart from this, by having a good holistic model of some critical machines we can verify and validate all the historical data on machine behaviour gathered previously. With the same model we can simulate the condition degradation during the modelled usage, and in this way we can build and then validate the models, and assess the machine conditions as summarized below:

- finding weak structural points where the early degradation occurs, so they can be modified in successive redesign. If this is not the case, they will be the first places for the installation of condition monitoring equipment.
- Finding and choosing the best damage-orientated symptom S (for detection, location and assessment), by simulating a number of possible measurable quantities, and by processing their signals accordingly.
- Choosing the best symptom operator $\phi(.)$ for the given case of wear, together with its shape factor γ, which allows us to adjust the evolution of mass, spring, damping coefficients in our holistic model, (see Table 2.2 and Eq. (2.3.30)). This will also serve for the prognosis of symptom behaviour in the given operational condition (see also the example on diesel engine condition monitoring in Sect. 2.3.6).
- Validating by response simulations the fault detection/assessment decisions obtained from previous statistical data.

Some of the enumerated tasks and cases are illustrated below.

6.2.1 The Choice of Symptom Operator Shape

In modelling the system condition evolution, either for simulation or in practical cases of condition monitoring, one of the important tasks is to choose the functional form of the condition degradation model. Table 2.2 will help us to choose a model. But how can the shape factors γ, α of the symptom distribution and the damage measure (see Chap. 2) be assessed?

As usual, we can proceed in two ways to choose heuristically the simplest possible model, for example a linear one: $S/S_0 = 1 + (1/\gamma)D, \quad D = \theta/\theta_b$, either by trial and error, or by the use of some additional data (corrosion speed in a given environment, for example) in order to assess the shape factors γ or α. Secondly, having some symptom data from the real case of condition monitoring, we can use a computer program (such as already discussed in Chap. 2) to make the job easy and more objective.

Let us consider this problem in connection with two particular machine diagnostics tasks: the monitoring of ball-bearing conditions in a small electric motor used in the textile industry, and the railroad diesel engines used for the main drive of trains.

In the first case a large number of units are considered under condition monitoring with the root mean square acceleration amplitude as the symptom (298 units). We can assume in a first approach that we have the case of a uniform distribution of damage evolution in a sample, i.e. $p(D) = const$, so in this way one can obtain the average symptom life curve. This was done with the computer program dem8.m already mentioned, and the results are shown in the first part of Table 6.5. There one can see that the Weibull and Fréchet models can be used, since they have almost the same determination coefficent[4] and χ^2—measure for evaluating the goodness of the fit, expressed by the ordering given in the last column. But due to the historical connotation of Weibull regarding ball-bearings, the latter was chosen (see the numbers with a circle) as the model for condition forecasting and residual life assessment (see, for example, Eq.(2.3.60)).

The second case concerns 56 symptom readings from 4 diesel engines in use. Here the average acceleration amplitude was used as the condition symptom. It was measured at the top of the first cylinder head [24] and each 10 thousand kilometres (56 readings). We can also process these data by the dem8.m program. The results are shown in the last part of Table 6.5. It is seen that the Weibull model of the symptom operator gives almost the optimum for the determination coefficient and χ^2-measure, although the uniform distribution is the better model here, not only due to its simplicity. Hence, when looking for the prediction of the next symptom value, or for the shape of the degradation function (see Eq. (2.3.30)) in a simulation, we will use the linear model of wear and symptom evolution.

[4] It is the square of the correlation coefficient.

Table 6.5. The choice of symptom models and their shape factors γ for the cases in machine condition monitoring

	SYMPTOM DISTRIBUT.	SHAPE FACTOR γ	SCALE FACTOR β	DETERMIN. COEFFIC. R^2[%]	χ^2	SYMPTOM MODEL ORDERING
SMALL ELECTRIC MOTORS	WEIBULL	0,715	11,483	98	4,13	①
	FRÉCHET	2,034	14,414	96,8	2,89	2
	PARETO	1,648	7,823	90,5	24,338	4
	UNIFORM	0,03	1	59,2	71,151	5
	EXPONENT.	0,318	1	89,3	14,661	3
	WHEN THE WEIBULL MODEL IS CHOSEN, SYMPTOM: $\frac{S}{S_0}=[-ln(1-D)]^{\frac{1}{\gamma}}, D=\frac{\theta}{\theta_b}, \gamma=0{,}715$ RELIABILITY: $R(S)=exp[-(\frac{S}{S_0})^{\gamma}]$					
DIESEL ENGINES	WEIBULL	4,117	2,001	96,7	11,476	2
	FRÉCHET	4,221	1,561	80,0	27,044	4
	PARETO	2,882	1,27	64,2	50,642	5
	UNIFORM	0,67	1	97,3	9,697	①
	EXPONENT.	1,066	1	93,5	15,777	3
	WHEN THE UNIFORM MODEL IS CHOSEN, SYMPTOM: $\frac{S}{S_0}=1+\frac{1}{\gamma}D, D=\frac{\theta}{\theta_b}, \gamma=0.67$ RELIABILITY: $R(S)=1+\gamma(1-\frac{S}{S_o})$					

6.2.2 Model-Based Diagnostics for Rotor Machinery Including Turbosets

Rotor machinery. Diagnostics is always understood as 'detection – localization – assessment', but in rotor machinery the localization problem is less important than in other engineering fields. This is due to the available experience and the large amount of statistics already gathered, and also to the much smaller physical dimensions compared with other systems, which gives another range of dynamic phenomena and wave behaviour in particular (see Fig. 3.2 on wave propagation). Hence, with respect to the model-based condition monitoring of machines, only the detection and the assessment problem have been studied extensively as yet. As one of the latest examples of such research let us refer briefly to the main idea and findings of paper [189].

This paper, with an extended literature survey, concerns the investigation of a simple supported rotating shaft loaded by concentrated mass (ring), which is the model of the rotor. The problem is to develop and/or improve some known methods of signal processing for the task of the detection and assessment of rotor unbalance and shaft crack. It was shown by analytical and experimental investigations that the extended Kalman filter and the modified instrumental variable method can be used succesfully for the detection and the assessment of both faults with acceptable accuracy when compared with the real condition introduced experimentally. The location problem is not investigated.

Turbosets. As was stated at the beginning of this section, the oldest application of condition monitoring and diagnostics in engineering began in power stations, and with steam turbines in particular. Initially (up to the Sixties), this produced the so-called Turbine Supervisory Equipment (TSE), based on transducers, analogue signal conditioning and processing. They mainly gave some alarm signals and sometimes a preselected emergency shut-off when the symptom limit value was exceeded. But the main work of the diagnostic interpretation of elaborated symptoms was carried out by diagnostic staff.

The rapid development in electronics and computer signal processing has transformed the TSE of turbosets into Turbine Monitoring Equipment (TME), with integrated tailor-made computers for signal processing especially devoted to turbosets (shaft trajectories, cascade spectrum of rundowns and run-ups, symptom alarm and limit values, etc). Since the Eighties this has been the standard equipment of most modern power stations, and due to the long application of both systems in practice many statistical data from turbine diagnostics have been gathered and elaborated. This currently makes possible the application of knowledge engineering and expert systems, in particular, to the diagnostics of turbosets. Due to the progress in identification theory and practice, and in computer-based systems of the FE-method, and to the successful modelling of some interaction effects typical to slide bearing machinery, it was possible to elaborate computer models for the simulation of about all dynamic phenomena important in diagnostics and in the supervision of operating turbosets. Another obstacle to the modelling of rotor machinery with journal bearings should be mentioned here: the highly nonlinear effects in journal bearings and shaft sleeves which were elaborated successfully quite recently.

Hence, according to some papers (see [196] and the refs. cited there) the possibility of elaborating some type of model-supported diagnostics of turbosets is now emerging. It is tentatively called the 'Intelligent Diagnostic System' which, together with the expert system, incorporates and integrates two knowledge bases (KB): KBI – all operational and diagnostic data already gathered through the traditional approach, KBII – validated dynamic model of the turboset, which can be used for checking the statements (assessment) previously made, for confirming the recent state conditions (symptom values) and for prediction.

Up to now no methods or computer programs exist for the location of faults, possibly due mainly to the three reasons already mentioned. This is the history and the state of development of model-based diagnostics for turbosets. It is illustrated in Fig. 6.7.

With regard to the current state of the development of dynamic models of turbosets [196] in general, it holds true for the:

- simulation of some dynamic phenomena (vibration, support reaction, temperature, etc.) in good agreement with the experiment,

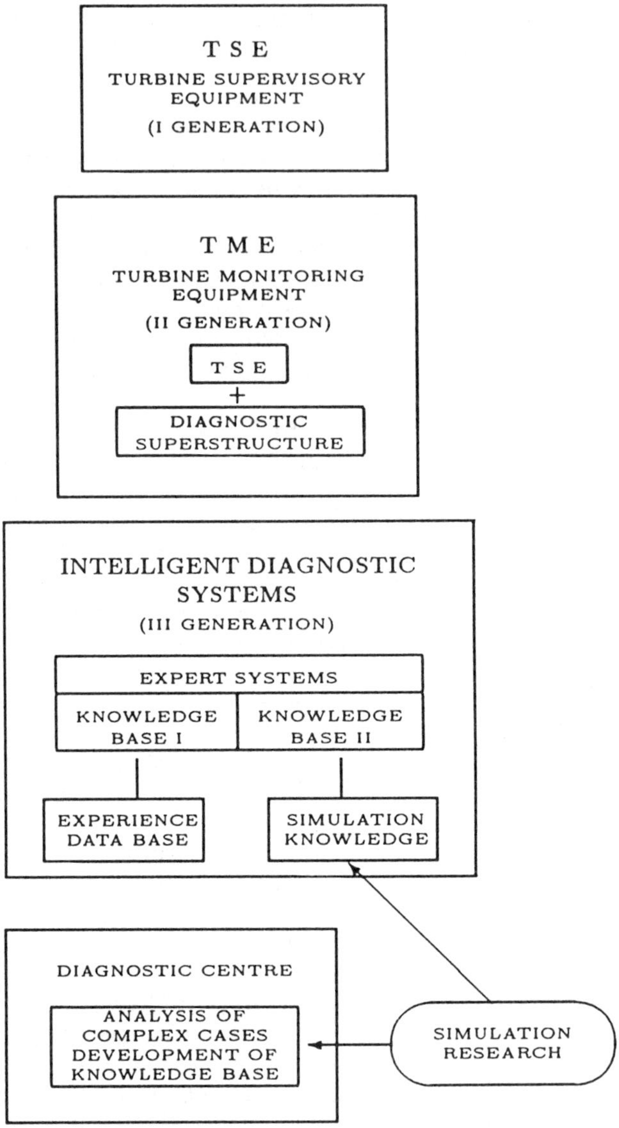

Fig. 6.7. Modern diagnostic systems of power-generating turbosets (after Kiciński 1995)

- modelling and simulation of some abnormal effects of operation, i.e. faults,
- description of the resulting simulated condition in terms of a set of symptoms, which already have a good diagnostic meaning.

In particular, the dynamic models mentioned in Fig. 6.7 permit one to simulate such faults as:

- unbalance at any point of the shaft and rotor,

- different types of oil instabilities in journal bearings,
- the evolution of the operational line of the shaft and the resulting position of bearing supports,
- cracks in a rotor and shaft and the evolution of the cross-section due to erosion,
- the change in the stiffness of bearing supports,
- abnormal electromagnetic forcings from the side of the generator,
- condition evolution in journal bearings,
- the misalignment of the shaft and couplings.

This can be checked under the given simulated operational conditions, and will validate the existence and intensity of faults in terms of diagnostic symptoms and operational parameters. The dynamic model also permits one to simulate the dynamic behaviour under measured operational conditions and to look for the causes of observed abnormalities. In addition, simulations with modified operational parameters dependent on the structural parameters lead to an assessment of future conditions. The computational expenditure is very high, as reported in [196]. For low-pressure casing the dynamic FE-model has 60,000 DOF, and it contains at least two nonlinear and non-stationary models of slide bearings.

Consequently, model-based diagnosis in its general sense can be a very powerful tool for application in mechanical engineering systems, and this field is also still at the research stage. There are reports that several turbine producers are working extensively on the appropriate developments.

6.3 Application in Civil Engineering: Norderelb Bridge

In connection with the expert system SAMBA already mentioned, the Norderelb bridge in Hamburg served as a benchmark test [111]. It is a real, existing system which is modelled by finite elements and tested using the traffic as excitation in order to identify the eigenfrequencies and normal modes in a certain frequency range. These results are taken as a reference. The damage artificially and only theoretically introduced follows the experience with the (first) FE-model, which did not fit the estimates from the test. Because the reason for the misfit between the calculated and the estimated eigenquantities was discovered afterwards, it is taken here as an example of a damage demonstration. It must be emphasized again that the damage discussed was not real damage to the bridge! Because the FE-model of this *Gedankenexperiment*[5] fits the measured quantities very well, the adjustment of this model is not presented here. However, the adjustment of FE-models has already been illustrated in Sect. 6.1.

[5] Thought-experiment

6.3.1 The System and its Model

The system. The main characteristics of the Norderelb motorway bridge are: it is a cable-stayed steel bridge consisting of five fields of 31 m – 64 m – 171 m – 64 m – 80 m, a total span of 410 m, and built as a support bridge. The superstructure consists of four main girders with load-distributing lateral disks. Together with the roadway and the ground plate, the two inner main girders form a stiff torsional box girder. In the middle span the main outside girders were included in the box girder by bracings. Two groups of four cables are led above each of the two pylons. The main girders are supported by the cables at four points of the main span. Their diameter is $11.2cm$ and their nominal strength is $1570N/mm^2$. They are fixed in the pylons. The remaining superstructure consists of steel (St 37 and St 52 according to DIN). The roadway girder is supported by two abutments, four (two × two) hinged supports and the two pylons. Figure 6.8 shows an overview drawing of the bridge. Figure 6.9 illustrates details of the cross-section of the middle span. The cross-section of the end field is an open one. Figure 6.10 is a photograph of the bridge.

The FE-model. The FE-modelling and the following calculations are performed by the program NASTRAN. QUAD4 and TRIA elements are used

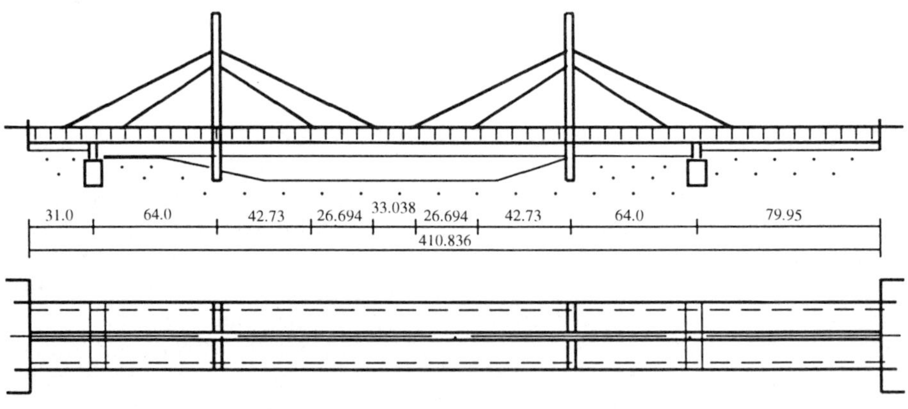

Fig. 6.8. Overview drawing

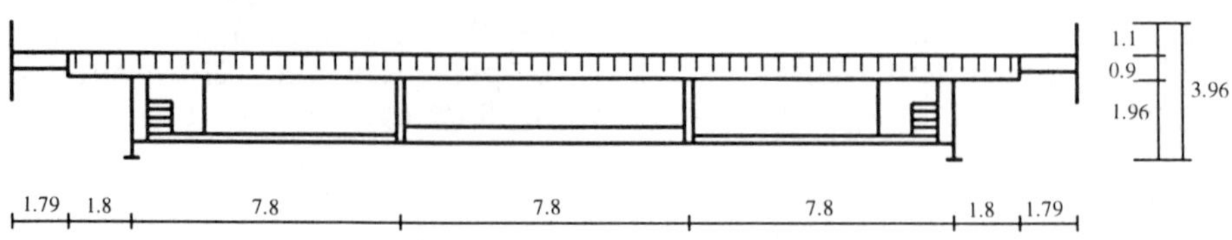

Fig. 6.9. Cross-section

Fig. 6.10. The Norderelb bridge

for roadway, main girders, ground plate and lateral disk. Bracings, ribs, and pylons are modelled by beam elements. Each of the 32 cables is represented by a rod element. Generally, cables are strongly nonlinear members. However, due to their high pre-stressing, the tension stresses will not change to compression during service[6]. Consequently, linear dynamics will represent their behaviour well. Some further details (Young's modulus etc.) are given in [202] and [203]. The associated undamped model will be considered. A total of about 2000 elements (approximately 7000 static DOF) are used.

Modal quantities are chosen and calculated, although they may not generally be the most sensitive quantities (symptoms) for indicating local damage. However,

- no forces are required for calculating eigenquantities
- "they are very popular in application"
- the eigenfrequencies have some spatial properties through their orders (ranking)
- the mode shapes (displacements) are insensitive with respect to local faults, but the 2nd and 3rd spatial derivatives (moments and shear forces of a beam element) are very sensitive to them [29] (differentiation of calculated or measured modes must not be performed; it can be substituted by integration using the equation of motion [9]).

In this example, that means with the assumed (and simulated) damage, it is proved that the eigenquantities are good symptoms, as will be shown in Sect. 6.3.4. The calculated eigenfrequencies and normal modes are shown

[6] It is emphasized that only forces due to the traffic are considered. For example, the self-excitation of the cables is excluded in these investigations.

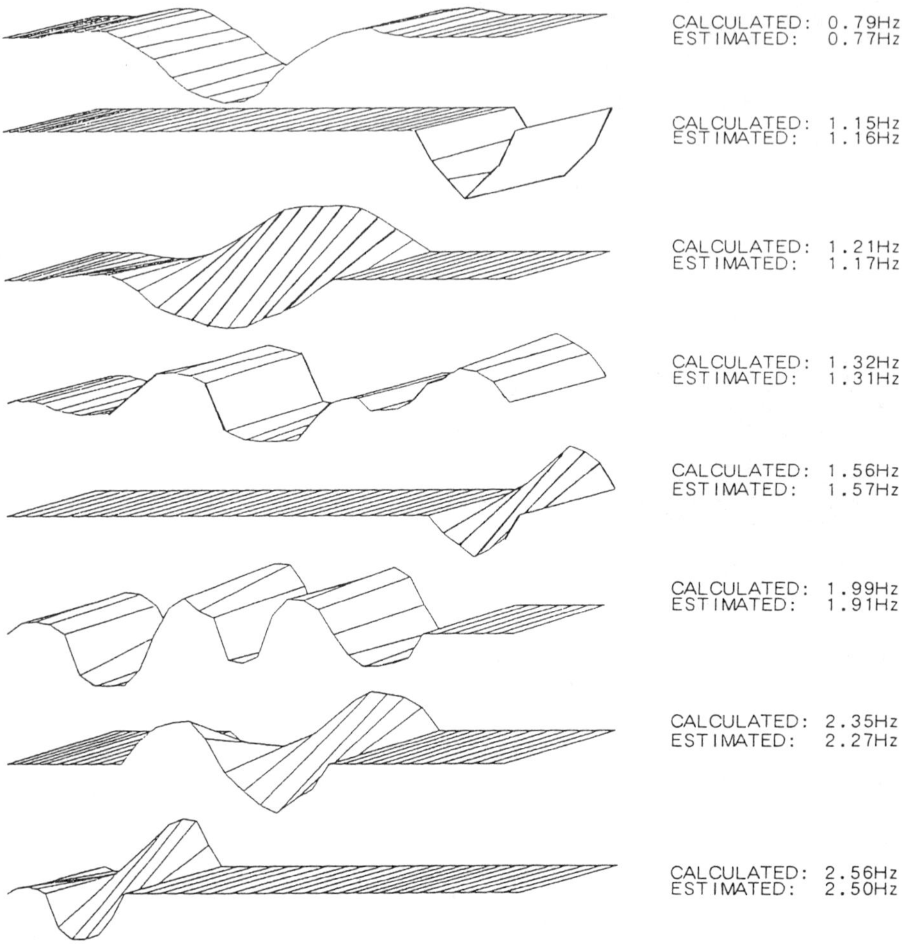

Fig. 6.11. Calculated eigenfrequencies and normal modes

in Fig. 6.11. The fieldwise vibration with the coupling of the fields is typical of this type of system. The reader's attention is drawn to the second mode, which shows a non-negligible deflection only in the end field. This is due to its span width and the change in the cross-section.

6.3.2 Identification of Eigenquantities

The test. Four reference geophones and, additionally, eight geophones are used blockwise. The latter are moved fieldwise during the measurements. A description of the modes was possible with this arrangement. Because of the importance of the motorway bridge investigated, stopping the traffic was impossible, and therefore it was taken as excitation. It was assumed as broadband noise, in the knowledge, of course, that the traffic, especially that of trucks, is an interaction problem with the bridge. For simplicity it is assumed, however, that data processing will reduce these introduced errors. Measurements were performed for a measuring time of 60 minutes.

Results. The frequency response functions had been estimated in the laboratory. The amplitude spectra at the reference points were performed in an initial step of the analysis process. Blocks of 8192 elements were therefore formed in the time domain, and this led to a frequency step of 0.0066 Hz at a sampling frequency of 54 Hz. Such a short frequency step was necessary due to the poorly damped steel system. 1400 amplitude spectra had to be calculated and averaged for each reference point. Random errors were thus reduced, and the curves were smoothed. Figure 6.12 gives an example of an averaged amplitude spectrum. SDOF models serve for the determination of the eigenfrequencies and mode components. Their estimates are obtained by averaging only 350 quantities in order to reduce the expenditure. Modal vector components, especially for the modes of the lower eigenfrequencies, were corrected by the use of the frequency response functions of the transducers (the lower limiting frequency of the geophones is 1 Hz) in order to obtain the correct phase relation which describes the relative sign of the modal components.

Because the absolute values of the estimated normal modes graphically show no quantitative difference from the calculated modes shown in Fig. 6.11, the quantitative comparison between the identified and calculated eigenquantities is given directly in the following.

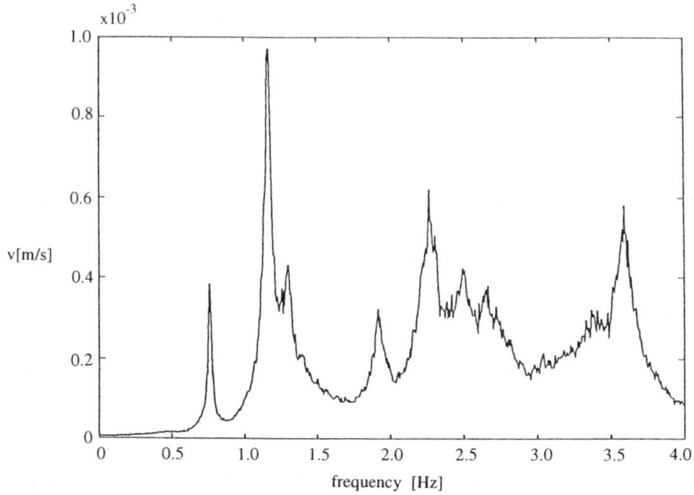

Fig. 6.12. Example of an averaged amplitude spectrum

6.3.3
Comparison of Calculated and Estimated Eigenquantities

Table 6.6 contains the calculated and identified results. The mode description characterizes the main deflections of the mode. The relative differences

Table 6.6. Comparison of identified and calculated eigenfrequencies and modes

DESCRIPTION	EIGENFREQUENCY [Hz]		DIFFERENCE	MAC
OF MODES	CALCULATED	MEASURED	%	
1.VERTICAL BENDING	0.79	0.756 ± 0.0021	3.3	0.997
2.VERTICAL BENDING	1.15	1.157 ± 0.0033	0.6	0.907
3.VERTICAL BENDING	1.32	1.307 ± 0.0051	1.3	0.887
4.VERTICAL BENDING	1.99	1.911 ± 0.0043	4.1	0.932
1.TORSION	1.21	1.173 ± 0.0054	3.2	0.996
2.TORSION	1.56	1.574 ± 0.0065	0.9	0.972
3.TORSION	2.35	2.274 ± 0.0069	3.3	0.832
4.TORSION	2.56	2.500 ± 0.0053	2.4	0.887

between the calculated and identified eigenfrequencies are relatively small. The last column contains the values of the MAC (see Sect. 2.2.2) for the calculated and measured eigenvectors. The MAC is the cosine of the angle between the calculated and identified eigenvectors. In the ideal case this value is identical to 1. As can be seen, the eigenvectors agree relatively well with each other.

Dependent on the catalogue of faults, it now has to be decided whether the used FE-model is usable. If not, it has to be adjusted in order to serve as a reference model. A correction of the FE-model will not be performed here, because, as will be seen next, the FE-model is usable for the detection and localization.

6.3.4
Simulated Damage

Damage description. The lower chord of the main girders is build as a bottom flange. The cross-section is closed in the middle field of the bridge, and in this area the stiffness is mainly determined by the ground plate. However, in the end field, defined by the large deflection of the 2nd vertical bending mode (see Fig. 6.11), the bottom flanges of the longitudinal girders are essential. It is assumed that these flanges are completely missing: the stiffnesses of the bottom flanges in the end field are removed in the related FE-model.

Effect of the damage on the eigenquantities. The damage simulated in this way leads to the simulated eigenquantities as shown in Table 6.7. Again, a comparison of the eigenquantities of the (simulated) damaged system with

Table 6.7. Comparison of the eigenquantities of the damaged system and the reference values

<table>
<tr><th rowspan="2">DESCRIPTION OF MODES</th><th colspan="2">EIGENFREQUENCY [Hz]</th><th rowspan="2">DIFFERENCE %</th><th rowspan="2">MAC</th></tr>
<tr><th>REFERENCE</th><th>DAMAGED SYSTEM (SIMULATED)</th></tr>
<tr><td>1.VERTICAL BENDING</td><td>0.765</td><td>0.79</td><td>3.3</td><td>0.985</td></tr>
<tr><td>2.VERTICAL BENDING</td><td>1.157</td><td>0.89</td><td>-23.0</td><td>0.860</td></tr>
<tr><td>3.VERTICAL BENDING</td><td>1.307</td><td>1.23</td><td>-5.9</td><td>0.562</td></tr>
<tr><td>4.VERTICAL BENDING</td><td>1.911</td><td>2.04</td><td>6.8</td><td>≈ 0</td></tr>
<tr><td>1.TORSION</td><td>1.173</td><td>1.23</td><td>4.9</td><td>0.994</td></tr>
<tr><td>2.TORSION</td><td>1.574</td><td>1.68</td><td>6.7</td><td>0.960</td></tr>
<tr><td>3.TORSION</td><td>2.274</td><td rowspan="2">2.37</td><td>4.2</td><td>0.818</td></tr>
<tr><td>4.TORSION</td><td>2.500</td><td>-5.2</td><td>0.120</td></tr>
<tr><td>5.TORSION</td><td>2.652</td><td>2.60</td><td>-2.0</td><td>0.938</td></tr>
<tr><td>6.TORSION</td><td>3.600</td><td>3.50</td><td>-2.8</td><td>0.922</td></tr>
</table>

the corresponding quantities of the reference model is now presented. The effect is enormous. The 2nd bending eigenfrequency shows a reduction of 23% as an effect only of the stiffness reduction in the end field. The related mode (see Fig.6.11) does not differ in its shape from the corresponding mode of the undamaged system. Most of the MAC values differ significantly from 1. This is mainly due to the changed signs of the modal vector components (not shown). By comparison with the reference model, the number of DOF (of normal modes) is now reduced. This is astonishing, when one looks at the stiffness modification in the end field. Here a sophisticated decision based on a significance test is unnecessary. This result shows the strong stiffness coupling of the various fields of the bridge.

The eigenquantities permit the detection and localization of the assumed damage, and therefore they are good symptoms in the example considered. The 2nd bending mode shows the main deflection in the end field (damaged as well as undamaged). The eigenfrequency of the damaged system is much smaller than that of the undamaged system; consequently, something has happened regarding stiffness reduction (or added masses) in this vibrating field. Inspection will confirm the result.

Consequently, the reverse method, namely the adjustment of the reference model, can easily be performed by looking at the used and known FE-models. The diagnosis is easy to perform with the available knowledge. Assessment will lead to action.

6.4 Conclusion

The academic example illustrates the difficulties arising from the ill-conditioning of the spatially discretized model in model adjustment, formulated as an inverse problem. However, this problem seems to have been overcome with the application of regularization methods.

The discussion of the state of science in diagnostics in the field of mechanical engineering emphasizes the importance of mathematical models of the dynamic behaviour. The difficulty here is to model the local nonlinearities which must be included in order to understand the system's dynamics. Additionally, adjustment of the model of the system condition and that for damage evolution can be provided.

The real world example of the bridge, except for the damage introduced (possibly unrealistically) and used, indicates that FE-modelling and model adjustment to the recent (measured) state enables the user, by means of a suitable choice of measured quantities and resulting estimates (symptoms, and information condensation!), to detect and localize faults of the system under investigation. An assessment results from the comparison of the recent model with the reference model, as described above. In the benchmark test discussed here the model-based diagnosis works well.

CHAPTER 7

Concluding Remarks

Signal and signature supported diagnosis are well known and have already been introduced for the monitoring and diagnosis of (vibrating) systems in operation. Higher demands with respect to the safety and lifetimes of systems require efficient and automatically elapsing tools for assessment and decision. From the economic point of view, the costs of development, and during operation/service of a system, can be lowered if the faults of a system are detected early and their evolution is known. Assessment of such faults then leads to action in time, and avoids subsequent costs. The costs of monitoring itself can also be lowered if the monitoring is performed state-dependent and event-dependent, and not periodically. With regard to machines the importance of monitoring and diagnostics is obvious, because condition monitoring was introduced in this field of application much earlier than in other technical fields. For civil engineering systems it is obvious that the deteriorating infrastructure and environment both pose challenging problems for engineering and diagnostics as well. It would be a great contribution to society if engineers could save even a fraction of the percentage of the cost by improving the present method of maintaining the infrastructure [204] and by designing in a new, more service-orientated manner. Efficient diagnosis procedures are the basis for economic renewal engineering.

Model-based diagnostics is theoretically an optimum method for damage detection, localization and assessment, because verified, validated and usable mathematical models at every state condition are the best knowledge base available. The expenditure is great. However, dependent on the complexity and criticality of the system and its societal and economic importance, the expenditure on monitoring, measurements and the subsequent computations can be much lower than any other approach required to achieve the necessary safety and performance, not to mention the breakdown costs.

The recent advancement in computational engineering (software, hardware, and numerical methods), the development in measuring techniques and in system identification encouraged the authors to propose and to discuss the system identification-based methodology for diagnosis.

As already stated, model-based diagnosis is discussed from a methodological point of view; this means that the intention is to provide the reader with a stimulus and an overview of some methods, so that he/she will be able to start the investigation of his/her particular problem.

To summarize (see Fig. 5.1):

- health monitoring has to use symptoms, and smart or intelligent monitoring systems with built-in sensors and a knowledge base are recommended;
- a measurement basis for model adjustment is required which demands an *appropriate* data set of measured values and/or identified quantities containing the information required (related measuring techniques should be known, the data processing background has to be known, too, and the available hardware and software should not be used as a black box in order to avoid misinterpretations and additional errors);
- after the decision concerning significant modifications between the model and measured quantities, the adjustment has to be performed using a global and/or a local model;
- a detailed investigation of the adjusted mathematical model(s) together with the previously theoretical studies will give a diagnostic result and assessment of the apparent system modifications (faults, damage) which will lead to a basis for actions.

Modified forces and environmental conditions have to be taken into account, of course. Environmental parameters and (resulting) special conditions, such as pre-stressing, have to be known and registered or taken as additional loads in the particular case.

The quality of the models used is decisive in the procedures discussed. The results will only be as informative as the models are able to describe the required properties. One has to distinguish between global identification and local identification. This is discussed through the distinction between dynamic models and static models when subsystem modelling is considered. Dependent on the criteria decisive for assessment, it can be necessary, after fault localization, to use an expanded static submodel for that part of the system where the damage affects safety, performance, or comfort. Or it may be necessary, too, to introduce a damage model in order to obtain detailed information about the extent of the damage. One will obtain what one introduces as information! These facts require adaptive models, but it may be a long path until a procedure which uses such adaptive models is achieved.

The future aim in diagnosis will be to use smart sensors, to incorporate self-learning, and to use adaptive models. Smart sensors are discussed in [205]; they include data processing and some knowledge base combined with conclusion rules. The next step can be smart systems, where the smart sensors are a part of the construction. Consideration may be given further to active systems, which allow a change of the load paths due to modified external conditions (adaptive elastomechanical systems, which can be realized, for example, with additional active elements: feedback control by sensed system modifications or external loading [206]). An initial and simple step in the direction of self-learning is already implemented here with the application of recursive estimation procedures. Adaptive models should take into account expected and predicted damage and the resulting changed external forces in order automatically to perform an assessment and trend prediction.

The effect of these future developments can be enormous with regard to the montoring, diagnosis, safety and integrity of systems, and, of course, the related decrease in total costs. Identification algorithms, for example, will be influenced by the application of neural networks (see, for example, [207]).

The content of the book is restricted to linear models, although non-linear system behaviour is discussed partially. In each application a check has to be made as to whether this assumption will be approximately valid (property-dependent and purpose-equivalent). For systems behaving non-linearly, the reader's attention is drawn to [208] and the references cited in it.

References

[1] Blanchard, B.S., Fabricky, W. J.,1990: *Systems Engineering and Analysis*, Prentice Hall, New York

[2] Blanchard, B.S., 1986, *Logistics Engineering and Management*, 3rd ed., Prentice Hall, Englewood Cliffs, N.J.

[3] Pardue, E.F., Piety, K.R., Moore, R., 1992, Elements of reliability-based machinery maintenance, Sound and Vibration, May, 14-20

[4] Eykhoff, P., 1974, *System Identification - Parameter and State Estimation*, John Wiley & Sons, London, New York, Sydney, Toronto

[5] Natke, H.G., 1992, *Einführung in Theorie und Praxis der Zeitreihen- und Modalanalyse - Identifikation schwingungsfähiger elstomechanischer Systeme*, 3rd edition, Vieweg, Braunschweig, Wiesbaden

[6] Natke, H.G. (Ed.), 1982, *Identification of Vibrating Structures*, CISM Courses and Lectures No. 272, Springer-Verlag, Wien, New York

[7] Natke, H.G. (Ed.), 1988, *Application of System Identification in Engineering*, CISM Courses and Lectures No. 296, Springer-Verlag, Wien, New York

[8] Natke, H.G., Yao, J.T.P. (Eds.), 1988, *Structural Safety Evaluation Based on System Identification Approaches*, Friedr. Vieweg & Sohn, Braunschweig, Wiesbaden

[9] Natke, H.G., 1989a, *Baudynamik - Einführung in die Dynamik mit Anwendungen aus dem Bauwesen*, LAMM 66, B.G. Teubner, Stuttgart

[10] Yao, J.T.P., 1985, *Safety and Reliability of Existing Structures*, Pitman Advanced Publ. Program, Boston, London, Melbourne

[11] Natke, H.G., 1994, Model verification and validation in engineering, Bulletin Polish Academy of Sciences, Techn. Scie., Appl. Mech., Vol. 42, No. 4, 505-519

[12] Soong, T.T., 1981, *Probabilistic Modeling and Analysis in Science and Engineering*, John Wiley & Sons, New York, Clichester, Brisbane, Toronto

[13] Elishakoff, I., 1983, *Probabilistic Methods in the Theory of Structures*, John Wiley & Sons, New York, Chichester, Brisbane, Toronto, Singapore

[14] Zadeh, L.A., 1965, Fuzzy sets, Information and Control, 8, 338-345

[15] Zimmermann, H.-J., 1991, *Fuzzy Set Theory - and its Applications*, Kluwer-Nijhoff Publ., Boston

[16] Teano, T., Asai, K., Sugeno, M., 1992, *Fuzzy Systems Theory*, Academic Press, Boston, MASS

[17] Kosko, B., 1992, *Neural Networks and Fuzzy Systems*, Prentice Hall, Inc., Englewood Cliffs, NJ

[18] Frank, P.M., 1994, Application of fuzzy logic to process supervision and fault diagnosis, Proc. SAFEPROCESS'94, Vol. 2, 531-538

[19] Isermann, R., 1994, Integration of fault detection and diagnosis methods, Proc. SAFEPROCESS'94, Vol. 2, 591-612

[20] Ben-Haim, Y., Elishakoff, I., 1990, *Convex Models of Uncertainty In Applied Mechanics*, Elsevier, Amsterdam, Holland

[21] Ben-Haim, Y., Chen, G., Soong, T.T., 1996, Maximum structural response using convex models, ASCE J. of Eng. Mech., April, 325-333

[22] Collacott, R.A., 1977, *Mechanical Fault Diagnosis and Condition Monitoring*, Chapman and Hall, London

[23] Braun, S. (Ed.), 1985, *Mechanical Signature Analysis - Theory and Applications*, Academic Press, New York

[24] Cempel, C., 1991, *Vibroacoustic Condition Monitoring*, Ellis Horwood, Chichester, UK

[25] Dieterle, R., Bachmann, H., 1981, *Experiments and models for the damping behaviour of vibrating reinforced concrete beams in the uncracked and cracked condition*, Institut für Baustatik und Konstruktion, ETH Zürich, Bericht Nr. 119, Birkhäuser Verlag Basel

[26] Natke, H.G., Cempel, C., 1991, Fault detection and localisation in structures: a discussion, Mechanical Systems and Signal Processing 5(5), 345-356

[27] Wallaschek, J., 1992, *Parameteridentifikation linearer ungedämpfter mechanischer Kontinua anhand von Eigenfrequenzen und Eigenschwingungsformen*, VDI Verlag, Reihe 11: Schwingungstechnik, Nr. 160, Düsseldorf

[28] Ben-Haim, Y., 1992, Adaptive diagnosis of faults in elastic structures by static displacement measurement: the method of selective sensitivity, Mechanical Systems and Signal Processing, 6(1), 85-96

[29] Cempel, C., Natke, H.G., Ziołkowski, A., 1992, Application of transformed normal modes for damage location in structures, in: *Structural Integrity Assessment*, Ed. P. Stanley, Elsevier Appl. Sci., London, New York, 246-255

[30] Courant, R., Hilbert, D., 1962, *Methods of Mathematical Physics*, Vol.1, Interscience, New York

[31] Ben-Haim, Y., Natke, H.G., 1993, Sequential adaptation in estimating elastic boundary-condition influence matrices, J. of Dynamic Systems, Measurement, and Control, Vol. 115, 370-378

[32] Chen, H.-S., Matzen, V.C., 1994, Detection of damage in a space truss using modal data; Proc. Smart Structures and Materials Conf., Orlando, FL, 13-18 Feb.

[33] Natke, H.G., Yao, J.T.P., 1988, Identification Approaches in Structural Safety, in [8]

[34] Hochweller, W., 1982, *Modifikationsproblem: Teilentwurfsproblem, Teiloptimierung*; Zwischenbericht zum Teilprojekt C4, Bericht des Curt-Risch-Instituts für Dynamik, Schall- und Meßtechnik der Universität Hannover, CRI-F 2/82

[35] Ben-Haim, Y., Prells, U., 1993, Selective sensitivity in the frequency domain, Part 1: Theory, Mechanical Systems and Signal Processing, Vol. 7, No. 5, 461-475

[36] Natke, H.G., 1993, *Subsystem Modelling and Synthesis - Contact Forces Expressed by Interacting Forces*, Report of the Curt-Risch-Institute for Dynamics, Acoustics and Measurements of the University of Hannover, CRI F-3/93

[37] Natke, H.G., Yao, J.T.P., 1993, Detection and location of damage causing non-linear system behaviour, in: *Safety Evaluation Based on Identification Approaches Related to Time-Variant and Non-Linear Structures*, ed. by H.G. Natke et al., Friedr. Vieweg & Sohn, Braunschweig, Wiesbaden, 188-203

[38] Böhling, S., 1993, *Theoretische und experimentelle Analyse eines nichtlinearen Einmasseschwingers*, Diplomarbeit am Curt-Risch-Institut für Dynamik, Schall- und Meßtechnik der Universität Hannover, Fachbereich Bauingenieur- und Vermessungswesen

[39] Rost, R., Brown, D.L., 1988, The use of spatial domain concepts in modal analysis, Proc. 13th Internat. Symposium on Modal Analysis, Leuven, Belgium, Part I, I-2

[40] Radeş, M., 1988, Modal analysis using frequency response measurement data, The Shock and Vibration Digest, Vol. 20, No. 1, 13-21

[41] Dat, R., 1975, L'essai de vibration d'une structure imperfaitement lineaire, La Rech. Aerospatiale, N0. 4, 223-227

[42] Natke, H.G., 1979, Fehlerbetrachtungen zur parametrischen Identifikation eines Systems mit kubischem Steifigkeits- und Dämpfungsterm, Czerwenka-Festschrift TU München (FRG), 109-121

[43] Mertens, M. et al., 1985, Basic rules of a reliable detection method for non-linear dynamic behaviour, Proc. 10th Internat. Seminar on Modal Analysis, K.U. Leuven, Pt. IV

[44] Sauer, R., Szabó, I., 1967, *Mathematische Hilfsmittel des Ingenieurs*, Teil 1, Grundlagen der Math. Wissenschaften in Einzeldarstellungen, Bd. 139, Springer-Verlag, Berlin, Heidelberg, New York

[45] Billings, S.A., Voon, W.S.F., 1982, *Structure detection and model validity test in the identification of non-linear systems*, Dep. of Control Eng., University of Sheffield, Sheffield Res. Rep. No. 196

[46] Newland, D.E., 1994, *An Introduction to Random Vibrations, Spectral & Wavelet Analysis*, 3rd ed., Longman Scientific & Technical, Copubl. in the US with John Wiley & Sons, Inc., New York

[47] Rajbman, N.S., Čadeev, V.M., 1980, *Identifikation - Modellierung industrieller Prozesse*, VEB-Verlag Technik, Berlin

[48] Bendat, J.S., Piersol, A.G., 1986, *Random Data - Analysis and Measurement Procedures*, John Wiley & Sons, New York, Chichester, Brisbane, Toronto, Singapore

[49] Goodwin, G.C., Payne, R.L., 1977, *Dynamic System Identification Experiment Design and Data Analysis*, Academic Press, New York

[50] Billings, S.A., Voon, W.S.F., 1984, Least squares parameter estimation algorithms for nonlinear systems, Int. J. Systems Sci., 15, 601-615

[51] Billings, S.A., Voon, W.S.F., 1983, Structure detection and model validity tests in the identification of nonlinear systems, Proc. Inst. Electr. Engrs., Tt. D. 130, 193-199

[52] Billings, S.A., Fadzil, M.B., 1985, The practical identification of systems with nonlinearities, 7th IFAC Symp. on Identification and System Parameter Estimation, New York, 155-160

[53] Hsin-Yi Lai, Su-Hua Shieh, 1987, Identification and quantification of system non-linearity through response data, IMAC, 1452-1459
[54] Natke, H.G., Zamirowski, M., 1990, On methods of structure identification for the class of polynomials within mechanical systems, ZAMM - Z. angew. Math. Mech. 70, 10, 415-420
[55] Argyris, J., Mlejnek, H.-P., 1991, *Dynamics of Structures*, North-Holland, Amsterdam, New York, Oxford, Tokyo
[56] Tomlinson, G.R., 1987, Developments in the use of the Hilbert transform for detecting and quantifying non-linearity associated with frequency response functions, Mechanical Systems and Signal Processing 1, 2, 151-171
[57] Tomlinson, G.R., 1987, Developments in the use of the Hilbert transform for detecting and quantifying non-linearity associated with frequency response functions, Mechanical Systems and Signal Processing 1, 2, 151-171
[58] Chui, C. K., 1992, *Wavelet Analysis and its Applications I: An Introduction to Wavelets*, Academic Press, New York
[59] Chui, C.K. (Ed.), 1992, *Wavelet Analysis and its Applications II: Wavelets: A Tutorial in Theory and Applications*, Academic Press, New York
[60] Zamirowski, M., 1992, *Einige zeitdiskrete Parameterschätzmethoden zur Identifikation nichtlinearer mechanischer Systeme*, Dissertation, Mitteilung des Curt-Risch-Instituts für Dynamik, Schall- und Meßtechnik der Universität Hannover, Fachbereich Bauingenieur- und Vermessungswesen
[61] Natke, H.G., Prells, U., 1992, A contribution to structure identification of nonlinear mechanical systems, in: *Symposium on Identification of Nonlinear Mechanical Systems from Dynamic Tests*, Proc. of EUROMECH 280, Eds. L. Jezquel, C.H. Lamarque, A.A. Balkema, Rotterdam, Brookfield, 85-89
[62] Bschorr,O., Mittmann, J., 1984, Qualitätskontrolle und Mustervergleich mittels differentieller Modalanalyse, VDI-Z., Bd. 126, Nr. 9, Mai(1), 308-310
[63] Armon, D., Ben-Haim, Y., Braun, S., 1994, Crack detection in beams by rank-ordering of eigenfrequency shifts; Mechanical Systems and Signal Processing, 8, 81-91
[64] Natke, H.G., and Yao, J.T.P., 1994, Applications of fuzzy sets in holistic dynamics; in preparation
[65] Rodden, W.P. 1967, A method for deriving structural influence coefficients from ground vibration tests, AIAA Journal 5, No. 5, 991-1000
[66] Natke, H.G., 1968, *Ein Verfahren zur rechnerischen Ermittlung der Eigenschwingungsgrößen aus den Ergebnissen eines Schwingungsversuches in einer Erregerkofiguration*, Dissertation TH München; Engl. translation: A method for computing natural oscillation magnitudes from the results of vibration testing in one exciter configuration, NASA-TT-F-12, 446, 1969, Ph.D. thesis
[67] Natke, H.G., 1978, Substrukturmethoden bei dynamischen Untersuchungen, in: *Aeroelastische Probleme außerhalb der Luft- und*

Raumfahrt, Hrg. H.G. Natke, O. Mahrenholtz, Mitteilung des Curt-Risch-Instituts der Technischen Universität Hannover, 17-46

[68] Cottin, N., 1993, On the use of reduced models in linear system identification, in: *Structural Dynamics - EURODYN '93*, Eds. T. Moan et al., A.A. Balkema, Rotterdam, Brookfield, 819-826

[69] Söderström, T., Stoica, P., 1989, *System Identification*, Prentice Hall, New York, London, Toronto, Sydney, Tokyo

[70] Baumeister, J., 1987, *Stable Solution of Inverse Problems*, Vieweg Advanced Lectures in Mathematics, Friedr. Vieweg & Sohn, Braunschweig, Wiesbaden

[71] Louis, A.K., 1989, *Inverse und schlecht gestellte Probleme*, Teubner Studienbücher Mathematik, Teubner, Stuttgart

[72] Natke, H.G., 1993, On regularization methods within system identification, in: M. Tanaka, H. Bui (eds.) *Inverse Problems in Engineering Mechanics*, IUTAM Symposium Tokyo, 1992, Springer-Verlag, Berlin, Heidelberg, New York, London, Paris, Tokyo, Hong Kong, Barcelona, Budapest, 3-20

[73] Heckl, M., 1993, Advantages and disadvantages of intensity measurements, in: *Safety Evaluation Based on Identification Approaches Related to Time-Variant and Non-Linear Structures*, ed. by H.G. Natke et al., Friedr. Vieweg & Sohn, Braunschweig, Wiesbaden, 12-27

[74] Elmer, K.-H., 1992, *Optimierung numerischer FE-Modelle zur Simulation der Wellenausbreitung mit Vektor- und Parallelrechnern*, 1. Zwischenbericht zum DFG-Forschungsvorhaben, Mitteilung des Curt-Risch-Instituts für Dynamik, Schall- und Meßtechnik der Universität Hannover

[75] Hagedorn, P., Sparschuh, S., 1988, Experimental modal analysis vs. power flow measurement. Some recent developments, 6th IMAC, 666-672

[76] Fitch, E.C., 1992, *Proactive Maintenance for Mechanical Systems*, Elsevier, Advanced Technology, Oxford

[77] Kliman, V., 1985, Fatigue life estimation under random loading using energy criterion, Int. Journ. Fatigue 1, 39-44

[78] Luczak, A., Mazur, T., 1981, *Physical wear of Machinery Elements*, (in Polish), WNT Warsaw

[79] Poltzer, G., Meissner F., 1983, *The Foundation of Friction and Wear*, (in German), VEB, Leipzig

[80] Engel, P.A., 1976, *Impact Wear of Materials*, Elsevier, New York

[81] Juvinal, R.C., 1967, *Stress, Strain and Strength*, Mc Graw Hill, New York

[82] Severdenko, V.P., 1976, *Ultrasound and Plasticity*, (in Russian), Nauka, Minsk

[83] Waelchli, F., 1992: Eleven Theses of General System Theory (GST), System Research, Vol. 9, No. 4, 3 - 8,

[84] Prigogine, I., 1971, *Thermodynamic Theory of Structure, Stability and Fluctuations*, (in French) Masson et Cie, Paris

[85] Cempel, C., Natke, H.G., 1990, *Damage Measures and Evolution in Machines and Structures*, Report of the Curt-Risch-Institute for Dy-

namics, Acoustics and Measurements of the University of Hannover, CRI-B-1/90

[86] Cempel, C., Natke, H.G., Yao, J.T.P., 1995, Damage capacity and symptom-based reliability as advanced signal processing procedures; Intern. Workshop on Structural Damage Assessment Using Advanced Signal Processing Procedures, Pescara, Italy, May 29-31, will be publ. in J. of Eng. Structures

[87] Cempel, C., Natke, H.G., 1992, Damage evolution and diagnosis in operating systems, in: *Safety Evaluation Based on Identification Approaches Related to Time-Variant and Non-Linear Structures*, ed. by H.G. Natke et al., Friedr. Vieweg & Sohn, Braunschweig, Wiesbaden, 44-61

[88] Natke, H.G., Cempel, C., 1996, Holistic modelling as a tool for the diagnosis of critical complex systems, Automatica 32, No. 1, 89-94

[89] Ben-Haim, Y., 1994, Fatigue lifetime with load-uncertainty represented by a convex model, ASCE Journal Eng. Mechanics, 120, 445-462

[90] Di Pascale, E., Cakmak, A.S., 1990, Detection of seismic structural damage using parameter-based global damage indices; Probabilistic Engineering Mechanics, Vol. 5, No. 2, 60-65

[91] Aven, T., 1992, *Reliability and Risk Analysis*, Elsevier, Applied Science, London, New York

[92] O'Connor, P.D.T., 1985, *Practical Reliability Engineering*, sec. ed., John Wiley, Chichester

[93] Franks, L.E., 1967, *Signal Theory*, Prentice Hall, New York

[94] Papoulis, A., 1965, *Probability, Random Variables, Stochastic Processes*, Mc Graw Hill, New York

[95] Winiwarter, P., 1985, Isodynamics of population size distribution in hierarchical systems; Proc. Society for General System Research, Los Angeles, May 27-31, Vol. 1, 87-95

[96] Winiwarter, P., Cempel, C., 1991, Life symptoms; Open energy transformation systems with limited internal structuring capacity, Systems Research, Vol. 9, No. 4, 9-34

[97] Cempel, C., 1992, Damage initiation and evolution in operating systems, Bull. of Polish Academy of Sciences, Techn. Sci. Vol. 40, No 3, 204-214

[98] Kalbfleisch, J.D., Prentice, R.L., 1980, *The Statistical Analysis of Failure Data*, John Wiley & Sons, New York, Chichester, Brisbane, Toronto, Singapore

[99] Birger, I. Ya, 1978, *Technical Diagnostics* (in Russian), Nauka, Moscow

[100] Cempel, C., 1993, Theory of energy transforming systems and their application in diagnostics of operating systems, Applied Mathematics and Computer Science, Vol 3, No 3, 533-548

[101] Cempel, C., 1993, Dynamics, life, diagnostics; holistic approach to modelling of operating systems, Proceedings of the Symposium 'Modellierung und Wirklichkeit', Hannover 7 -8 October 1993, Curt-

Risch-Institute für Dynamik, Schall- und Meßtechnik der Universität Hannover, Eds. H.G. Natke et al., CRI-K 3/93, 1-14

[102] Soong, T.T., Grigoriu, M., 1993, *Random Vibration of Mechanical and Structural Systems*, P T R Prentice Hall, Englewood Cliffs, New Jersey

[103] Bertalanffy, L. von, 1973, *General System Theory*, G. Braziler, New York

[104] Cempel, C., 1993, Machine degradation model - Its generalization and diagnostic application, Proceedings of IUTAM Symposium, Wupertal August 1993, to be publ.

[105] Cempel, C., Ziołkowski, A., 1995, New method of symptom data base processing for machinery condition recognition and forecasting, Mechanical Systems and Signal Processing (9) 2, 129-137

[106] Reid, J.G., 1985, *Linear System Fundamentals - Continuous and Discrete, Classic and Modern*, McGraw-Hill, Auckland, Bogotá, Hamburg, London, Tokyo

[107] Collacott, R.A., 1985, *Structural Integrity Monitoring*, Chapman and Hall, London

[108] Müller, P.C., Schiehlen, W.O., 1976, *Lineare Schwingungen*, Akademische Verlagsgesellschaft, Wiesbaden

[109] Patton, R., Frank, P., Clark, R., 1989, *Fault Diagnosis in Dynamical Systems - Theory and Application*, Prentice Hall, New York

[110] Cook, R.K. et al., 1955, Measurement of correlation coefficients in reverberant sound fields, J. Acoustical Soc. of America, Vol.27, No. 6, 1072-1077

[111] Natke, H.G., Doll, H., Hildebrandt, P., Schütze, M., 1995, Bridge condition assessment using an expert system; Structural Engineering Review, Vol. 7, No.3, 165-180

[112] Broch, J.T., 1970, *Messungen von mechanischen Schwingungen und Stößen*, Brüel & Kjaer, Naevum, Danemark

[113] Buzdugan, G., Mihăilescu, E., Radeş, M., 1986, *Vibration Measurement*, Martinus Nijhoff Publ., Dordrecht, Boston, Lancaster

[114] Juckenack, D. (Ed.), 1989, *Handbuch der Sensortechnik. Messen mechanischer Größen*, Verlag Moderne Industrie, Landsberg a. Lech

[115] Harris, C.M., Crede, C. (Eds.), 1976, *Shock and Vibration Handbook*, McGraw-Hill, New York, Düsseldorf

[116] Baseville, M., Nikiforov, I.V., 1993, *Detection of Abrupt Changes, Theory and Application*, Prentice Hall, New York

[117] Kolerus, J., 1990, Merkmalgewinnung zur Interpretation von Schwingungen; in: *Diagnoseverfahren, GMA-Aussprachetag Baden-Baden*, 17. Sept. 1990, VDI-Berichte 854, VDI-Verlag, Düsseldorf, 93-105

[118] Candy, J.V., 1988, *Signal Processing - The Modern Approach*, McGraw-Hill, New York

[119] Cottin, N., 1991, On the optimum experimental design for the parametric identification of linear elastic systems, in: *Structural Dynamics, EURODYN'90*, Eds. W.B. Krätzig et al., A. A. Balkema, Rotterdam Brookfield

[120] Jain, V.K., Dobeck, G.J., 1979, System identification techniques, A tutorial review, Winter Annual Meeting New York, N.Y., Dec. 2-7, ASME paper No. 79-WA/DSC-20, 1-12

[121] Isermann, R., 1988, *Identifikation dynamischer Systeme*, Bd. I, Springer, Berlin, Heidelberg New York, London, Paris, Tokyo

[122] Cottin, N., Natke, H.G., 1986, On the parameter identification of elastomechanical systems using weighted input and modal residuals, Ing.-Archiv 56, 106-113

[123] Cottin, N., Prells, U., Natke, H.G., 1992, A parameter identification technique for elastomechanical systems using modal quantities, The Int. J. of Analytical and Experimental Modal Analysis, Vol. 7, No. 3, 197-212

[124] Natke, H.G., Lallement, G., Cottin, N., Prells, U., 1995, Properties of various residuals within updating of mathematical models, Inverse Problems in Engineering, Vol. 1, 329-348

[125] Natke, H.G., 1979, Vergleich von Algorithmen für die Anpassung des Rechenmodells einer schwingungsfähigen elastomechanischen Struktur an Versuchswerte, ZAMM 59, 257-268

[126] Bard, Y., 1984, *Nonlinear Parameter Estimation*, Academic Press, Orlando, San Diego, New York, Austin, Boston, London, Sidney, Tokyo, Toronto

[127] Natke, H.G., Dellinger, E., 1972, Funktionalanalytische Behandlung der Flattergleichung unter Verwendung des Newtonverfahrens, Z. Flugwiss. 20, Heft 8, 300-306

[128] Golub, G.H., van Loan, C.F.,1983, *Matrix Computations*, North Oxford Academic, Oxford

[129] Gill, P.E., Murray, W., Wright, M.H., 1981, *Practical Optimization*, Academic Press, London, New York, Toronto, Sydney, San Francisco

[130] Unbehauen, H., Göhring, B., Bauer, B., 1974, *Parameterschätzverfahren zur Systemidentifikation*, R. Oldenbourg-Verlag, München, Wien

[131] Kalman, R.E., 1960, A new approach to linear filtering and prediction, Trans. ASME Ser.D.J. Basic Eng., Vol. 82, 35-45

[132] Kalman, R.E., Bucy, R.E., 1961, New results in linear filtering and prediction theory, Trans. ASME J. Basic Eng., Vol. 83, 95-109

[133] Ruymgaart, P.A., Soong, T.T., 1985, *Mathematics of Kalman Filtering*, Springer-Verlag, Berlin

[134] Vetter, W.J., 1973, Matrix calculus operations and Taylor expansions, SIAM Review, April, Vol. 15, No. 2, 352-369.

[135] Cottin, N., 1992, *Reduktion der Ordnung und Festlegung der zu identifizierenden Parameter bei der Korrektur von großen Rechenmodellen für elastomechanische Systeme*, Report CRI - F-4/92, Curt-Risch-Institute for Dynamics, Acoustics and Measurements of the University of Hannover

[136] Oeljeklaus, M., Natke, H.G., 1995, Parallel interval algorithm for the parameter identification of linear elastomechanical systems, Proc. 15th Biennal Conf. on Vibration and Noise of the ASME Design Eng. Technical Conf., Boston MA, Vol. 3, Part C, 1329-1336

[137] Oeljeklaus, M., 1995, *Das lineare Ausgangsgrößenverfahren mit rechnerisch adaptiver Erregung im Frequenzbereich*, Ph.D. thesis, University of Hannover, Curt-Risch-Institute for Dynamics, Acoustics, and Measurements, Report CRI-F 3/95

[138] Gerhardt, L., 1993, *Der Einsatz von Zustandsbeobachtern zur Parameteridentifikation an mechanischen Systemen*, Dissertation, Technische Universität Dresden, Fakultät für Maschinenwesen

[139] Natke, H.G., Zamirowski, M., 1992, ARMAX modelling in structural dynamics - Approximations of the differential equations of motion, ZAMM, Z. angew. Math. Mech. 72, 12, 631-637

[140] Krämer, P., 1983, Rekonstruktion des Zustandes linearer zeitinvarianter eingangserregter dynamischer Systeme mit Hilfe des vollständigen Zustandsbeobachters, msr, Berlin, 26, Heft 10, 562-565

[141] Ackermann, J., 1976, Einführung in die Theorie der Beobachter, Regelungstechnik 24, 7, 217-226

[142] Oreta, A.W.C., 1994, *Development of a Method of Identification of Local Structural Parameters by Kalman Filter*, P.H.D. thesis, Dep. of Civil Eng., The Graduate School of Eng., Nagoya University, Japan

[143] Frank, P.M., 1987, Advanced fault detection and isolation schemes using nonlinear and robust observers, Proc. 10th IFAC World Congr. München, Vol. 3, 63-68

[144] Wünnenberg, J., 1990, *Observer-based Fault Detection in Dynamic Systems*, Diss. Universität-GH-Duisburg, Fortschrittsberichte VDI 8, Nr. 222, VDI-Verlag Düsseldorf

[145] Natke, H.G., Zamirowski, M., 1993, ARMAX modelling in structural dynamics - Numerical integration of the integral representations, ZAMM, Z. angew. Math. Mech. 73, 9, 217-221

[146] Hoen, C., 1991, *System Identification of Structures Excited by Stochastic Load Processes*, Ph.D. thesis, University of Trondheim, MTA-rapport 1991:79

[147] Gawronski, W., Natke, H.G., 1986, On ARMA models of vibrating systems, Probabil. Eng. Mechanics, 1, 1, 150-156

[148] Zamirowski, M., 1990, *ARMA Modelling and Identification of Structural Systems, A Review*, Report of the Curt-Risch-Institute, University of Hannover, CRI-F1/90

[149] Gawronski, W., Natke, H.G., 1988, Order estimation of AR and ARMA models, Int. J. Systems Sci., 19, No. 7, 1143-1148

[150] Kozin, F., Natke, H.G., 1986, System identification techniques, Structural Safety 3, 269-316

[151] Renker, G., 1989, *Zur Identifikation nichtlinearer strukturmechanischer Systeme*, Report of the Institute for Mechanics No. 68, Ruhr University Bochum

[152] Natke, H.G., 1991, Recent trends in system identification, in: *Structural Dynamics*, Krätzig et al. (Eds.), Balkema, Rotterdam, 283-289

[153] Natke, H.G., 1988, Updating computational models in the frequency domain based on measured data: a survey, Probabilistic Engineering Mechanics, 3, 1, 28-35

[154] Baruch, M., 1995, Parametric methods of reference basis for the identification of linear dynamic structures, James H. Belfer Memorial Symposium "Modelling of Structures and Mechanical Systems", Technion, Faculty of Mechanical Engineering, Haifa, Israel, May 1995

[155] Baruch, M., Bar Itzhack, I.Y., 1978, Optimum weighted orthogonalization of measured modes, AIAA Journal 16, 4, 346-351

[156] Baruch, M., 1995, *Damage Detection Based on Reduced Measurements*, Technion-Israel Institute of Technology, Faculty of Aerospace Engineering, Report TAE No. 754

[157] Schmidt, R., 1988, *Die Anwendung von Zustandsbeobachtern zur Schwingungsüberwachung und Schadensfrüherkennung auf mechanische Konstruktionen*, Mitteilungen aus dem Institut für Mechanik der Ruhr-Universität Bochum, Nr. 60

[158] Lin, C.C., Soong, T.T., Natke, H.G., 1990, Real-time system identification of degrading structures, J. Eng. Mechanics 116, 10, 2258-2274

[159] Brandon, J.A., Cremona, C.F., 1990, Singular value decomposition: Sufficient but not necessary, IMAC '90, 1376-1380

[160] Schoukens, J., Pintelon, R., 1991, *Identification of Linear Systems*, Pergamon Press, Oxford, New York, Beijing, Frankfurt, São Paulo, Sydney, Tokyo, Toronto

[161] Schwetlick, H., 1991, Nichtlineare Parameterschätzung: Modelle, Schätzkriterien und numerische Algorithmen, GAMM-Mitteilungen 2, 13-51

[162] Prells, U., Natke, H.G., Cottin, N., 1993, On regularized error localizations, in: Proc. of the IUTAM Symposium on Identification of Mechanical Systems, 23-27 Aug. Wuppertal, Germany, P.C. Müller (ed.), to appear

[163] DeMichele, D.J. (ed.), 1993, *Proceedings of the 11th Internat. Modal Analysis Conference (IMAC'93)*, 1-4 Feb. 1993, Kissimmee, FL, USA

[164] Lallement, G., Cogan, S., Alves-Rade, D., 1993, Parametric updating in elastodynamics: enlargement the knowledge space of the structure, in: H.G. Natke (ed.) *Proc. of the Internat. Workshop on Inverse Problems*, University of Hannover, Germany, 8-9 Oct. 1992, Mitteilungen des Curt-Risch-Instituts CRI-K1/93, 46-66

[165] Tanaka, M., Bui, H.D. (eds.), 1993, *Inverse Problems in Engineering Mechanics*, IUTAM-Symposium Tokyo 1992, Springer-Verlag, Berlin, Heidelberg, New York, London, Paris, Tokyo, Hong Kong, Barcelona, Budapest

[166] Prells, U., 1995, *Eine Regularisierungsmethode für die lineare Fehlerlokalisierung von Modellen elastomechanischer Systeme*, Ph.D. thesis University of Hannover, Curt-Risch-Institute, CRI-F-1/95

[167] Natke, H.G., 1991, Error localization within spatially finite-dimensional mathematical models, a review of methods and the application of regularization techniques, Computational Mechanics, 8, 153-160

[168] Prells, U., 1992, *Regularisierte Modellfehlerlokalisierung*, Report CRI-F-2/92 of the Curt-Risch-Institute for Dynamics, Acoustics and Measurements of the University of Hannover

[169] Zhang, J.H., Natke, H.G., 1991, A two-level updating procedure of the component mode synthesis model, Mech. Systems and Signal Processing, 5, 501-514

[170] Ben-Haim, Y., Prells, U., 1993, Selective sensitivity in the frequency domain, Part 1: Theory, Mech. Systems and Signal Processing, 4, 3, 233-242

[171] Prells, U., Ben-Haim, Y., 1993, Selective sensitivity in the frequency domain, Part 2: Applications, Mech. Systems and Signal Processing, 7, 6, 551-574

[172] Natke, H.G., 1987, Input and damping identification within the frequency domain, J. of Modal Analysis 4, 92-98

[173] Natke, H.G., Zhang, J.H., 1989, The application of modal transformation as a tool for handling large models in updating procedures, in: G.M. Carlomagno et al. (eds.), *Computer and Experiments in Stress Analysis*, Proc. 4th Internat. Conf. on Computational Methods and Experimental Measurements, Capri, Italy, Computational Mech. Publ., Springer-Verlag, Berlin, Heidelberg, New York, London, Paris, Tokyo, 335-346

[174] Cottin, N., 1983, *Parameterschätzung mit Hilfe des Bayesschen Ansatzes bei linearen elastomechanischen Systemen*, Report of the Curt-Risch-Institute of the University of Hannover, Report CRI - F - 2/1983

[175] Gantmacher, F.R., Krein, M.G., 1960, *Oszillationsmatrizen, Oszillationskerne und kleine Schwingungen mechanischer Systeme*, Akademie-Verlag, Berlin

[176] Gladwell, G.M.L., 1986, *Inverse Problems in Vibration*, Martinus Nijhoff Publ. Dordrecht, Boston, Lancaster

[177] Rapoport, A., 1989, *Decision Theory and Decision Behaviour - Normative and Descriptive Approaches*, Kluwer Academic Publ., Dordrecht, Boston, London

[178] Cempel, C., 1990, Symptom limit values in practice of condition monitoring, Mech. Systems and Signal Processing, Vol. 4, No. 6, 483,493

[179] Williams, J.H., Davies, A., 1992, System condition monitoring - an overview, Noise & Vibration Worldwide, Sept., 25-29

[180] Therrien, C.W., 1989, *Decision, Estimation, and Classification - An Introduction to Pattern Recognition and Related Topics*, John Wiley & Sons, New York, Chichester, Brisbane, Toronto, Singapore

[181] Zadeh, L.A., 1973, Outline of a new approach to the ananlysis of complex systems and decision processes, IEEE Transactions on Systems, Man, and Cybernetics, Vol. 3, 28-44

[182] Freudenthal, A.M., Garrelts. J.M., Shinozuka, M., 1966, The analysis of structural safety, J. of the Structural Div. ASCE, Vol. 92, No. ST1, 267-325

[183] Wehling, H.-J., 1993, Schwingungsüberwachung an Reaktorkühlsystemen und Dampfturbosätzen von Druckwasserreaktoren; in: *Dynamische Probleme – Modellierung und Wirklichkeit-*, ed. by

H.G. Natke et al, Mitteilung des Curt-Risch-Instituts der Universität Hannover, CRI-K 3/93, 15-34

[184] Fritzen, C.-P., Seibold, S., Buchen, D., 1995, Application of filter techniques for damage detection in linear and nonlinear mechanical structures, Proc. 13 IMAC, Nashville, TN, USA, Feb.13-16, 1874-1881

[185] Tabaszewski, M., Cempel, C., Wojsznis, J., 1995, Optymalizacja Symptomowych Modeli Prognostycznych dla Celów Diagnostyki Technicznej (Optimization of symptom prognostic models for vibration condition monitoring), Proc. III Diagnostic Conference, Szczyrk, Poland, Oct., 4, 63-68

[186] Natke, H.G., 1989, Minimum modifications due to dynamic requirements: comparison and application, in: *Computer Aided Optimum Design of Structures: Recent Advances*, Eds. C.A. Brebbia, S. Hernandez, Proc. of the Internat. Conf., Southampton, UK, June 1989, Computational Mech. Publ., Springer-Verlag, Boston, Heidelberg, New York, London, Paris, Tokyo, 31-40

[187] Natke, H.G., 1987, Minimaländerungen an Teilsystemen aufgrund von dynamischen Anforderungen, Ing.-Archiv 57, 197-208

[188] Kirsch, U., 1981, *Optimum Structural Design Concepts, Methods and Applications*, McGraw-Hill, New York

[189] Seibold, S., 1995, *Ein Beitrag zur modellgestützten Schadensdiagnose bei rotierenden Maschinen*, Fortschrittberichte VDI, Reihe 11: Schwingungstechnik, Nr. 219, VDI-Verlag, Düsseldorf

[190] Morassi, A., 1993, Crack-induced changes in the eigenparameters of beam structures, ASCE J. Eng. Mech. Vol. 119, No.9, 1798-1803

[191] Davini, C., Gatti, F., Morassi, A., 1993, A damage analysis of steel beams, Meccanica 28, 27-37

[192] Hamamoto, T., Kondo, I., 1992, Damage detection of existing building structures using two-stage system identification, Theoretic and Appl. Mech., Vol. 41, 147-157

[193] Hamamoto, T., Kondo, I., 1994, Global and local damage detection of existing structures, in: *Structural Safety and Reliability* (ICOSSAR '93), Eds. G.I. Schuëller et al., A.A. Balkema, Rotterdam, Brookfield, 789-797

[194] Sheinman, I., 1994, Damage detection in framed structures (TN), AIAA Journal, Vol.31, No. 5, 1103-1105

[195] Sheinman, I., Damage detection and updating of stiffness and mass matrices with the aid of test data, prepared for publ.

[196] Kiciński, J., Drozdowski, R., Materny, P., 1995, The influence of selected parameters on the non-linear vibrations of power turbogenerators, Proc. 2nd Conf. on Acoustical and Vibratory Surveillance, Methods and Diagnostic Techniques, Senelis, France, Oct., 111-123

[197] Rytter, A., 1993, *Vibrational Based Inspection of Civil Engineering Structures*, Ph.D. thesis, Fracture and Dynamics, Paper No. 44, University of Aalborg, Danmark

[198] Zimmerman, D.C., Kaouk, M., 1994, Structural damage detection using a minimum rank update theory, ASME J. of Vibration and

Acoustics, Vol. 116, 222-231

[199] Link, M., Santiago, F.O., 1991, Updating and localizing structural errors based on minimization of equation errors, Proc. Intern. Conf. on Spacecraft Structures and Mechanical Testing, (ESA/ESTEC), Noordwijk

[200] Blake, M P., Mitchel, W.S., 1972, *Vibration and Acoustic Measurements Handbook*, Spartan Books, New York

[201] NN, 1984, *Allianz-Handbuch der Schadensverhütung*, 3rd ed., VDI-Verlag, Düsseldorf, Engl. translation: *Handbook of Loss Prevention*, 1978, Allianz Versicherungs-AG, Berlin, München

[202] Doll, H., 1994, Eigenfrequencies and normal modes of the Norderelb bridge near Hamburg: Numerical and measuring investigations, Proc. 12th IMAC, Honolulu, Hawai, 449-455

[203] Doll, H., Hildebrandt, Kuhlmann, H., Schütze, M., 1995, *Wissensbasiertes System ‚Meßtechnik im Bauwesen'*, Endbericht, Curt-Risch-Institut für Dynamik, Schall- und Meßtechnik der Universität Hannover

[204] Yao, J.T.P., Natke, H.G., 1994, Structural control and civil infrastructure systems education, 1st World Conf. on Structural Control, 3-5 Aug. 1994, Los Angeles, CA, USA, FA3-13 - FA3-18

[205] Tzou, H.S., Anderson, G.L., 1992, *Intelligent Structural Systems*, Kluwer Academic Publ., Dordrecht, Boston, London

[206] Soong, T.T., Natke, H.G., 1988, From active control to active structures, VDI-Schwingungstagung, Köln, VDI-Berichte Nr. 695, 1-18

[207] NN, 1995, Proc. 15th Biennal Conf. on Vibration and Noise of the ASME Design Eng. Technical Conf., Boston MA, Vol. 3, Part C

[208] Natke, H.G., Tomlinson, G.R., Yao, J.T.P., Eds., 1993, *Safety Evaluation Based on Identification Approaches Related to Time-Variant and Nonlinear Structures*, Vieweg, Braunschweig, Wiesbaden

Subject Index

Printing: COLOR-DRUCK DORFI GmbH, Berlin
Binding: Buchbinderei Lüderitz & Bauer, Berlin